WORLD FOOD PRODUCTION, DEMAND, AND TRADE

Iowa State University
Center for Agriculture
and Rural
Development

AGRICULTURAL
AND ECONOMIC
DEVELOPMENT

World Food Production, Demand, and Trade

Leroy L. Blakeslee, *Assistant Professor of Agricultural Economics.* WASHINGTON STATE UNIVERSITY

Earl O. Heady, *C. F. Curtiss Distinguished Professor in Agriculture; Professor of Economics; Director, Center for Agricultural and Rural Development.* IOWA STATE UNIVERSITY

Charles F. Framingham, *Associate Professor, Department of Agricultural Economics and Farm Management, Faculty of Agriculture.* UNIVERSITY OF MANITOBA

IOWA STATE UNIVERSITY PRESS / *Ames, Iowa* / 1973

This project was partly financed by the
U.S. AGENCY FOR INTERNATIONAL DEVELOPMENT,
Contract No. AID/csd-2163.

Composed and printed by
The Iowa State University Press

First edition, 1973

Library of Congress Cataloging in Publication Data

Blakeslee, Leroy L 1935–
    World food production, demand, and trade.

    1. Food supply. 2. Agriculture—Statistics.
I. Heady, Earl Orel, 1916– joint author.
II. Framingham, Charles F., joint author. III. Title.
HD9000.6.B57   338.1'9   72-2785
ISBN 0-8138-1800-1

CONTENTS

v

PREFACE

OPINIONS OF EXPERTS and farm leaders fluctuate through fairly regular cycles in assessment of the world food situation. Now, however, these cycles seem to be overlapping. While one school of experts has dampened its notions that world food requirements will outrun supply and now suggests that optimism is in prospect because of improved technology or other reasons, another group extends its visions and sees the calamities predicted by Thomas Malthus to be near at hand and exploding in nature. In recent years, we have had a few experts who express belief that the green revolution finally promises to overthrow that depressing ruler of the world, hunger. But during the same years, other writers have predicted the imminence of a world food crisis, so much so that developed countries might have to decide by 1975 which food-short countries to save and which to let starve.

More substantial evidence, including that of this study, indicates that the potentials in supply expansion and in demand constraints are such that the world could attain a favorable food production-demand balance in the next three decades. It could do so to the extent that broad problems of hunger might be erased by simultaneous war on the distribution and equity problems of economic development. Those who will guarantee that this end can be attained, or that it will be forfeited, with subsequent decades of misery for people, are government leaders in countries of large populations with high birthrates or in countries of large unexploited agricultural resources. Leaders in the former countries have the long-run solution of population in their hands and will deliver it or let it slip through humanity's hands; leaders in the latter countries can provide short-run solutions but, try as hard as they may, cannot extend food supplies to match population growth that other countries do not control.

This volume further substantiates the potential of solving world food problems in the next three decades. The reality is there. Of course, it will not be attained without vigorous population policies, particularly in the less developed countries. To say that the possibility exists does not cause its realization. If, because the possibility seems to prevail, pol-

vii

iticians and government administrators relax and do little or nothing to constrain births, the possibility will breed an even more intense problem in a few decades.

This study was made to help further sift the possibilities and prospects of matching food supply and demand in future decades. It represents an attempt which is as empirical and quantitatively detailed as possible in terms of methodology justifiable by the amount and nature of data available in many countries and the resources available for the study. It does provide important insights, on the basis of conservative projections in production possibilities, on the prospects of meshing future population and world food demand with supplies. It does so even without incorporating shift variables to express potential effects of the green revolution. Too few observations, in both geographic and time dimensions, prevailed for incorporation of these variables into the types of quantitative methods used. However, modifications potentially generated by the green revolution are discussed.

The realization of potentials in meshing food production in some parts of the world with food demands in others will depend on trade even under the best outcome wherein country leaders effectively restrain population growth. Accordingly, a trade model has been applied to determine "to where" and "from where" foodstuffs might move at future points in time. While the model applied is not nearly so detailed and sophisticated as those we have applied at Iowa State University in the analysis of interregional production patterns and trade among 200 U.S. regions, it is among the first fairly elaborate attempts that have been made through a formal optimizing model. The model also is used to examine potential world and interregional trade in fertilizers in both the short run and the long run.

LEROY L. BLAKESLEE
EARL O. HEADY
CHARLES F. FRAMINGHAM

PART ONE

Background and Methods

CHAPTER 1

Historical Perceptions of the World Food Problem

EVEN IN THE ABSENCE of recorded history, food problems were among the first concerns engaging the physical and mental capacities of man. The problem was encountered initially in its most elemental form and man's response was direct and unsophisticated. Later perceptions of the problem were more esoteric and varied, and general awareness of the problem rose and subsided as new theses were put forward and as experience became history. For centuries, in some form and in some part of the world, the problem remained as a perplexing dilemma, never fully resolved. Today, in the 1970s, we again find scientists, philosophers, and men of public affairs reinterpreting the past, dissecting the present, and probing the future in search of new insights about the nature and significance of the world food problem and man's response to it.

By any reckoning it is clear that past attempts to make agriculture more productive have been highly successful. Some segments of the world's population now are virtually assured of an adequate food supply for the foreseeable future. All this, of course, has occurred concurrently with the general increase in man's productivity, which we term economic development.

At the same time, world population has continued to grow, and the world's agricultural output now meets the needs of a population far larger than earlier analysts could imagine. Although documentation is difficult, severe famine appears to be a much less common occurrence in recent decades than in earlier times. Most recently, many specialists suggest that major breakthroughs in agricultural technology and in public support of agricultural improvement have signaled the end of food problems even in South Asia, an area which has been the focus of food scarcity concerns over the last decade.

Man's persistent fixation with the food problem attests to its importance and its complexity. However, unfulfilled past prophecies of widespread calamity, together with recent agricultural advances, suggest the possibility that ongoing predictions of food scarcity in the underdeveloped countries are illusory. Obviously, the consequences of concluding wrongly that future food needs will be met as part of the normal course of events are immense. On the other hand, man's capacity for directing events is greater than at any time in history. Thus two central questions deserve consideration: Are there new and significant elements in the present or near future situation which make past experience largely irrelevant? If forces tending toward a food crisis are at work, what policies seem desirable and feasible to meet future needs?

These and other questions have stimulated a large and expanding literature in recent years, and the present work is a contribution to this literature. In broad terms, we undertake an analysis of ongoing trends and future prospects for food production and demand in most major countries of the world. Recent trends in area, yield, and production of important crops are estimated and projected, and income and population trends are used to project food demand. Based on these projections, prospects for balancing future food production and demand are evaluated. In several respects, particularly in its focus on *de facto* production trends and in its commodity detail, we believe that this analysis contains significant elements new to the literature on the subject.

Four Food Crises of the Past

M. K. Bennett, writing in 1949, noted three periods in the history of the English-speaking world when pessimism about the pressure of population on food supply was especially widespread.[1] He wrote in the midst of the fourth such period. Each of these waves of concern had a character of its own, and the fifth wave, that of the last decade, is again somewhat different from its predecessors.

Bennett, like many others, points to the 1789 writings of Malthus as the earliest systematic statement of the relationship between population and growth in food supplies.[2] The characterization of economics as "the dismal science" was in no small measure the result of Malthus's conclusion that the normal state of affairs is one where food supplies are barely adequate for the population.

---

1. M. K. Bennett. "Population and Food Supply: The Current Scare." *Scientific Monthly* 68 (Jan. 1949): 17–26.

2. T. R. Malthus. *An Essay on the Principle of Population*, 9th ed. London: Reeves and Turner, 1888. Tracing the genealogy of ideas is always a difficult task. During the eighteenth and nineteenth centuries a pro-populationist attitude prevailed in many quarters as population (and labor supply) growth came to be viewed as a major symptom, or even a cause, of growth in economic wealth. To the extent that this view still persists today, it is probably most prevalent in the countries of Eastern Europe. The anti-populationist thesis which reemerged in the nineteenth century is most commonly associated with the name of Malthus who was most successful in popularizing the idea. Schumpeter, however, finds the Malthusian principle fully developed in its essentials as early as 1589 in the writings of Botero. Schumpeter also cites several other seventeenth- and eighteenth-century writers who, like Botero and Malthus, adopted the central premise that human fecundity could be compared to a "spring that is held down by a weight and is certain to respond to any decrease in this weight." (J. A. Schumpeter. *History of Economic Analysis*. New York: Oxford University Press, 1954, p. 255.)

Reasoning from basic demographic truisms, Malthus determined that man's potential reproductive ability implied an exponential population growth curve over any significant length of time; he also suggested that food supplies for a region could grow no more vigorously than by an arithmetic progression. This belief stemmed from observation of a bounded total quantity of cultivable land and diminishing returns with a fixed land input. His conclusion that food supply potential is no match for population potential follows immediately.

But Malthus did not infer widespread famine as a result. Instead, he concluded that population numbers will almost always increase to a point near the limits of sustenance, but only rarely will this limit be reached so abruptly that actual starvation results.[3] He proposed two kinds of restraints, positive and preventive, to limit population. Positive checks encompass most maladies of man which increase death rates, including ". . . all unwholesome occupations, severe labor, and exposure to the seasons, extreme poverty, bad nursery of children, large towns, excesses of all kinds, the whole train of common diseases and epidemics, wars, plague and famine."[4] The preventive checks are those limiting the birthrate, but Malthus had only limited hopes for widespread voluntary actions to limit births.

Some facets of Malthus's thesis have withstood the test of human experience remarkably well; others have been found lacking. Observations on *potential* population growth, based on reproductivity alone, go unchallenged. His positive and preventive checks, taken together as factors containing population to sustenance limits, are qualitatively correct whether or not the moral values he attaches to these checks are accepted. Certainly the power of the positive checks is evident in the African statistics of life expectancy and, to some extent, in the vivid news films which came from Asia in the 1960s. If dominance of the positive checks corresponds to the early phases of the demographic transition, then Malthus can probably be faulted for neglecting the later phases. Clearly, he grants the power of preventive checks to contain population numbers to the bounds set by food supplies (productive resources), but he doubts the power of these checks to bring the rate of population growth below the rate of growth in sustenance (productive resources).

As generalizations, his judgments on yield potentials were conservative. He specifically states his doubts as to whether a country such as Japan could ever double food production, even with the best directed efforts and an infinite time span.[5] Yet recent statistics show that between 1954 and 1967 Japan's net agricultural output increased by 60

---

3. Malthus admits the possibility of population remaining below the limits set by food supplies, but he views this as an unlikely happening.
4. Malthus, p. 8.
5. Malthus.

percent.[6] Other examples could be cited, but this one is Malthus's own reference term and should suffice. Still, while agricultural productivity far exceeded Malthus's expectations in many cases, there are numerous countries where historical evidence has not contradicted him.

What, then, is the relevance of the Malthusian model to the present world food situation? First, and most important, it is quite consistent with circumstances presently existing in countries with low incomes and poor diets. Life expectancy is low, and the dominance of the positive checks in limiting population growth seems clear. Until the recent emergence of the "green revolution," expansion of food supplies in these countries has often been closely tied to cropland expansion and only modest yield increases. While yield experience does not reflect diminishing returns to technically advanced agricultural inputs, there seems little doubt that additional returns from further application of conventional inputs would be small indeed. Nevertheless, widespread famine has been rare.

Little will be said now of food supply trends in the economically developed countries, except to note that average diets have been fully adequate by any nutritional standard. It is the experience of these countries that provides evidence for challenging certain of Malthus's conclusions. On the production side, the successful application of science in the development of new agricultural inputs and new farming practices was a factor far outside Malthus's reckoning. Furthermore, technical and sociological changes accompanying economic development in Europe, Japan, Oceania, and North America have resulted in circumstances where preventive checks now restrain population growth rates to a fraction of the biological potential. Together, these forces have resulted in diets near the saturation level, and even major food surplus problems.

Developments discussed above are recent phenomena. This is underscored vividly as we move to the second period identified by Bennett when the topic of population pressure on food supplies was again a subject of widespread debate. A single source also can be identified as the keystone of the discussion: the 1898 presidential address, "The Wheat Problem," of Sir William Crookes to his colleagues in the British Association for the Advancement of Science.[7] His concern was not, as Malthus's, with the global problem of population and food supply; or even with the food problems of underdeveloped countries, the large concern

---

6. U.S. Department of Agriculture, Economic Research Service. *The Agricultural Data Book for the Far East and Oceania.* U.S. Department of Agriculture, ERS–Foreign 189, 1967.

U.S. Department of Agriculture, Economic Research Service. *The 1968 Agricultural Data Book for the Far East and Oceania.* U.S. Department of Agriculture, ERS–Foreign 219, 1968.

7. Sir William Crookes. *The Wheat Problem.* New York: Knickerbocker Press, 1899.

today. Rather, his concern was of food supply for the "bread-eaters" of the world, or the Caucasian race. Thus, Crookes's main concern was the welfare of present economically developed countries.

Crookes cited evidence of above average world wheat crops during the period 1889–96, of consumption levels above production in 1897 and 1898, and of rising population in countries where bread was a staple food. He inventoried possible additions to the world's wheat growing area and the likely rate of increase among the number of bread-eaters. From these calculations he concluded that at the most, only about three decades remained before all available wheat acreage would be required for current needs. Crookes was particularly alarmed because the period of rapid and easy expansion of U.S. wheatlands, so important in fulfilling England's import needs at that time, appeared to be drawing to a close. In the period 1870–88, U.S. cropland expanded to 230 percent of its 1870 value, while in the next ten years further expansion amounted to less than 4 percent.

Crookes's American contemporaries took issue with his assessment of U.S. acreage trends, but Crookes was essentially correct. Morrison and Commager record the 20-year period following 1870 as the time of greatest westward expansion in the United States, and also as the time of the disappearance of the American frontier. It may seem ironic today that the establishment and settling of what is undoubtedly one of the world's most productive agricultural regions should have been regarded as a calamitous event, but it was so regarded by Crookes.

Although Crookes's remarks precipitated great alarm among his colleagues, he also pointed the way out of the seeming calamity. He focused on the potential for yield improvement through increased application of fertilizer, particularly nitrogen. Crookes's tentative optimism was based not on potentials for applying known and proven inputs but on his assessment of the commercial potentials for fixing atmospheric nitrogen and using it in the production of chemical fertilizers. Seen from Crookes's time, the prospects for obtaining needed fertilizer from traditional sources seemed poor indeed; and fixation of atmospheric nitrogen was a process which had not yet left the confines of chemists' laboratories.

As was true of Malthus's ideas, the circumstances of the 1898 "food scare" have some interesting counterparts in recent events. Potentials for food shortages were apparent enough, and history alone suggested few solutions to the problem. Certain promising technical developments were visible on the horizon, but their ultimate practical impact was not assured. The final outcome was only a matter for conjecture. One need only shift the time scale forward 65 years, redirect the object of these comments from the bread-eating countries to today's underdeveloped countries, and we have a fair description of recent writings on the world food problem.

The demise of the world's bread-eaters never materialized and over-production of wheat has even generated serious farm income problems. Many scientific developments other than fixation of atmospheric nitrogen were involved, but Crookes clearly was right in foreseeing the potential applications of science in agriculture. However, in the setting of the 1890s, his projections of wheat supply and demand under assumptions of continuation of trends, current at that time, were by no means a redundant exercise in the arithmetic of consumers and acres. In fact, they showed quite clearly that contemporary patterns of change, if continued, would lead to disastrous results.

Bennett refers to the period immediately following World War I as the third time of widespread interest in food supply problems. However, the discussion of this period seems largely peripheral to our interests, and we choose not to take it up in detail. The main stimulus was apparently the devastation brought about by the war itself, rather than the economic and demographic considerations raised in earlier and subsequent periods.

In the late 1940s and early 1950s, the theme of population-food imbalance was again revitalized. The combined influence of three new elements provided the critical mass leading to an explosion of the topic in the realm of public debate. First among these was the growing impact of the Food and Agriculture Organization of the United Nations. The mere existence of FAO was a significant fact in itself, quite apart from its programs, for implicit in the formation of the organization was acceptance of the idea that food problems in all parts of the world should, to some extent, be common interests of all members of the world community. To embody this principle in a permanent major institution having international sanction, membership, and support at the governmental level was to give far greater credibility to the principle than had ever existed before.[8] Once in existence, a major part of FAO's early program was to dramatize the poor state of human nutrition existing over major portions of the globe. There was much to dramatize, especially for millions of Asians, Africans, and Latin Americans. Also, the destruction of World War II had increased the severity of conditions, as had

8. Sir John Boyd-Orr, first Director-General of FAO, in recounting the history of attempts to internationalize the attack on world hunger, notes that in the late 1940s efforts were made to endow UN agencies dealing with food and developmental problems with much more autonomy than was actually granted. Speaking of an unsuccessful proposal for a World Food Board put forth by the Director-General in 1946, Orr states, "It would be a step toward the evolution of the United Nations Organization into a World Government without which there is little hope of permanent world peace." (Sir John Boyd-Orr. *The White Man's Dilemma.* London: George Allen and Unwin, Ltd., 1953.) Thus, in some respects it appears that the unification of efforts to improve human nutrition worldwide may be seen as an extension of the trend toward collective action among nations which was so prevalent at the end of World War II, particularly in defense matters.

poor weather and relatively low crop production in Western Europe. The thrust of FAO's information program brought the attention of wealthy nations to the long-standing food scarcity conditions among the world's poor.

From this information base, FAO nutrition specialists established goals, or target consumption levels, for major world regions. Per capita food consumption targets designed to erase food shortages worldwide were incorporated into estimates of future world food demand. Such projections, reflecting both population growth and greatly improved per capita consumption levels, appeared to make staggering demands upon world agriculture, especially in view of (then) current growth of agricultural productivity in underdeveloped countries.

This same time period saw what was probably the high point of public concern, at least until the environmental concern of late, over the conservation of natural resources. The writings on this topic were the second inflammatory factor contributing to the food scare. The dissipation of land and water resources resulting from reckless mismanagement was perceived by some to be proceeding at such an alarming rate that even present food production levels might soon be threatened.[9] Or, as Bennett put it, "The soil conservation school adds to the old concept that the food-producing land of the world is strictly limited in extent, the new concept that the land is actually being destroyed, and at a rapid rate."[10]

The third component, recognition of the effects of lifesaving techniques on death rates in the underdeveloped countries, which contributed to the fourth period of concern over world food supplies, has been magnified greatly in recent years. Practices of medicine, public health, and sanitation were cheap and easy to apply on a mass scale in such countries. Full recognition of the importance of these developments did not come until later, and the topic will be treated more fully in our discussion of current perceptions of the world food problem. Nevertheless, it is instructive to briefly contrast the demographic outlook of two decades back to that of the present. Bennett, in 1949, reviewed authoritative projections which placed the probable world population in the year 2000 at about 3.3 billion—a projected increase of 900 million people in the last half of the century. However, data available at the present indicate that world population probably passed the 3.3 billion mark in the late 1960s and UN demographers now estimate that world population in 2000 may be between 5.3 and 7.4 billion. This striking change in the outlook for future population growth has become an important, if not dominant, factor in current thinking on the world food problem.

---

9. William Vogt. *Road to Survival.* New York: Sloane Associates, 1948. Fairfield Osborne. *Our Plundered Planet.* Boston: Little, Brown, 1948.

10. Bennett, p. 19.

RECENT PERSPECTIVES ON THE WORLD FOOD PROBLEM

We now consider the events of the past decade which again triggered a wave of pessimism over the world's future food supplies. Four major factors appear to have been involved.

Throughout most of the 1950s, modest gains were made in both total and per capita food supplies in most underdeveloped countries. These had much to do with limiting the duration of the fourth food scare. But by the late 1950s and early 1960s, small but highly significant changes were observed in per capita food supply trends of the underdeveloped countries. They first appeared to slow, then nearly to stop. The empirical fact that the momentum of the 1950s had been lost was the first factor.

But a second factor, two consecutive years of monsoon failure in South Asia (1965 and 1966), made the records of progress appear worse and reduced food production to dangerously low levels in one of the world's most populous and poorly fed regions. The third factor, mentioned earlier, was falling death rates and the accompanying rapid spurts in population growth and slowing of per capita food supply trends throughout the underdeveloped world.

Finally, a new component was added to earlier discussions. Following World War II, the goal of general economic development was embraced by the governments of the world's poorer nations. Typically, countries most lagging in economic development are those with large agricultural sectors and low per capita food supplies. Current thinking emphasizes that the agricultural sectors of such countries can play pivotal roles in realizing broad development goals. Thus the "food problem" of earlier times has come to be seen as only one aspect of the broader problem of lagging economic development coupled with rapid population growth, and the goals of increasing agricultural productivity and dietary levels in the underdeveloped countries acquire additional significance in the pursuit of the expanded aspirations so evident in today's world.[11] Thus earlier expressions of the food problem primarily in terms of demographic, agronomic, and nutritional terms have phased into recognition of economic considerations. This is most evident in concepts of food demand where it is now recognized that demand estimates based on nutritional needs of a population may bear little resemblance to actual demand in the absence of sufficient income. Any realistic assess-

---

11. The work of Coale and Hoover provides an excellent analysis of the relationship between the rate of population growth and the rate of accumulation of productive resources in determining the rate of general economic development among poor nations. (Ansley J. Coale and Edgar M. Hoover. *Population Growth and Economic Development in Low-Income Countries.* Princeton, N.J.: Princeton University Press, 1958.) Their conclusion that rapid population growth inhibits the accumulation of productive capital provides an instructive counterpoint to the Malthusian theme that growth in food supplies (or productive resources) limits population growth.

ment of the present and future world food situation must recognize these basic economic variables.

Having introduced the four principal aspects of the world food problem as seen in the 1960s, we now move to consider them in fuller detail.

POSTWAR FOOD SUPPLY TRENDS

Table 1.1 presents indices of total and per capita food production for eight world regions covering post-World War II years. As the data show, food supplies in most underdeveloped countries moved upward at modest rates, in many cases, from a base which permitted only marginal or even substandard diets. These trends, apart from the concepts of green revolutions, seemed to offer hope that underdeveloped nations were departing from the Malthusian model as developed nations had done earlier. Besides gains in production there were gains in the application of medical and sanitation knowledge to eliminate diseases and pestilence, which formerly had killed or debilitated large numbers of people, particularly children.

While the aggregations in Table 1.1 hide considerable detail, evidence of truly stagnant agricultural productivity is rare. An element of war recovery is in these data, but the gains in food production are still impressive. Apparent increases in total food production during the postwar years, both in the developed and underdeveloped regions, were approximately 50 percent.

However, trends in per capita food production exhibit divergent patterns. In Latin America and Africa, the largest increases since the 1948–52 period amounted to 7 and 8 percent, respectively, but in no year since 1960 have these peaks been exceeded. Indices of per capita food production in the Near and Far East have risen more rapidly, but again, most of the gains occurred in the earlier part of the period. By 1960, per capita food production in the Far East had risen 15 percent above the level of 1948–52, but since that time, the largest recorded per capita food production exceeded the 1960 level by only 2 percent. Similarly, the 1959 per capita production index for the Near East was 16 percent greater than that of 1948–52, and the largest gain since that time was the 2 percent increment estimated for 1963.

Due to the varying quality of agricultural production statistics of underdeveloped countries, the figures cited here are subject to qualification. The significance of the rather small year-to-year percentage changes should not be overly stressed. However, we believe that the broad patterns of change evident in the 1948–70 period are significant insofar as they portray approximately 12 years of growth in per capita food production followed by approximately 11 years of virtual stagnation in less developed countries.

TABLE 1.1. Indices of Total and Per Capita Food Production by World Regions
(1952–56 = 100)

| | Production in Developed Regions | | | | | | | |
| | W. Europe | | E. Europe, USSR | | N. America | | Oceania | |
Years	Total	Per Capita	Total	Per Capita	Total	Per Capita	Total	Per Capita
1948–52	84	86	83	87	92	99	92	102
1953	101	102	94	96	98	100	100	103
1954	101	101	96	96	97	97	98	98
1955	102	101	104	103	101	100	104	101
1956	103	102	114	111	104	101	101	97
1957	106	104	118	113	101	96	99	92
1958	110	106	129	122	109	101	117	107
1959	113	108	131	122	109	100	115	103
1960	119	114	133	122	111	100	123	107
1961	119	112	137	123	110	98	123	105
1962	126	118	140	124	114	99	135	113
1963	128	118	134	118	121	104	138	113
1964	129	118	146	127	120	101	145	116
1965	130	118	149	128	122	102	137	107
1966	133	120	167	142	127	104	158	122
1967	142	127	168	142	132	107	148	109
1968	146	129	176	147	133	107	175	130
1969	146	129	171	141	132	105	167	121
1970[a]	147	128	180	147	130	103	166	118

| | Production in Underdeveloped Regions | | | | | | | |
| | Latin America | | Far East[b] | | Near East | | Africa | |
Years	Total	Per Capita	Total	Per Capita	Total	Per Capita	Total	Per Capita
1948–52	88	98	87	94	84	93	87	95
1953	95	98	98	100	100	103	98	100
1954	100	100	100	100	98	98	102	102
1955	102	100	104	102	100	98	101	99
1956	109	103	107	103	110	105	106	101
1957	112	103	108	102	115	107	106	99
1958	117	105	113	104	119	108	108	98
1959	116	101	118	107	122	108	113	100
1960	118	100	123	108	123	105	120	103
1961	125	103	127	109	124	104	117	99
1962	126	101	129	109	134	109	124	102
1963	132	102	132	109	138	110	128	103
1964	137	103	136	110	139	107	130	102
1965	140	103	134	105	141	106	130	100
1966	141	101	135	104	145	107	130	97
1967	151	105	141	103	150	107	132	97
1968	152	102	149	107	155	108	137	97
1969	157	103	156	108	159	107	140	97
1970[a]	166	105	162	110	159	104	141	95

SOURCE: United Nations, Food and Agriculture Organization. *The State of Food and Agri-culture, 1968 and 1971.*

[a] Preliminary.

[b] Excluding Mainland China.

TABLE 1.2. 1957–59 Average Per Capita Daily Nutritive Values by World Regions (retail level)

Region	Calories	All Protein	Animal Protein	Calories as % of Requirement	% Calories from Cereals, Roots, and Sugar
		(grams)	*(grams)*		
E. Europe, USSR	3,180	94	33	122	71
W. Europe	2,910	83	39	113	55
N. America	3,110	93	66	120	40
Oceania	3,250	94	62	125	48
Latin America	2,510	67	24	104	63
Far East[a]	2,030	53	9	90	79
Near East	2,470	76	14	103	72
Africa	2,360	61	11	101	74

SOURCE: United Nations, Food and Agriculture Organization. *Third World Food Survey.* United Nations Food and Agriculture Organization Freedom from Hunger Campaign Basic Study 11, 1963.
[a] Excluding Mainland China.

Data on absolute levels of food consumption in various parts of the world provide further information necessary for a full understanding of the recent world food situation. The FAO of the UN has, on three occasions, analyzed data on population, food production, inventory changes, trade, waste, and extraction rates to make comprehensive estimates of food consumption around the world. The most recent study was based on data for 1957–59, and results were reported as part of the *Third World Food Survey.* Selected findings from this study are presented in Table 1.2. The four underdeveloped regions achieved little more than two-thirds of the levels of calorie consumption found in the developed regions. Furthermore, daily average calorie consumption levels were barely meeting FAO's recommended minimum standards in Latin America, the Near East, and Africa, while the reported consumption level for the Far East was only 90 percent of the standard. Supplies in the remaining four regions were well above minimum physiological requirements.

Calorie consumption requirements estimated by FAO and incorporated in Table 1.2 allow for normal growth and physical activity and reflect interregional variations in needs arising from differences in average physical stature, age and sex distribution of the population, and mean environmental temperature.

Figures presented in Table 1.2 are averages for regions. Within a region, some countries have greater or lesser supplies than the regional average, and some groups in individual countries are nutritionally better off than others. Thus, in regions such as Latin America, the Far East, Near East, and Africa where average calorie supplies are at or below the physiological requirement, it is certain that there are significant numbers of people for whom hunger and the array of maladies set forth by Malthus are everyday experiences.

Protein consumption estimates show a similar contrast between developed and underdeveloped regions, but the most striking differences appear in estimates of animal protein consumption. Specialists in human nutrition are not in full agreement on the importance of animal protein to human diets. A proper combination of plant proteins can provide the essential nutrients contained in animal products. However, incomplete evidence available suggests that the proper combination has not been achieved in many cases since presence of diets low in animal protein is highly correlated with presence of protein-deficiency diseases. Moreover, diets rich in animal protein will almost certainly provide the amino acids essential to good health.

Figures presented in Table 1.2, showing the percent of calories derived from cereals, starchy roots, and sugar, indicate a general pattern of heavy reliance on these foods in underdeveloped regions. Average diets in the Far East, Near East, and Africa all derived over 70 percent of total calories from these foods. Countries in Eastern Europe and the USSR also rely heavily on these foods, but this fact must be judged against a background of higher levels of consumption for all kinds of food than in underdeveloped regions.

Seen against this background of substandard consumption, it is not difficult to understand the great urgency which was attached to the food crisis following two consecutive years of monsoon failure in the Far East during 1965 and 1966. Per capita food production declined 5 to 6 percent from the immediately preceding years. Net imports in 1965–67 were about 50 percent above the level of 1962–64, but still, FAO estimates that the effective per capita supply fell about 5 percent.[12] The situation was further aggravated by the low levels of food reserves in the region and by the lack of effective food distribution channels. In many Western countries, a production decline of this magnitude would result in little hardship, but in a region where undernourishment and malnutrition were already common, the situation posed a genuine threat to survival for millions.

Data available for the period 1967 through 1971 indicate that the immediate crisis in South Asia has passed. Indeed, the introduction of improved, nonnative wheat and rice varieties, together with increased governmental emphasis on improving agricultural output, perhaps marks the beginning of a period when old evolutionary patterns will be altered drastically. While evidence on this point is still not firm, more urgent implications are those of the third factor entering discussions over the world food problem: the present growth rate of world population, the so-called population explosion.

---

12. United Nations, Food and Agriculture Organization. *State of Food and Agriculture, 1968.*

POPULATION GROWTH IN UNDERDEVELOPED COUNTRIES

The current population growth rate for the world is presently at an all-time high. Also, many underdeveloped countries now have high built-in growth-rate potentials, because of the past high birthrates and a large proportion of populations in younger age brackets. Likelihood of death for this generation has been most affected by application of health knowledge in underdeveloped countries; therefore, their chances of surviving through childbearing age have been very much enhanced. Even if fertility were to decline quite markedly, the potential for substantial population growth is still large.

We turn now to mortality changes which have dominated recent demographic considerations for underdeveloped countries. Stolnitz prefaces his excellent review and interpretation of their mortality experience with the following summary statement:[13]

> The most personal of the post-war revolutions pursues an everwidening course. Headlong mortality declines in the underdeveloped areas began only a decade or two ago, in isolated fashion. Today they have become commonplace and near-continental. Large parts of Latin America and Asia have already experienced such declines and much of Africa seems likely to do so in the near future. With amazing regularity the nations of these regions which provide reasonably reliable information show recent 10- to 20-year trends which match or exceed the maximum declines ever found in the industrialized, low mortality parts of the world.

Due to the scarcity of reliable data, Stolnitz used expectation of life at birth, rather than the crude death rate, as his primary indicator of mortality. A second reason for using life expectancy at birth is its independence of the population age distribution.[14] Stolnitz's findings for Latin America were particularly striking. In tropical South America, he found life expectancy had risen from 44.0 years in 1945–50 to 51.6 years in 1955–60. The corresponding change in Central America and Antilles was 46.0 to 52.0 years. In Ceylon, expectation of life at birth rose from 52.6 years in 1947 to 60.3 years in 1954 for males, while for females the change was from 51.0 to 59.4 years. Taiwan males gained in life expectancy from 41.1 years in 1936–41 to 61.3 years in 1959–60. The change for females' life expectancy was from 45.7 to 65.6 years. In India, the most populous country in non-Communist Asia, life expectancy for

13. George J. Stolnitz. "Recent Mortality Trends in Latin America, Asia, and Africa: Review and Reinterpretation." *Population Studies* 19 (1965): 117–38.
14. The crude death rate, the number of deaths per thousand population, must be interpreted with caution when applied to underdeveloped countries. They are characterized by high past and present birthrates, and thus exhibit broadly based age distribution pyramids. Thus a lowering of the age-specific death rate for the young has an exaggerated effect on the crude death rate simply because there are relatively few people in the higher-aged cohorts where age-specific death rates may have been less affected.

males rose from 32.4 years to 41.7 years and for females from 31.7 to 42.1 years from 1941–50 to 1956–61. Data on the African countries are very meager, but available evidence suggests a similar, though somewhat smaller, change.

Stolnitz compared the rates of decline in mortality now being experienced in the underdeveloped countries with historical rates of decline in the now-developed countries. Japan's mortality levels in the interwar period approximated those of the West at a time 50 years earlier, while the lag by 1965 was only 10 years. Taiwan, in the last 30 years, achieved mortality reductions which took 50 years in the West; and Mauritius accomplished mortality reductions in 10 years which matched the gains made by the Western nations during the full nineteenth century. The quick and easy advances in mortality reduction have, of course, been accomplished in many areas, and future advances will come more slowly. But any casual examination of recent mortality trends in the underdeveloped countries, and life expectancy differences still remaining between these and the developed countries, surely suggests that further significant declines are possible and likely.

The reason for a nearly complete omission of birthrates from the preceding discussion of population growth in underdeveloped countries is that until very recently birthrates have remained nearly constant at relatively high levels and have not been actively associated with the important changes in the aggregative world food situation during recent years. The active agent, insofar as population growth is concerned, has been the death rate. Birthrate trends are a matter of great importance for the future, and this subject will be given attention in a later chapter when other prospects for the future are being examined.

In Malthusian terms, it can be said that man already has interdicted to suppress a subset of the positive checks. This was done with intent, and provisionally, the result must be judged favorably. Although it may lack analytical content, the requirement that population numbers remain below the bounds of sustenance is a rigid inequality. The more important question to be asked is: How do we ensure against the aforementioned relation becoming an equality for part of the world's populace? Undoubtedly this must come through a combination of effective preventive checks and increased agricultural output (in both cases, beyond Malthus's expectations). But ultimately, man must address himself to a more important and difficult question, for answering the first, and no more, ensures nothing but misery. That question, of course, is how can underdeveloped nations raise their populace above the subsistence level and allow their efforts to be applied to tasks more rewarding than simply keeping alive.

CHAPTER 2

Purpose and Nature
of the Investigation

MAN DOES POSSESS the means to alter the supply of food products and both the birth and mortality rates of people. The manner and extent to which these options are exercised over the next three decades will determine whether, for the greatest mass of the world's population, the possibility for man is to exercise his abilities in positive enjoyment of life or whether it is only to fight off the burdens which face all animals and to minimize his misery. Certainly the outlook is positive. Investment in and communication of new technology have recently given rise to renewed hope on the food supply side. Agricultural surpluses in developed countries are a reflection of this possibility. The wide geographic applicability of new wheat and rice varieties and their successful application with other technologies, especially in underdeveloped countries such as India and Pakistan, are even more substantial evidence. In the short run, ability to increase food supplies beyond a narrow margin above subsistence levels rests more on food production possibilities than on population controls in the less developed countries. Over the longer run, however, solutions obviously must rest on the latter means. The possibilities here are no less than on the food supply side, even though the prospect is that countries of rapidly growing populations are likely to give first emphasis to the latter.

Obviously the course of future events will depend on the policies various countries formulate and implement in respect to both (a) food supply as reflected in advanced farm technologies, and (b) food demand as reflected especially in population magnitudes. The options are numerous and the outcome will depend on the imagination and leadership of governments in attaining improved nutrition and human welfare through either or both enlarged food supplies and restrained food demands. In one sense, these are the important considerations relating to future food policies.

Our purpose is not, however, to detail these policy possibilities or to outline the instruments needed to attain them. Instead, we seek to project food production and demand in terms of major variables and trends

17

reflected in the recent past. In doing so, we make no attempt to incorporate the "shift variables" which characterize the green revolution in a few countries or the fairly rapid decline in birthrates in a very limited number of less developed nations. This study is directed toward projection of food production and demand up to the year 2000 in terms of trends expressed, especially in the postwar period. Further, based on certain of these projections, a trade model is developed and applied to estimate the potential future movement of food and fertilizer among surplus and deficit producing countries.

RELATED STUDIES

Future food balances have been estimated previously. Important studies in these respects were those of FAO and the USDA. The *Third World Food Survey* by FAO described worldwide food consumption levels in the period 1957–59 and estimated quantities of additional food production required to achieve specified nutrition targets throughout the world in the years 1975 and 2000.[1] Food balance sheets were prepared for over 80 countries, including Mainland China, and covered about 95 percent of the world's population. The balances were used to estimate nutritional adequacy of average diets in countries under study. Nutritional targets were established for the future, and needed future food supplies were estimated, based on the targets, population projections, and estimated nutritional values of individual foods. The projected 1975 world food needs were 35 percent above 1958 levels when no improvements in diets were assumed and 50 percent above to attain specified short-term nutritional targets. Underdeveloped countries required a 79 percent increase in total food and a 121 percent increase in supplies of livestock products to achieve 1975 targets. By the year 2000, total world food needs were estimated to be 174 percent greater than 1958, while needed increases in livestock output were estimated to be 208 percent greater.

Under the authority of Public Law 480, the USDA initiated a series of country studies in the early 1960s. Supply-demand projections were completed for a fairly large number of countries, particularly those related to U.S. export potentials or facing domestic food problems. The objectives of these studies were to project the future growth of the economy through projections of population and the major components of national accounts, to project domestic demand for major agricultural products consumed, to project domestic supply for major products produced, and to estimate future agricultural trade potentials for the country under study. Ultimately, the goal of the USDA studies was to

---

1. United Nations, Food and Agriculture Organization. *Third World Food Survey.* United Nations Food and Agriculture Organization Freedom from Hunger Campaign Basic Study 11, 1963.

assess future market prospects for U.S. agricultural exports. In general, these studies were used by the USDA in preparing its study, *The World Food Budget, 1970*.[2] The objectives of this investigation were to study supply and utilization of food commodities for the world in 1970; to evaluate problems and possibilities of closing the food gap; and to arrive at consistent and balancing estimates of production, utilization, and trade.

First approximation demand estimates were made for 1970 using the country studies, or using population and income forecasts for countries not included in this set. In some cases, past per capita consumption trends were simply extrapolated. Production was provisionally estimated by extrapolating past trends estimated from 1954–63 data. Areas and yields were extrapolated for some important commodities as a check. Trade and nonfood uses were also projected by extrapolating trends. The provisional estimates were then examined for consistency and adjusted, where necessary, by USDA country specialists in consultation with agricultural attachés. The numerical results of this study show individual countries and the world to be in balance in the sense that demand equals production plus imports minus exports, in all but a few cases. Diet-deficient countries, as a group, are shown to be net importers of grain at a level of about 24 million tons annually. They also imported over 5 million tons of milk products in 1970. However, for the remaining six commodity classes considered, the diet-deficient countries are net exporters.

Somewhat parallel to the USDA's *World Food Budget* were the two extensive studies by FAO: *Agricultural Commodities: Projections for 1970*[3] and *Agricultural Commodities: Projections for 1975 and 1985*.[4]

The two studies share essentially the same primary objectives: to provide estimates of trends and future prospects for world commodity markets in the context of world economic development, to assess the scale of worldwide demand for food and of the prospective food gaps, to aid in government planning in a developmental context, and to determine the interrelations between trade and aid policies. The 1970 projections study was oriented strongly toward gauging commercial trade prospects, while the later study was broadened substantially. Production and demand projections were made for 99 countries and for most major agricultural crops. Demand projections were made under two income and two population assumptions. The two population assumptions corre-

2. U.S. Department of Agriculture, Economic Research Service. *The World Food Budget, 1970*. U.S. Department of Agriculture, Economic Research Service, Foreign Agricultural Economic Report 19, 1964.
3. United Nations, Food and Agriculture Organization. *Agricultural Commodities: Projections to 1970; Special Supplement to Commodity Review*. Rome, Italy, 1962.
4. United Nations, Food and Agriculture Organization. *Agricultural Commodities: Projections for 1975 and 1985*, 2 vols. Rome, Italy, 1966.

sponded to UN's low and medium variant projections, while the income assumptions were determined on an essentially ad hoc basis after analyzing past trends, national development plans, past productivity changes, and natural resource bases. Both studies based their per capita demand projections on analyses of household consumption surveys in about 80 countries. In terms of the later study, 1975 production estimates for the world are very close to demand estimates for most commodities. This is true even though the projections imply that substantial excess agricultural capacity will remain in the developed countries in 1975. (Only demand estimates were prepared for the 1985 projections.) However, as in other studies, there are large deficiencies between domestic production and demand in the underdeveloped countries—deficits so large as to raise serious doubts about the possibilities for erasing them through international food shipments.

OECD ESTIMATES

The Organization for Economic Cooperation and Development (OECD) prepared two sets of food production and consumption estimates for major foodstuffs. One set was mainly for the OECD member countries, or roughly the most developed countries of the world.[5] The other set of projections was for the developing countries.[6] A main concern of these studies, initiated during the most recent "food scare" of the 1960s, was the potential and means of food aid furnished by developed countries to the developing nations.

The study for developed countries assumed that ongoing agricultural policies would continue, and that land in the United States would be returned to production. Population projections were the medium variant of the UN, except in a few cases where the birthrate assumption was modified. Consumption was projected on the basis of population and income elasticities of demand. Land for agriculture was adjusted to account for some shift to urban uses and to less extensive crops as farms adjust in size. Yields were projected mainly as linear trends based on observations of the previous 15 years. A different set of slope coefficients was sometimes used for 1975–85 and 1962–75, under the assumption that yields might approach an asymptote. The balances between projected production and consumption showed the OECD area and Oceania with capacity to increase output more rapidly than consumption by both 1975 and 1985. The area could export quite large quantities of additional grain and some greater amount of dairy products. While it would require some more grain for expanded beef

5. Organization for Economic Cooperation and Development. *Agricultural Projections for 1975 and 1985: Europe, North America, Japan and Oceania.* Paris, 1968.
6. Organization for Economic Cooperation and Development. *The Food Problem of Developing Countries.* Paris, 1967.

and veal consumption, the growing capacity of the total area to export grain elsewhere in the world would still be vast by 1985.

For the comparisons in developing countries, OECD used the UN's medium population variant and two levels of growth in agricultural productivity, 2.60 and 3.10 percent per year. Under the first productivity assumption, food demand was projected to exceed production by $8 billion per year in 1980. Under these assumptions, food demand was projected to grow at 3.25 percent and food production at 2.60 percent per annum in the developing countries. For the higher level of food growth, 3.10 percent, demand was projected to exceed production in the developing countries by $4 billion per year in 1980 under a food demand growth rate of 3.40 percent per annum. While developing countries are simply grouped in one aggregate (status estimates are not derived for individual countries and the dates are not the same), the two sets of OECD projections pose the possibility that the food import needs could be met through the excess supplies of developed countries up to the 1980s.

USDA ESTIMATES

Abel and Rojko made a further set of USDA estimates.[7] Under their estimates, grain import requirements of the less developed countries could be met from a U.S. grain acreage of 158 million acres (including that to meet domestic demand requirements) in 1970. This acreage compares to 150 million acres of grain for the United States in the 1964–65 crop year when approximately 55 million acres were idled under cropland retirement or supply control programs. Abel and Rojko indicate that if less developed countries increased production at only historical rates to 1980, their grain requirements could be fully met if the United States harvested 186 million acres of grain, compared to the 165 million acres of grain harvested in 1967. They project that the less developed countries will need to import 54.3 metric tons of grain by 1980. While they do not provide data for individual countries, a summary of their projections is included in Table 2.1.

A recent USDA study of the likely production, demand, and trade in grains in 1980 has used an internal equilibrium consistency-type model to predict ex post demands, supplies, trade, and the attendant price levels simultaneously.[8] The forecast is made, assuming that the world would seek to satisfy its requirements at least cost, that grain markets would be

7. Martin E. Abel and Anthony S. Rojko. *World Food Situation: Prospects for World Grain Production, Consumption and Trade.* U.S. Department of Agriculture, FAER-35, 1967.

8. Anthony S. Rojko, Francis S. Urban, and James S. Naive. *World Demand Prospects for Grain in 1980, with Emphasis on Trade by Less Developed Countries.* U.S. Department of Agriculture, FAER-75, 1971.

TABLE 2.1. World Grain Production and Trade, 1959–61 Average, 1964–65, and Projections to 1970 and 1980 (million metric tons)

Country or Region	1959–61 Average		1964–65		1970 Projection II[a]		1980 Projection II[a]	
	Production[b]	Net Imports	Production[b]	Net Imports	Production[c]	Net Imports	Production[c]	Net Imports
Less Developed Countries:								
India	67.6	4.0	73.6	6.6	87.0	6.0	106.1	16.5
Pakistan	15.5	1.2	17.5	1.6	20.6	2.2	27.4	3.3
Other Less Developed Countries, excluding Grain Exporters	105.7	15.5	118.0	20.8	138.0	22.5	180.0	34.5
Subtotal	188.8	20.7	209.1	29.0	245.6	30.7	313.5	54.3
Net Grain Exporters[d]	33.1	−9.4	45.2	−15.3	50.2	−14.9	68.0	−20.0
Total, Less Developed Countries	221.9	11.3	254.3	13.7	295.8	15.8	381.5	34.3
Developed Countries:								
United States[e]	170.4	−27.5	159.7	−38.2	217.1	−54.8	315.0	−109.5
Developed Exporters (less U.S.)[f]	59.5	−18.0	73.4	−26.6	92.5	−36.5	115.0	−42.5
Other Developed Free World	83.1	32.8	89.2	37.0	92.5	58.5	106.8	73.2
Eastern Europe (incl. USSR)	156.2	−0.1	171.6	7.5	192.6	4.3	230.2	1.2
Total, Developed Countries	469.2	−12.8	493.9	−20.3	594.7	−28.5	767.0	−77.6
Communist Asia	117.6	0.7	130.8	4.9	150.0	5.7	183.5	9.0
World Total	808.7	−0.8	879.0	−1.7	1,040.5	−7.0	1,332.0	−34.3

SOURCE: Martin E. Abel and Anthony S. Rojko. World Food Situation: Prospects for World Grain Production, Consumption and Trade. USDA, FAER-35, 1967.

[a] Calendar year basis.

[b] Year beginning July 1 for wheat and coarse grains, following calendar year for rice; negative numbers mean either a world surplus or an increase in stocks.

[c] For 1970, assumes 1954–66 rate of growth in production with normative evaluation of the impact of agricultural policies and development plans. For 1980, assumes production growth rates are 2.5 percent per year for grain importers, 3.1 percent per year for grain exporters, and 2.6 percent per year for less developed countries, using the above 1970 estimates as a base level of production.

[d] Argentina, Mexico, Burma, Cambodia, and Thailand.

[e] Grain production in the United States is based on harvested acreages of 150 million in 1964, 158 million in 1970, and 186 million in 1980.

[f] Canada, Australia, France, and Republic of South Africa.

cleared, and that excess demands would be satisfied out of trade and rising domestic production.

On the demand side of the analysis, the investigators develop demand *functions* in which the own price, per capita income, and population explain the quantity demanded. In addition, there are relationships describing the demands for carry-over stocks and a demand for feeding purposes for wheat and coarse grain (in which the substitution of wheat for coarse grain is allowed if relative wheat/coarse grain prices warrant).

On the supply side, the characteristic yield-area formulation of gap models is subsumed into a true supply function, in which quantity produced is determined by the own price and the prices of the competing grains. A trend variable allows for the inclusion of a systematic shift in the entire supply function under technological change.

The demand and supply relationships for all countries and regions are linked through relationships equating quantities supplied with quantities demanded and through the requirement that wholesale prices must be the same in a given region whether explained from the demand or supply side. In each of the 96 countries involved in the 22 areas (including Communist Asia), wholesale price–export price relationships are specified, with an added requirement that import prices and export prices must be equivalents up to the cost of shipping commodities both in and out of regions.

The system is solved simultaneously and normatively for the year 1980 after projections of population (one projection taken from UN-FAO 1966 estimates) and income (one projection taken from OECD and FAO modified for assumed goal attainment in some countries) are specified for the demand equations. The value of the trend variable is specified for the supply equations. The solution is summarized in Table 2.2.

The solutions arrived at permit the price-quantity-trade implications of five "policy" sets to be observed, including a "no-change-from-the-present" (case I) policy, a failure of remedial policy in the face of lower growth in productivity than projected in LDCs (case III), and a policy of aggressive agricultural development in LDCs (case II). Two policy alternatives are specified for developed countries as well as options to the case I solution (cases IIa and IIb).

The absence of net trade in the projections for World Total indicates that the international market for grain is cleared. The excess demands are met but at rising prices, if case III comes about, or falling prices, if the other cases occur. Thus, unlike the gap models, which would show the equivalent of case III as a wide deficit for LDCs compared with 1964–66, the consistency model shows LDCs' consumption rising relative to the base year (but not as high as if case II obtained) and imports meeting the excess consumption over supply at an interna-

TABLE 2.2. Projected Production, Consumption, and Trade of Grains, 1980, under Differing Assumptions (million metric tons)

Region	1964-66		1980 Case I[a]		1980 Case II[a]		1980 Case IIa[a]		1980 Case IIb[a]		1980 Case III[a]	
	Consumption	Net Trade	Consumption	Net Trade	Consumption	Net Trade	Consumption	Net Trade	Consumption	Net Trade	Consumption	Net Trade
Less Developed Countries												
—Traditional Importers	237.5	28.1	389.7	51.6	431.8	80.7	438.0	40.1	435.6	36.3	364.8	65.8
—Traditional Exporters	24.2	13.8	36.6	15.8	37.7	20.2	38.6	17.0	38.2	18.2	34.3	13.7
—Net total	261.7	−14.3	426.3	−35.8	469.5	−10.5	476.6	−23.1	473.8	−18.1	399.1	−52.1
Centrally Planned Economies (net)	327.2	−13.5	442.4	−1.7	442.7	−2.3	441.5	−5.9	440.8	−4.5	442.3	−1.3
Developed Countries												
—United States	144.1	44.5	203.6	51.4	210.4	36.0	226.3	57.4	218.5	60.2	199.1	61.4
—Net total	326.4	30.0	440.4	37.5	451.4	12.9	480.3	29.1	475.9	22.6	433.1	53.5
World Total	915.3	2.3	1,309.2	...	1,363.6	...	1,398.5	...	1,390.6	...	1,274.5	...

SOURCE: Anthony S. Rojko, Francis S. Urban, and James S. Naive. *World Demand Prospects for Grain in 1980, with Emphasis on Trade by Less Developed Countries.* USDA, FAER-75, Dec. 1971, pp. 78–79.

[a] Set I assumes a continuation of present food and fiber policies, allowing for moderate gains in productivity in the less developed countries. Set II assumes that agricultural productivity and economic growth in the less developed countries would be higher than projected in Set I. Set II-A assumes that major developed exporters would maintain their traditional share of the world market. Set II-B assumes that the major developed importers would become more sensitive to world grain prices and adjust their high internal prices to changes in world prices. Set III assumes that agricultural productivity and economic growth in the less developed countries would be lower than projected in Set I.

tional price that exceeds the 1964–66 world price. While the gap model would give cause for alarm because of the prospects of food shortages, the equilibrium model would give cause for alarm because of the rising cost of mere survival and the shortage of wealth available for investment in development that such a situation portends.

Numerous other summaries have been made of agricultural production capacity relative to food demand or requirements. A 1963 study by Pawley was a companion study to the *Third World Food Survey*.[9] This study was descriptive in nature and sought to determine in a qualitative way whether the natural resources of the world, if properly used, can provide an adequate level of nutrition for a greatly expanded population in coming decades. Pawley concludes that a favorable future balance is assured in the developed regions; that resources are more than ample in Latin America and Africa, but that additional efforts will be needed to mobilize them; that twentieth-century needs in the Near East will approach resource availabilities; and that a favorable balance between resources and future needs in the Far East is at least questionable.

Using the OECD estimates as a basis for his summary, Kristensen painted a deary outlook for the developing countries up to the year 2000.[10] He projected a mammoth food gap, with a burden on the developed countries in filling it. He argued that the developed nations have the comparative advantage as well as the greatest area of arable land in food production and that developmental assistance and world trade should be oriented accordingly. In this context, the developed countries should produce food and trade it for nonfood and labor intensive products of developing countries.

While other published analyses prevail, they generally are of a qualitative nature. During the last decade, and especially in the last few years, a number of books and articles urging alarm over the combined and complex set of variables represented by population growth, food needs, and the environment have been published. However, our selected review has been restricted to the more detailed and quantitative studies of projected food production and demand, and space has prevented review of some of these.

NATURE OF STUDIES

The studies reported here represent a detailed and extended analysis of food production, demand, and trade over future time periods. The basic projections were initiated concurrently with some of the studies

9. Walter H. Pawley. *Possibilities of Increasing World Food Production.* United Nations Food and Agriculture Organization Freedom from Hunger Campaign Basic Study 10, 1963.
10. Thorkil Kristensen. "The Approaches and Findings of Economists." *Proceedings of the International Conference of Agricultural Economists.* London: Oxford University Press, 1969, pp. 65–88.

reviewed above. However, since we incorporate considerably more detail and extend the analysis through a quantitative trade model, a greater burden of analysis and interpretation was involved.

In addition, the study in this analysis involves a detailed soils investigation. Whereas previous studies have considered additional land for cultivation on a more aggregate and trend basis, a specific soils study was made prior to the research reported in this book. As explained in a later chapter, one foundation for our research was employment of a soils specialist for nearly two years to measure, evaluate, and correlate all land in Africa, Latin America, and non-Communist Asia (except for the few excluded countries) that is not cropped, is cropped only under shifting cultivation, or handled in a similar manner. Then, soils which are not now cropped, or only intermittently cropped, were related to rainfall, water supplies, location, transportation, and roads. The potential yield of adapted crops under these circumstances was then estimated to determine (a) additional land which could be brought into production or be subjected to multiple cropping and (b) the upper bounds on cropped area (under two assumptions) in each individual country of the study.

The foundation soils study was made possible through Dr. Charles Kellogg, who arranged for access to restricted classified data for the world in the possession of the U.S. Department of Defense. A large amount of data and information was evaluated and correlated and included aerial surveys, climatological data, soil characteristics, transportation networks, human settlements, and other information. We believe that no other study incorporates such great detail in estimation of soils inventories and land restraints.

Also, as compared to some of the more aggregative projections of world food supplies and demands, our study incorporates considerable detail in the projection of production and demand (although other studies include an equal degree of specification detail in relating consumption to major demand variables and parameter estimates). Because of the nature of data available in individual countries, especially in the less developed countries but also in a large number of developed nations, it is impossible to apply an "econometric perfect" model to estimating supply and demand relations as a means for projecting world food demand and balances.

DESIRED AND ACTUAL METHODS

If data were available in sufficient quantities and desired forms for all countries of the world, we could imagine the development of an appropriate econometric or regression model based on time series and cross-sectional data. It would be formulated in a manner to allow equilibrium predictions of future food supplies and demands, both on a world and individual country basis. Prices and quantities for agricultural commodities and resources would be determined endogenously to the system. Supply and demand relationships for both the commodities and inputs of agriculture would be estimated simultaneously in a set of appropriate equations. Then, manipulating the exogenous and policy variables determined outside the system, and allowing endogenous variables (including export quantities and prices) to be jointly determined within the system, we could generate a system of endogenous quantities and prices to characterize demand and supply for the inputs and outputs and trade by countries, among countries, and for the world. But not only are data lacking for this approach in the majority of the countries we treat but also we even have difficulty in identifying the base and observations for projecting the exogenous variables in some. To be certain, we are forced to do so as we make our projections, but we doubt that the quality of the data, even that representing conventional exogenous variables, in some of the underdeveloped countries justifies methods of greater sophistication and predictional power than we use.

This study is devoted to projection of production and demand in 96 countries of the world. We do not project production in the sense of some system of equilibrium crop returns for farmers. Production is projected only on an individual crop basis and thus must be considered in this light. We use the term *surplus* or *deficit* in comparing production and demand for individual countries. The *surplus* or *deficit* so generated results only from the system of projection used and suggests the aggregative potential in a country. For example, in a particular country, we may show a large surplus of cereals and a small deficit of sugar and vegetable crops or vice versa. Quite obviously, we would expect farmers to shift land between the so-designated surplus and deficit crops. However, we do not aggregate further than the crop categories used later (which would thus "aggregate out" a surplus or deficit) because (a) we are only concerned with broad comparisons of projected production potentials and demand, and (b) this arithmetic can be accomplished by interested readers on the basis of commodity and county groupings of interest to them.

Similar qualification should be made with respect to our use of the term *demand*. We use the term as an abbreviation for our projection of consumption based on the two major variables, population and per capita income, which determine total demand. Hence, we do not use *demand* in terms of estimation of demand functions. Neither do we use it in terms of an estimated demand-supply equilibrium in which price elasticities are considered. The *demand* is the quantity of food commodities for human consumption projected in terms of population and income changes in the individual countries. Synonyms for the term *de-*

mand as we use it might be *requirements* or *consumption*. However, neither of these terms is any more correct than demand. Hence, for purposes of brevity and equal accuracy, we stick by the term demand.

OBJECTIVES OF ANALYSIS

The study has two major overall objectives. One is to project production and consumption for the majority of the world's population and agricultural producing areas. The other is to apply a trade or distribution model to certain production-demand comparisons generated for groups of countries. However, the trade model also applies to fertilizers as well as farm commodities. The projections of production and demand presented are for the years 1980 and 2000. The data generated for 1975 also are used in the trade or distribution phase.

OBJECTIVES IN PROJECTIONS

The present investigation seeks to encompass projections of area, yield, and production for most major crops; estimates of cropland expansion potentials; and demands estimated, using population and income projections together with consumption functions to assess future production-demand comparisons which might result if present trends continue. Detailed estimates such as these are needed if developing problem situations are to be recognized in time for remedial programs to be designed and implemented. Moreover, projections reflecting continued recent trends in production are especially important since they provide a means of gauging the amount by which present efforts must be stepped up to achieve future goals. This study was initiated to meet these needs. In meeting this objective, these steps were required: (a) Estimating recent time trends in crop area and yield, or production, for all major crops grown over the world, including in the coverage as many as possible of those countries which are significant producers of agricultural commodities. (b) Projecting estimated area and yield trends through the year 2000, subject to estimated upper bounds on cropland expansion in the underdeveloped countries; but otherwise assuming that the factors affecting recent trends will continue to affect them in the same way in the future. (c) Estimating possible future demand through the year 2000 for certain food commodity aggregates in each country under three alternative sets of population estimates, and under three alternative sets of future total income estimates. (d) Computing the surpluses or deficits implied in the future production and demand estimates for each commodity class, for each country, and for certain aggregates of countries. (e) Analyzing these future food production-demand comparisons in relation to their determinants to identify those factors which are critical in determining the outcome. (f) Interpreting the comparisons in terms of the magnitudes of adjustments in agricultural productivity, demand, or trade which would be required to bring about acceptable production-demand balances.

To attain the major study objectives, broad geographic coverage of world agriculture was important. The degree of coverage in the study was determined primarily by data considerations. No country was included unless past trends in either area and yield or production for a significant share of its crops could be documented with data of minimally acceptable quality. On this basis, 96 countries are included in the study. The total number of crops and agricultural commodities represented is 73, although the most which appear in the analysis for any individual country is 41. The only country omitted from the basic projections and which is of real significance to world totals is Communist China. Surely, omission of a country with one-fourth of the world's population is damaging to the global aspects of the results, but data on Chinese agriculture are inadequate for certain facets of this study. The analysis for many countries in Africa was carried out in less detail for similar reasons, but results are included. However, estimates for China are included in one part of the trade analysis.

No attempt has been made to evaluate nutritional aspects of the diets which might result if the projected production and demand figures were to be realized. Neither has any attempt been made to determine future levels of production, domestic demand, and trade which would reduce excess demand to zero for every commodity, country, and time. A balance will exist, of course. Moreover, if such a grand plan could be specified on grounds which were normative in some universal sense, it would certainly be invaluable. The aim of this study is much more modest and relates only to evaluating the consequences of a continuation of recent trends, and some variants about them, to identify possible future disparities; determining their causes; and judging the magnitudes of some possible adjustments which might resolve them. Comparisons are made only for primary agricultural products or crops. Demand for livestock products is projected as a basis for estimating feed use of the crops. However, no judgment is made as to whether these livestock products will be produced domestically or imported (or whether the demand will even be fulfilled); but the derived feed demand is a component of domestic demand in any case. Food and industrial demands are estimated directly, as is allowance for seed and waste.

OBJECTIVES OF TRADE MODEL

The term *trade model* is used for that part of the study that relates to the distribution of major food commodities and fertilizer materials among surplus and deficit producing regions of the world. The distribution results from a least-cost optimization by means of a linear programming model. Costs involved in the optimization process are those of transportation and commodity acquisition in the various countries. The linear programming model is applied for surpluses and deficits among regions and countries for the years 1975, 1980, and 2000. It is applied, however, only to a specific combination of population, income, and land variants which allow feasible solutions.

In one sense the programming model can be considered a simulation of trade patterns under competition and lack of trade barriers. The principal reason for its application is to examine the pattern of trade which might be optimal in this context in meeting future world food requirements. We recognize, of course, that the policies of individual countries and international organizations such as trading blocks and common markets are not likely to allow short-run prevalence of these conditions.

AUXILIARY OBJECTIVES

We are interested in the above objectives in terms of evolving problems and potential policies of countries concerned with increasing food production and trade to meet future world food demands and requirements. Hence, in later chapters, we interpret projected production trends in terms of recent developments in new varieties and auxiliary technologies. Too, in closing, we summarize some policy needs to attain these potentials.

CHAPTER 3

Methodology for
Demand Projections

THE METHODS USED in projecting food demand by countries necessarily must conform to the data available. Were ideal data available, logically consistent econometric models, based on time series and family budget sources, might have been used. While appropriate data in these forms are available in certain developed countries, generally they are lacking in the less developed nations. Hence, if demand projections are to be made, as certainly they must if enlightened outlook and intelligence are to be available for these important variables, we are required to use methods which conform to the types of statistical observations available and which allow acceptable estimates. We realize that the models which follow hardly conform to those theoretically desired under a regime of ideal data. However, those we use can be applied uniformly over all nations, represent perhaps the maximum quantitative sophistication justified by the data of many countries, and appear to be acceptably efficient for the analysis at hand.

BACKGROUND OF PROJECTIONS

Demand projections are made for the following nine product classes: cereals, sugar (raw value), starchy roots, pulses, vegetables and fruits, oil crops, meat, milk, and eggs. Corresponding production figures also are projected. For both production and demand estimates, it was necessary to project certain basic or "exogenous" variables for the future. These projections are based on time series observations in most cases and are estimated by conventional statistical methods. As nearly as possible, the projections for production represent an extension of past trends. For reasons discussed later, we do not hold so closely to "continued recent trends" in the case of demand projections, although trends in income variables are rather closely related to relationships of the recent past.

Least-square estimation techniques are used throughout, but primarily as a means of extracting existing information from the series of discrete historical data and to incorporate it into a single, manageable

31

parametric function of time. Only limited emphasis is placed on the statistical properties of the resulting functions. Examination of a few confidence intervals for predicted values of the dependent variables ten or twenty years in the future suggests that the resulting least-square estimates cannot be justified alone on their statistical merits.

Only functions which are monotonic over the projection period were considered. While others may provide better fits in some cases, they typically are inappropriate for projection purposes and in documenting agricultural development in underdeveloped countries under the paucity of published statistics which exists. The specific set of functions used and the criteria for their selection are discussed later. The most important consideration in the selection of methodology was the quality of available data. Necessary data often do not exist. Limited data-gathering capabilities in underdeveloped countries tend to concentrate on commodities important in foreign trade or on those that move through domestic commercial markets. Important subsistence crops are omitted from statistical records or are estimated with extremely crude methods. Two or more agencies in a country frequently publish quite different official statistics on the same variable. If these data estimated by questionable techniques simply had large measurement errors with zero means, their effect would be of less importance. However, the methods used tend to produce errors systematic through time and caution must be exercised in interpreting the estimates which result.

Growing realization of the need for improved data has resulted in an expanding geographic coverage of crop production statistics in many underdeveloped countries. This is a favorable tendency. However, when two discrete sets of historical statistics on crop area are combined without adjustments for the different coverage, a completely spurious time trend results. Taxation is sometimes based on agricultural production while at the same time output is calculated from questionnaires sent to farmers or village leaders. The resulting estimates sometimes are only aggregate guesses of the village leader. Still other countries take national pride in showing high production and consumption, or a definite upward trend in these quantities. These and other factors require extreme care in selection and analysis of much data. However, to meet the objectives stated earlier, maintaining broad geographic and crop coverage, it is necessary to use such data with whatever correction is possible.

COUNTRIES USED IN THE ANALYSIS

The countries included in the study are listed in Table 3.1 (column 2). They are grouped under the regional aggregates indicated in the first column. The same countries also are aggregated into three groups according to average per capita income levels (low, medium, and high).

TABLE 3.1. Countries Included by Region and Income Class

Region	Country	Income Class
United States	United States	high
Canada	Canada	high
Mexico	Mexico	medium
Central America	British Honduras	medium
and Caribbean	Costa Rica	medium
	Cuba	medium
	Dominican Republic	low
	El Salvador	low
	Guatemala	low
	Haiti	low
	Honduras	low
	Jamaica	medium
	Nicaragua	low
	Panama	medium
	Trinidad and Tobago	medium
Brazil	Brazil	low
Argentina and Uruguay	Argentina	medium
	Uruguay	medium
Other South America	Bolivia	low
	Chile	medium
	Colombia	low
	Ecuador	low
	Paraguay	low
	Peru	low
	Venezuela	medium
Northern Europe	Austria	high
	Belgium and Luxembourg	high
	Denmark	high
	Finland	high
	France	high
	Ireland	medium
	Netherlands	high
	Norway	high
	Sweden	high
	Switzerland	high
	United Kingdom	high
	West Germany	high
Southern Europe	Greece	medium
	Italy	medium
	Portugal	medium
	Spain	medium
Eastern Europe	Bulgaria	medium
	Czechoslovakia	medium
	East Germany	high
	Hungary	medium
	Poland	medium
	Rumania	medium
	Yugoslavia	medium
USSR	USSR	high
North Africa	Algeria	low
	Ethiopia	low
	Libya	low
	Morocco	low
	Sudan	low
	Tunisia	low
	United Arab Republic	low

TABLE 3.1 (cont.)

Region	Country	Income Class
West Central Africa	Angola	low
	Camaroun	low
	Congo (Kinshasa), Rwanda, and Burundi	low
	Ghana	low
	Guinea	low
	Ivory Coast	low
	Liberia	low
	Nigeria	low
	Senegal	low
	Sierra Leone	low
	Togo	low
East Africa	Kenya	low
	Malagasy Republic	low
	Malawi, Rhodesia, and Zambia	low
	Tanganyika[a]	low
	Uganda	low
Republic of South Africa	Republic of South Africa	medium
West Asia	Cyprus	medium
	Iran	low
	Iraq	low
	Israel	high
	Jordan	low
	Lebanon	low
	Syria	low
	Turkey	low
India	India	low
Other South Asia	Ceylon	low
	Pakistan	low
Japan	Japan	medium
Other East Asia	Burma	low
	Cambodia	low
	China (Taiwan)	low
	Federation of Malaya	low
	Indonesia	low
	Philippines	low
	South Korea	low
	South Vietnam	low
	Thailand	low
Oceania	Australia	high
	New Zealand	high

[a] Now part of Tanzania.

The classification of countries by income levels is given in column 3. Certain entries under the "country" heading are groups of countries. The later analysis is by aggregates of countries grouped under these regional and income designations.

POPULATION PROJECTIONS

In the analysis which follows, three alternative population projections are derived from estimates published by the UN.[1] Recent trends

1. United Nations, Department of Economic and Social Affairs. *Provisional Report on World Population Prospets as Assessed in 1963.* New York, 1964.

in total national real income for each country are estimated from historical data and two alternative future trends are established. Population and total income projections are combined to estimate future per capita income by countries. Using these per capita income and population projections, projected demand for nine aggregates of food commodities is estimated via a consumption function and a base period income elasticity estimate specific to the commodity class. Prices are assumed constant throughout. Future demands for nonforage livestock feeds are then derived from estimated future demand for livestock products. No demand projections are made for agricultural commodities which are not used for food or livestock feed.

The UN estimates of population were the result of two sets of analyses. First, all available recent projections of populations for individual countries were assembled and analyzed. A standard set of assumptions, or a demographic model, was developed and individual country projections through 1980 were adjusted, where necessary, to conform to the norm. In the second phase, the UN model was used to estimate low, medium, and high variant projections for 24 regional totals encompassing the world.[2] Such estimates were prepared for the overall period 1960–2000 by 10-year intervals. The model also was used for country projections in cases where other sources were lacking. Individual country estimates through 1980 correspond to the medium variant.

Similar mortality assumptions were used under all variants of the projections. A gain in life expectancy of one-half year per year is assumed until life expectancy reaches 55 years. Thereafter, it is assumed to rise more rapidly until a level of 65 years is reached and then asymptotically to approach 73.9 years.

United Nations projections for the medium level of population in individual countries for 1975 were used directly. Low and high population projections for 1975 were estimated by decomposing the published regional totals into country components. The regional total for the forward year was generated as a proportion equal to the percent of the medium variant regional total allocated in individual countries (in terms of their base period population). This procedure also was applied to 1980 population projections under each assumption for individual countries. The 1980 estimates were not used directly, but they were required for further computations.

The UN study did not provide individual country estimates for years after 1980. Furthermore, country populations within the same region frequently grow at quite different rates; each country's population is a changing proportion of its regional total in successive time

2. Actually, a fourth variant titled "continued recent trends" also was estimated, though it is not included here. The continued recent trends variant assumes constant fertility at recent levels for all countries through 2000 and mortality declining throughout the world at recent rates. For comparison, the world population estimates in 2000 under the low, medium, high, and continued recent trends variants are 5,296 million, 5,965 million, 6,828 million, and 7,410 million, respectively.

periods. It is necessary to decompose the regional totals in a way which
accounts for these differing growth rates. The procedure used assumes
that the rates of population growth for countries within a region maintain the same proportions as those implied between the published 1975
and 1980 medium variant projections. This concept allows a computational process which yields country estimates but maintains consistency
with UN regional totals. The method used can be explained by means
of these symbols and equations:

$N_{i80}{}^k$ = 1980 population estimate for country i in region k.
$N_{\Sigma 90}{}^k$ = 1990 total population for region k.
$N_{\Sigma 100}{}^k$ = 2000 total population for region k.
n^k = number of countries in the kth region.
$r_i{}^k$ = percentage change in population between 1975 and 1980 for
country i in region k under the medium variant projection.
$c_{85}{}^k$ = proportionality constant unique to region k over the period
1980–90.

Equation 3.1 is solved for $c_{85}{}^k$ and individual country estimates for
1985 and 1990, N_{i85} and $N_{i90}{}^k$, are computed as in equations 3.2 and 3.3.[3]

$$N_{\Sigma 90}{}^k = \sum_{i=1}^{n_k} N_{i80}{}^k(1 + c_{85}{}^k r_i{}^k)^2 \qquad (3.1)$$

$$N_{i85}{}^k = N_{i80}{}^k(1 + c_{85}{}^k r_i{}^k) \qquad (3.2)$$

$$N_{i90}{}^k = N_{i80}{}^k(1 + c_{85}{}^k r_i{}^k)^2 \qquad (3.3)$$

Estimates for 2000 are computed by substituting $N_{\Sigma 100}{}^k$ for $N_{\Sigma 90}{}^k$, and
$N_{i90}{}^k$ for $N_{i80}{}^k$ in equation 3.1, determining a new proportionality constant, $c_{95}{}^k$, and similarly "updating" equation 3.3. This procedure is re-
peated for all three population variants. The same $r_i{}^k$ is used for the ith
country in all calculations, but values of $N_{i80}{}^k$, $N_{\Sigma 90}{}^k$, and $N_{\Sigma 100}{}^k$ appropriate for each variant are used to compute two values of the propor-
tionality constant.

In some cases the UN study provided population projections for individual countries through 2000 and it was possible to use them directly.
Korea, China (Taiwan), India, and Pakistan are examples, although
the estimates beyond 1980 are for only 10-year intervals. In such cases,
a geometric mean of 1980 and 1990 estimates was used to derive the
1985 population. The UN projections do not distinguish between the
southern and northern portions of Korea and Vietnam. However, since
only the southern regions are included, the country totals have been

---

3. Equation 3.1 is a quadratic in $c_{85}{}^k$, and for nonnegative $r_i{}^k$, the roots are real
and of opposite signs. The positive one is selected.

decomposed by assuming in each case that the two component regions will grow at the same rate as the total. Results from these calculations and other basic results will be presented after other methodological approaches have been discussed.

INCOME PROJECTION

The projections of total income upon which estimates of food demand are partially based represent extrapolations of past trends. For all but 12 out of the 96 countries studied, it has been possible to assemble historical time series data on an income variable, and to base estimates of future trends on such data. The 12 for which data were lacking included British Honduras, Angola, Cameroun, Ethiopia, Guinea, Ivory Coast, Liberia, Libya, Malagasy Republic, Sierra Leone, Togo, and Jordan. Trends for these countries are based on income growth in nearby countries judged to be similar, and in a few cases, some inferences from scattered historical data. Data required for most other analyses in these countries are available. Hence, we assume that a crude income indicator, in light of other data available, justifies inclusion of the 12 countries.

Time series on total personal consumption expenditures were assembled wherever possible. However, since these data were not always available, personal consumption expenditure is the income variable used for 57 countries. Other income variables used include national income (9 countries), gross domestic product (9 countries), individual consumption (4 countries), net material product (3 countries), and net domestic product (2 countries).

Time series collected were of varying length, but all were relatively short.[4] The commonest beginning year is 1950, and the year of the most recent observations used is 1964. For some countries, longer series of useful data could have been assembled, but because of World War II disruption, and overall poor quality in the earlier data for the underdeveloped countries, the series were limited to the postwar years.

Income series were measured in local currencies. These were not transformed into standard units since the introduction of income into the demand projections requires only that relative changes in income be specified. The data series were deflated with cost of living indices before proceeding.

---

4. The following data sources were used:
United Nations, Department of Economic and Social Affairs. *Yearbook of National Accounts Statistics, 1958, 1959; Yearbook of National Accounts Statistics, 1964, 1965; Yearbook of National Accounts Statistics, 1965, 1966; International Monetary Fund; Supplement to 1965/66 International Financial Statistics.* Washington, D.C., 1966.
United Nations, Statistical Office. *Monthly Bulletin of Statistics,* Jan., 1966.
Eugene A. Brady. "1950–64 Revised Data on Private Consumption Expenditures in Peru." Unpublished. Iowa State University, Department of Economics, Ames.

Income and price measurements have numerous deficiencies. A large amount of economic activity in underdeveloped countries falls outside market channels, and income measurements may be biased accordingly. Prices pose similar problems. In some cases, deflators available reflect only prices of urban wage goods.

REGRESSION MODELS FOR INCOME PROJECTIONS

Least-square regression techniques were used to estimate time trends in total income. The selection of appropriate functional forms to represent income paths is more difficult than for variables underlying agricultural production. After income data of each country were plotted and the variety of patterns observed, it was apparent that growth paths which increase at accelerating rates are common.

Projections based on estimated functions, $f(t)$, for which $d^2f/dt^2 > 0$, have an "explosive" time path, and where estimates are based on only a few time series data, projected values are very sensitive to observations off the trend line. In view of the uncertainty about the appropriateness of individual forms (and about the future path of income), two alternative income growth paths are specified for each country.

Wherever possible, the two variants have been specified as (a) a low variant represented by a function linear over the projection period and (b) a high variant based on a function which is exponential over the period. Exceptions were made, however. In a few cases, historical income observations appear to increase at decreasing rates. For these countries, a function for which $d^2f/dt^2 < 0$ is used for one of the two alternative growth paths. In still other cases, the estimated functions are judged to be unduly influenced by wars, political disturbances, and extraordinary economic experiences during the years when the observations were generated. Both of the two income trends then are specified on a priori grounds. The functional forms attempted are:

$$Y_t = a + bt + \varepsilon \tag{3.4}$$
$$Y_t = a + bt + ct_d + \varepsilon \tag{3.5}$$
$$\ln(Y_t) = \ln(a) + bt + \varepsilon \tag{3.6}$$
$$\ln(Y_t) = \ln(a) + bt + ct_d + \varepsilon \tag{3.7}$$
$$Y_t = a + b\sqrt{t} + \varepsilon \tag{3.8}$$

In these equations, t is the year minus 1900; Y_t is deflated total income in year t; t_d in a "dummy" time variable assigned a value 0 for $t < d$, and $t - d$ for $t \geq d$; ε is an error term; and a, b, and c are constants to be estimated. Equations 3.5 and 3.7 appear as curves composed of two segments joined at year d. Three alternative values of d were prespecified: 53, 55, and 57. In equation 3.5, the two segments are linear in form, and throughout the projection period the curve appears as a continuos linear function with intercept $a - cd$ and slope $b + c$. The two segments in equation 3.7 are exponential, but the curve is a continuous exponential function throughout the projection period with a multiplicative constant ae^{-cd}, and (constant) cumulative growth rate $b + c$. Functions described by equations 3.5 and 3.7 are used in a limited number of cases where income growth in recent years (that is, in years after d) is influenced by forces unlike those of earlier years but where it is desired to maintain the basic linear or exponential function over the total projection period. An example of a discrete underlying change is recovery during a postwar period.

The final selection of two functions to represent alternative future income growth requires a judgment based on the statistical properties of estimated functions, inspection of plotted observations, known discrete historical occurrences which have affected the data, and values projected by each function. Obviously no single empirically objective method can be used in selection, nor does any truly objective procedure exist for the selection. However, even though two parametrically different functions may be used for each country, and though the resulting projections may be quite different, fundamentally both are determined by trends revealed in the historical observations.

PER CAPITA INCOME PROJECTIONS

A total of nine different food demand constellations were estimated for each of the 96 countries at each of three future time periods. The nine varied according to population and income growth assumptions. Three were estimated assuming constant per capita income, but with population alternatively based on low, medium, and high growth rates. Food demand growth then is proportional alone to population. Six additional demand patterns were estimated by combining first a low growth rate for total real income and then a high growth rate for this variable each with the three above population rates.

Total income and population are projected independently and per capita income is computed as the ratio between the two. Hence, the methodology used excludes possible dependence between population and income variables, reflecting especially population growth which expands the labor supply and influences output and income accordingly. This assumption is used by Coale and Hoover, though they are more specific in claiming that rapid population growth lowers saving and diverts a high proportion of investment to nonproductive forms.[5] But this linkage is tenuous in countries where underemployment is high and labor productivity is low.

5. Ansley J. Coale and Edgar M. Hoover. *Population Growth and Economic Development in Low-income Countries.* Princeton, N.J.: Princeton University Press, 1958.

CONSUMPTION FUNCTIONS

In the demand projections based on population and income variables, changes in consumer preference are ignored. Demand for forage and some industrial crops is not included. Industrial crops used for food and nonforage livestock feeds are included in the projections.

Projection of demand for each commodity is carried out in two steps: per capita demand is first projected on the basis of estimated per capita real income; the resulting figure then is multiplied by projected population to provide a total direct demand estimate. The relation between per capita income and food demand has been studied by FAO and OECD, based on a large number of household surveys and time series data from different countries. The FAO analysis investigated the appropriateness of several types of consumption functions for different food groups.[6] Two findings of the FAO study were adapted for use: (a) a set of estimates of income elasticity of demand for various food groups in most countries of the world, and (b) a specification of consumption function forms judged appropriate for long-term demand projections in the same countries and for the same food groups. The elasticity estimates were not all for the same base year, but the dates range over only three years, 1959–62. Projected income changes are often large, and the methodology for projecting per capita demand must consider the declining marginal propensity of consumption for individual foods. All but one of the consumption functions used have declining income elasticities at higher income levels.

The projection procedure used here assumes the following: (a) At some base period, observed per capita consumption and income values constitute a point on a consumption function. (b) The consumption function is one of the mathematical forms specified in FAO's investigation. Given estimates of per capita income and consumption and estimated income elasticities at the same income-consumption point, it is possible to estimate future consumption for other income levels. Admissible consumption functions are those with two parameters. With specification of the functional form, the algebraic equation for the income elasticity can be derived and set equal to the estimated base period elasticity. This equation, together with base period income and consumption levels entered in the consumption function, constitutes a set of two equations in two unknown consumption function parameters. The set is then solved and future estimates consistent with the above assumptions are made.

The four alternative forms of consumption functions used are

6. United Nations, Food and Agriculture Organization. *Agricultural Commodities: Projections to 1970; Supplement to Commodity Review.* Rome, Italy, 1962.
United Nations, Food and Agriculture Organization. *Agricultural Commodities: Projections for 1975 and 1985,* 2 vols. Rome, Italy, 1966.

shown below. The corresponding expressions for income elasticity, μ, and projected consumption follow each consumption function.

$$C = aY^b \tag{3.9}$$
$$\mu = b$$
$$C_t = C_0 Y_t^{\mu} Y_0^{-\mu}$$

$$C = a + b\ln(Y) \tag{3.10}$$
$$\mu = bC^{-1}$$
$$C_t = C_0[1 + \mu\ln(Y_t Y_0^{-1})]$$

$$C = \exp(a - bY^{-1}) \tag{3.11}$$
$$\mu = bY^{-1}$$
$$C_t = C_0 \exp[\mu(1 - Y_0 Y_t^{-1})]$$

$$C = a - bY^{-1} \tag{3.12}$$
$$\mu = bC^{-1}Y^{-1}$$
$$C_t = C_0[\mu(1 - Y_0 Y_t^{-1}) + 1]$$

C and Y, respectively, are per capita consumption and income. Variables with a zero subscript denote base period values and those with t denote a future value. The symbols a and b are consumption function parameters.

ESTIMATES OF OTHER CONSUMPTIVE USES OF CROPS

The computational procedure discussed above requires fairly complete data on domestic utilization of agricultural products in a base period. Such data were available as an average for the 1959–61 period in the form of the U.S. Department of Agriculture food balance sheets for 91 countries or groups of countries included in the present study.[7] Five others (Cambodia, Senegal, South Korea, South Vietnam, and Uganda) have been added. Estimates of total domestic disappearance of food commodities were made for these five countries, using available production and trade data.

For projections which follow, it was necessary to convert the USDA food balances into corresponding aggregates and to adjust them to re-

---

7. U.S. Department of Agriculture, Economic Research Service. *Food Balance for 24 Countries of the Western Hemisphere, 1959–61.* U.S. Department of Agriculture, ERS-Foreign 86, 1964; *Food Balance for 16 Countries of Western Europe, 1959–61.* U.S. Department of Agriculture, ERS-Foreign 87, 1964; *Food Balance for 12 Countries of the Far East and Oceania, 1959–61.* U.S. Department of Agriculture, ERS-Foreign 88, 1964; *Food Balance for 30 Countries in Africa and West Asia, 1959–61.* U.S. Department of Agriculture, ERS-Foreign 119, 1965; *Food Balance for 8 Countries in East Europe, 1959–61.* U.S. Department of Agriculture, ERS-Foreign 124, 1965.
Charles A. Gibbons. "Food Balance for the United States, 1959–61." Unpublished. U.S. Department of Agriculture, Economic Research Service, 1967.

flect total domestic disappearance of all foods in terms of primary agricultural products. Except for sugar and meat, all commodities are expressed in 1,000 metric tons of unprocessed products. Three categories of end use are identified for each product class: food and industrial, feed, and seed and waste.[8] No attempt was made to project industrial uses separately, and industrial demand is assumed to grow at the same rate as food demand.

It was necessary to estimate the oil crop category separately since USDA balances include only the vegetable oil portion of such commodities, while other uses are included in this study. The allocation to feed and food uses presents conceptual problems since most such products are physically fractioned and the components are subject to dissimilar demand structures. The procedure used was to allocate the total, less estimated seed and waste, to food and feed uses according to the oil and cake fractions, respectively, after accounting for any direct food consumption. Grapes used for wine (in grapes equivalent) also are included.

In adjusting base period balances to reflect total domestic disappearance measured as primary product equivalents, seed and waste were considered as part of domestic disappearance. In an ex post sense they actually arise in the process of meeting a demand which may or may not be wholly of domestic origin. Thus for countries which are net food exporters in the base period, seed and waste are adjusted downward while importers are charged for more seed and waste than are realized domestically. The computation procedure is based on the assumption that seed and waste are associated with production only. Adjusted seed and waste for the rth produce class in the base period, SW'_{r0}, is computed as:

$$SW'_{r0} = SW_{r0}DD_{r0}P_{r0}^{-1} \qquad (3.13)$$

The symbols DD_{r0}, P_{r0}, and SW_{r0} refer to domestic disappearance, production, and seed and waste, respectively, all for commodity r in the base period.

Projected demand for the rth commodity contains only two components: food and industrial demand, and feed demand. However, the projected values are defined to include an allowance for the seed and waste indirectly incurred in meeting the domestic demand (i.e., a fixed proportion of deliveries to demand). No reductions are made for waste and seed requirements as yields are improved. Consistent with this

---

8. The USDA food balances record an entry in a category of "non-food industrial use" when frequently the commodity actually reappears in the balance sheet as "production" of a processed food product. At other times, quantities so designated are actually destined for nonfood uses. Appropriate adjustments were made in this study to avoid double counting whenever the need was apparent.

specification of demand, base period values of feed use, F_{r0}, and food and industrial use, FI_{r0}, are adjusted as in equations 3.14 and 3.15.

$$F'_{r0} = SW'_{r0}F_{r0}/ (F_{r0} + FI_{r0}) + F_{r0} \qquad (3.14)$$

$$FI'_{r0} = SW'_{r0}FI_{r0}/ (F_{r0} + FI_{r0}) + FI_{r0} \qquad (3.15)$$

FEED DEMAND

Certain parameters developed for projecting feed demand required to complete the adjustments on base period balances now are explained. Future feed demand is based on projected demand for livestock products. An index of total feed concentrate demand is computed for each country and at each time when a production-demand comparison is to be made. The index, I_t, is computed as a weighted sum of projected demand for the three livestock product aggregates in year t:

$$I_t = w_7 A_{7t} + w_8 A_{8t} + w_9 A_{9t} \qquad (3.16)$$

Subscript values 7, 8, and 9 represent meat, milk, and eggs, respectively. A_{rt} is the estimated future domestic demand for the rth class of livestock product in year t. The weights, w_r, are proportional to concentrate requirements per 1,000 metric tons production of the rth livestock product aggregate and are scaled so that $I_0 = 1.0$.

Weights for countries of Northern and Mediterranean Europe and for Canada were based on UN data.[9] Weights for Eastern European countries and for the USSR were assumed proportional to those for Mediterranean Europe, and those for Oceania were assumed proportional to weights for Canada. Analogous estimates for the United States were derived from USDA data.[10] For underdeveloped countries where few usable estimates were available, feed requirements per unit of production were assumed proportional to calories for human consumption per 1,000 metric tons for each livestock product aggregate (in farm weight).

As is evident from the procedure, internal consistency requires only that the weights, w_r, for a given country maintain proper proportionality among themselves; absolute values are not crucial. A constant mix of feeds in the livestock "diet" is assumed and future feed conversion rates are implicitly assumed constant at base period levels. Finally, no explicit measures are taken to segregate quantities of feed fed to draft animals. Needed data generally are not available and draft animals fre-

---

9. United Nations, Food and Agriculture Organization. *Agricultural Commodities: Projections to 1970; Special Supplement to Commodity Review.* Rome, Italy, 1962.

10. Earl F. Hodges. *Livestock-Feed Relationships, 1909–1963.* U.S. Department of Agriculture, Economic Research Service Statistical Bulletin 337, 1963.

quently serve other purposes in underdeveloped countries. Some are milked, and ultimately many contribute to meat supplies.

Returning to adjustments on base period food balances, the livestock aggregation weights are used to inflate base period feed estimates, F'_{r0}, of countries which are net importers of livestock products, and to deflate the feed estimates of countries which are net exporters of such products. A multiplicative adjustment factor, k, is computed for each country as illustrated in equation 3.17. Final adjusted feed-use figures for the base period, F''_{r0}, are computed as in equation 3.18 and the final value for adjusted total domestic disappearance is computed as in equation 3.19.

$$k = \sum_{r=7}^{9} w_r A_{r0} \Big/ \sum_{r=7}^{9} w_r P_{r0} \tag{3.17}$$

$$F''_{r0} = kF'_{r0} \tag{3.18}$$

$$DD'_{r0} = FI'_{r0} + F''_{r0} \tag{3.19}$$

When these adjustments are completed, the resulting balances differ considerably from the original data. Major exporters of crops exhibit reduced domestic disappearance by virtue of seed and waste "exports." Food, industrial, and feed demands of all countries are inflated by their pro rata shares of adjusted seed and waste. Finally, countries which import (export) livestock products show increased (reduced) feed allocations.

Using base period food and industrial use estimates in their adjusted forms, base period and projected population estimates, and base year and projected income estimates, future food and industrial demands are projected for all nine commodity aggregates, using the procedure described above.

The feed component of demand must also be projected for estimates of total future demand. As a step in this direction, consider again the aggregation weights. For meat and eggs, all future demand arises from food and industrial uses, and these are projected in the manner just described. The process is illustrated in equation 3.20 where Y_t and N_t, respectively, are income and population in year t. However, milk is

$$A_{rt} = FI'_{rt} = f_r(Y_t, N_t); \quad r = 7,9 \tag{3.20}$$

used for both food and feed in many countries and total demand must be expressed as in equation 3.21.

$$A_{8t} = FI'_{8t} + F''_{8t} = f_8(Y_t, N_t) + F''_{8t} \tag{3.21}$$

At any time t, feed demand for the rth class of food (including milk) will be estimated as in equation 3.22 (where $I_0 = 1.0$).

$$F''_{rt} = I_t F''_{r0}; \quad r = 1, 2, \ldots, 9 \tag{3.22}$$

Substitution of equations 3.20, 3.21, and 3.22 into equation 3.16 forms equation 3.23.

$$I_t = w_7 f_7(Y_t, N_t) + w_8 f_8(Y_t, N_t) + w_8 I_t F''_{80} + w_9 f_9(Y_t, N_t) \tag{3.23}$$

Finally, we can collect terms and rewrite equation 3.23 as equation 3.24,

$$I_t = w'_7 f_7(Y_t, N_t) + w'_8 f_8(Y_t, N_t) + w'_9 f_9(Y_t, N_t) \tag{3.24}$$

where $w'_r = w_r/(1 - w_8 F''_{80})$. The resulting weights, w'_r, account for functional circularity in the role of milk in the same way that interdependence coefficients in an input-output model account for the dual role of all commodities in that context. Milk is required to produce feed and feed is required to produce milk.

All feed demands, then, are projected using I_t as computed from equation 3.24 and then F''_{r0} as illustrated in equation 3.22. Total domestic demand is estimated by adding the result to food and industrial demand.

CHAPTER 4

Methodology for
Production Estimates

THE METHODS USED in projecting production are less complex, even though involving a greater computational burden, than those for consumption. As in the case of demand, theoretically optimal projections would be based on a set of supply and demand relationships in which quantities and prices would be taken as jointly determined variables. As well as demand and supply relations for food commodities, demand and supply functions also would be included for farm resource inputs. Obviously, however, these types and forms of data are not available for the less developed countries and we must resort to other methodology if consistent methods are used across all countries. Production projections made from time series data and presented in the next chapter, but qualified later in the chapter relating to such breakthroughs in technology as the green revolution, entail only extrapolation from past trends in yields and an adjustment considering additional land which might be made cultivable. Projections of production thus are in terms of potentials, if past trends extend to the future and cultivable land is devoted to crops.

Area and yield trends are estimated where data are available. In other cases, projections are based on trends estimated from production data only. The area trends so estimated are extrapolated subject to a constraint on total cropland in each country. Preparatory to making production-demand comparisons, the resulting production projections are aggregated into six commodity classes corresponding to the crop classes defined in Chapter 3.

ESTIMATES OF TIME TRENDS FOR AREA, YIELD, AND PRODUCTION

Estimation of production trends was initiated with assembly of data for area, yield, and output of all individual crops in all countries of the world for which usable data were available. The next chapter reports results of this analysis for over 3,000 individual sets of such data. Because of their numbers, it is impractical to include precise documentation of sources for each data set. Instead, a general description is given

46

of the priorities applied in selecting data sources and of the overall sources.

All basic data on which estimates are based were drawn from sources published by either USDA or FAO.[1] The FAO data are broader in both geographic coverage and number of crops. Data from statistical reporting services of individual countries were not used because the same data, with very few omissions of significant crops, are available in FAO publications. However, for certain countries of the world, the USDA estimates were considered more reliable. A priority is thereby established: with other considerations equal, USDA data were used whenever available.

With the exception of the United Arab Republic, projections for African countries are based on production trends only. An effort was made to project production on an area and yield basis, but it was determined that available data were generally inadequate. In addition to the basic FAO sources, supplemental data on the USSR were used.[2] Other sources were used occasionally for production data.[3]

The production series collected, like those for income, were of variable length. However, 1946 is the earliest year considered, and 1964 the most recent. All data are transformed into standard units of measurement: 1,000 hectares for area, 100 kilograms per hectare for yield, and 1,000 metric tons for production. Each data series was then plotted against time.

The various functional forms used to represent time trends are displayed in equations 4.1 through 4.8. According to the estimate being made, the variable Z_t indicates either area, yield, or production in year t. Again, a, b, and c are constants to be estimated:

$$Z_t = a + bt + \varepsilon \qquad (4.1)$$
$$Z_t = a + b\sqrt{t} + \varepsilon \qquad (4.2)$$
$$Z_t = a + b\log(t) + \varepsilon \qquad (4.3)$$
$$Z_t = a + bt + ct_d + \varepsilon \qquad (4.4)$$
$$Z_t = a + bt + c\sqrt{t} + \varepsilon \qquad (4.5)$$

1. United Nations, Food and Agriculture Organization. *Production Yearbook,* 1958 through 1965, and 1959 through 1966; *Yearbook of Food and Agricultural Statistics,* Part 1, Production, 1949 through 1957, and 1950 through 1958.

U.S. Department of Agriculture. *Agricultural Statistics,* 1949 through 1965, and 1950 through 1966.

U.S. Department of Agriculture, Economic Research Service. *Indices of Agricultural Production in 29 African Countries.* Mimeographed. Washington, D.C., 1965.

2. Harry E. Walters and Richard W. Judy. "Soviet Agricultural Output by 1970." Unpublished paper presented at Conference on Soviet and East European Agriculture, University of California, Santa Barbara, California. August 1965. Washington, D.C. U.S. Department of Agriculture, Economic Research Service, 1965.

3. U.S. Department of Agriculture, Economic Research Service. *Indices of Agricultural Production in 29 African Countries.* Mimeographed. Washington, D.C., 1965; *Indices of Agricultural Production in 10 Near East Countries.* Mimeographed. Washington, D.C., 1965.

U.S. Department of Agriculture. *Indices of Agricultural Production for the 20 Latin American Countries.* U.S. Department of Agriculture, ERS-Foreign 44, 1966.

$$Z_t = a + bt + c \log(t) + \varepsilon \qquad (4.6)$$
$$Z_t = \bar{Z} + \varepsilon \qquad (4.7)$$
$$Z_t = \overset{*}{\bar{Z}} + \varepsilon \qquad (4.8)$$

and ε is a disturbance term. The variable t_d is defined as in equation 3.5, except that values of d equal to 52 and 58 are considered. The form of equation 4.4 was used for several purposes. In general, it may be used to represent any trend which appears to possess two distinct, more or less linear phases, if the two phases are connected. Certain series seem to be characterized by a period of rapid war recovery, followed by a markedly dampened trend. A second purpose in its use was to detect any leveling-off of production trends near the end of the 1950s when earlier aggregate per capita production gains in the underdeveloped countries seemed to lessen. However, only limited success was achieved with this approach. Such trends (to the extent that they are present) often can be approximated as well or better with equations 4.2 or 4.3. By any rigorous statistical criteria, it is difficult to verify a distinct "break" in area, yield, or production trends on the basis of time series of the quality and length available.

Observed time paths of the variables related to production infrequently appear to increase at accelerating rates. However, in a few cases, forms of the type in equation 4.4 are used to (a) represent such trends and permit a more rapid rise than would be possible with equation 4.1 through 4.3 and (b) maintain a form less "explosive" in character than an exponential function.

Equations 4.5 and 4.6 are used only rarely. These forms are introduced as representatives of the class of function for which $d^2Z/dt^2 > 0$, but which have linear limiting forms. Thus, using equation 4.6 as an example, $dZ/dt = b + c \log(e)\, t^{-1}$, and $\lim_{t \to \infty} dZ/dt = b$. This "damping" trend causes these forms to be more useful in representing data which appear to increase at an increasing rate. However, for the projection period under study, the damping effect is small and, as with the exponential functions, projected values are highly sensitive to observations away from the trend line.

An overall mean of the observed data, $\bar{Z}$, is used in equation 4.7 if no discernible trend is evident and the entire series appears to have only random fluctuations about a constant value. The symbol $\overset{*}{\bar{Z}}$ in equation 4.8 refers to a mean of only a recent portion of the data set. This expression is used when an apparent trend in early years appears to have flattened out in recent years, or when a distinct shift is observed from one apparent constant level to another. The above are monotonic functions of time throughout the projection period. (Equations 4.5 and 4.6 are used only when the fitted curve rises at an increasing rate.)

ESTIMATION TECHNIQUES

Least-squares techniques were used to estimate all equations. Several criteria were applied in selecting functions for individual time trends. Among these were the appearance of the plotted observations and statistical properties of the fitted functions. Trial projections of dependent variables were made by evaluating each fitted function at future time values. Comparative growth rates between two or more forms fitted in the same data series also were considered in selecting among them. More recent, often provisional data, not used in fitting regressions, were also considered in relation to the data which are included and the resulting estimated functions. Again, extraordinary historical events which are known to have affected the observed data were considered. Sometimes, data associated with these extraordinary events were omitted and functions were estimated without them. In other cases, such events only condition the selection of particular functional forms estimated from the full data set. Examples of such historical events are major agricultural policy changes, extreme weather, war or political disturbances, severe market disruptions, etc. Before specifying that any function had a non-zero trend, a general statistical selection criterion was used requiring an F value significant at the 90 percent confidence level. Equations 4.7 or 4.8 sometimes were used even though statistically significant trends can be measured. For example, when a series moves from one more or less constant level to another, equation 4.8 was used. For two variables, a priori linear trends were specified because past observations are determined to a large degree by government agricultural policies. Also, evidence is available to suggest the likely future continuance and impact of these policies. These variables are the area of wheat and corn in the United States. As in the case of income trends, the selection could not always be based on empirically objective criteria. However, every effort has been made to select functions which allow projections representing extension of recent underlying trends.

ESTIMATES OF MAXIMUM POTENTIALS FOR CROPLAND EXPANSION

While recent trends of variables are used wherever possible, this procedure is not possible in all cases. For example, while land used for crops has expanded, this trend cannot go unbounded forever. Many persons express concern about the future world food situation on grounds that present rates of cropland expansion in the underdeveloped countries will soon exhaust available land supplies.

MEASUREMENT OF LAND AVAILABILITY AND CROP POTENTIALS

After surveying estimates available, it was decided that the objectives of the investigation could be served best by undertaking a special analysis of land resources in the underdeveloped countries of the world.[4] A soils specialist, William G. Harper, was employed to measure land availability for crops in various countries of the world. Land available for crops was designated in terms of its topography or slope, rainfall or water availability, and absence of serious problems such as alkalinity. The soils analysis was conducted under an agreement between the Soil Conservation Service of the U.S. Department of Agriculture and the Center for Agricultural and Economic Development, Iowa State University of Science and Technology. Full-time work by the soil scientist was required over most of a two-year period in measuring, from aerial maps and available soil classifications, land available for cultivation, crops which could grow on it, water availability, possibilities in multiple cropping, market location, and other important characteristics.

The soils analysis was limited to less developed countries of the world due to resource limitations.[5] The necessary research is expensive in both time and money. Also, the most pressing food problems and the most rapidly rising total land trends are in the less developed countries, and knowledge of their agricultural land resources is least adequate. Finally, food problems generally are not pressing in the developed countries, knowledge is more adequate, and total cropland trends are modest. Thus it was believed that continued trends would not result in large cropland increases in the developed countries.

UPPER AND LOWER BOUNDS ON EXPANSION OF LAND

Projections of crop areas in most countries of Central and South America, the Near East, and non-Communist Asia are made subject to upper bounds on cropland expansion derived from the special soils study. Since production projections for Africa are not based on area and yield trends, land limits were not used for these countries. However, as indicated later, the food outlook would be unchanged even if limits on crop area had been incorporated for these countries. In the special soils analysis, land for crop use was classified according to suitability under either fallow, irrigated, or rain-fed conditions. In many, but not all cases, it was possible to distinguish between presently irrigated and potentially irrigable land. The rain-fed and irrigated lands were placed in three classes on the basis of potentials for producing reasonable continuous (a) high, (b) moderate, or (c) low yields of adapted

---

4. This analysis was made possible through the aid and cooperation of Charles E. Kellogg, Deputy Administrator for Soil Survey, and Arnold Orvedal, Chief, World Soil Geography Unit, Soil Conservation Service, USDA; arrangements were made for access to unpublished materials on world soil resources and the library facilities of the Soil Conservation Service. The study of land resources was based on these working materials.

5. However, Communist China was excluded, since it was not to be included in the portion of the study on projections.

crops. Crop yields were estimated accordingly. However, land of all three classes was grouped together for establishing upper bounds on total area trends.

Some cropland is designated as suitable for only a limited number of specified crops, generally such as fruit, coffee, and cocoa. Cropland suitable for growing either a country's major or minor crops was assigned to the general cropland categories. A location criterion also was used in the classification system, and land more than fifty miles from water, rail, or road transportation was differentiated from land nearer these facilities.

The special soils analysis also provided a basis for estimating multiple cropping potentials. Areas were identified where multiple cropping is possible from the standpoint of soil characteristics. Multiple cropping potentials were estimated for irrigated land on the basis of water availability and climate. For areas judged to have multiple cropping potentials on upland soils, climatological data were summarized to estimate length of growing season and quantity and distribution of rainfall.

Underlying all classification decisions and projected potentials is an assumption of relatively high management. It is assumed that management and cultural practices are analogous to those of commercial agriculture in North America and Western Europe. Potential productivity of soils also is considered relative to costs of required practices. Implied in this assumption is development of adapted technology, and in many cases, lowering of farm input prices relative to farm product prices to levels where economic incentives for adoption of practices are parallel to those in North America and Western Europe.

From the basic data described above, upper boundaries on cropland expansion were established for each country at two alternative levels. For each level, two kinds of cropland were distinguished: land suitable for minor crops only and a general cropland class. A hectare of land designated as suitable for multiple cropping was considered as two hectares of cropland, whereas a hectare of cropland for dry fallow management was added as one-half hectare. Wherever cropland is designated as suitable for long-term crops such as coffee, sugarcane, etc., no multiple cropping is assumed in any case.

The basis for establishing low and high upper bound levels on land availability differs between Latin America and the Near and Far East. For Latin American countries, the low level is computed as the sum of all cropland located within fifty miles of rail, water, or road transportation. Land for both general crops and minor crops is so measured. Little cropland is presently planted more than once annually in Latin America and the low level or variant here assumes no multiple cropping. The high variant or upper bound includes all cropland regardless of location. In addition, multiple cropping is assumed possible on irrigated land where sufficient water is available and the length of the growing season is sufficient. It is also assumed in upland areas having favorable soil characteristics and acceptable temperature and moisture for eight months or longer.

For Near East countries, multiple cropping is assumed possible only on presently or potentially irrigable land. No potential cropland is identified at distances greater than fifty miles from transportation facilities. The low variant upper bound reflects all cropland identified in upland areas and all irrigated land presently in use. Multiple cropping is assumed to be at estimated present levels on irrigated land. Under the high level of land availability, potentially irrigable land is added and multiple cropping is increased to reflect potential increases in supplies of irrigation water and better management of existing and future flows.

Beginning with the eastern part of India and moving farther eastward, annual precipitation is judged adequate to permit multiple cropping on upland soils in some cases. Thus, in these countries, the conceptual basis for differentiating between low and high variant land levels on upper bounds is the same in the Near East, except for (a) multiple cropping assumptions on nonirrigated cropland and (b) the distance criterion in the case of Indonesia. Under the high variant, multiple cropping is assumed possible on nonirrigated cropland having favorable soil characteristics and acceptable temperature and precipitation for eight months or longer annually. Under the low variant, a minimum ten-month growing season is assumed needed. Certain portions of Indonesia's cropland are designated as more than fifty miles from transportation and were omitted from the low variant land constraint. In no other case is the location criterion operative in Asia.

It was next necessary to make allowance for the fact that the crops for which area trends were estimated do not include all crops grown in any country. Hence, the sum of the area trends cannot be interpreted as the trend in total cropland. Total potential cropland was reduced to allow for crops not included in analysis of area trends: each country's land constraints (low and high variants and land for minor crops and general crops) were multiplied by the estimated percentage of 1960 cropland included in crops for which area trends were computed. For the 39 countries with land constraints determined by the special soils analysis, crops whose area trends have been estimated accounted for 81 percent of the total 1960 estimated cropped area. The percentage coverage varies among countries and may be subject to errors, since estimates of undercoverage were sometimes based on inferences drawn from fragmentary data rather than estimates.

For most countries not included in the special soils analysis, constraints on land expansion routinely were set at levels only slightly above the 1960 level. Most estimated total area trends show only modest changes. Aside from Finland and Norway, no European country's data show rates of total cropland expansion in excess of 0.4 percent per year. Trends for Japan and 14 countries in Europe were negative to 1960. Norway, with an estimated annual growth in total cropland of 1.6 percent, was constrained to 115 percent of 1960 crop area, and Denmark, with an 0.8 percent annual growth rate was limited to 110 percent of 1960 area. Finland and the USSR were permitted a 10 percent increase in cropland under the land constraint on the basis of their respective 0.4 and 0.3 percent annual cropland growth rates, and France was allotted 10 percent increase on the basis of other estimates.[6]

Australia and New Zealand present special problems because both exhibit rapid growth in total cropland. Evidence suggests that substantially more expansion is possible and neither was designated for the separate detailed study in the special soils analysis. In the case of New Zealand, estimated available area suggests that ample underdeveloped land is available to permit the expansion implied in the projections.[7] Hence, the land constraint was set at a level which left the trend unbounded. For Australia, where the 1960 growth rate in cropland was 3.4 percent per year, a land constraint 45 percent above the 1960 level was used.[8]

PRODUCTION PROJECTIONS

Where area and yield trends could not be estimated but acceptable production data were available, the estimated production trend was extrapolated to project future production of individual crops. Nearly all fruit crops were projected in this manner and the projections do not consider the possible constraint of total cropland (where land is limited for other crops in the same country).

Where future production was estimated as the product of projected area times projected yield, it was necessary to project individual crop areas subject to overall land constraints. All crops were included in this process regardless of whether they were later incorporated into the food production-demand comparisons. By including them, it was possible to allow competition for food-producing resources arising from industrial and beverage crops.

All functions depicting area of individual crops in a country were first extrapolated to 1985 and 2000 to estimate "unconstrained" area values. The assumption was made that any crop present may be grown on land classified in the general cropland categories, but that land designated for minor crops could be used only for those crops as specified

6. United Nations, Food and Agriculture Organization. *Production Yearbook,* 1964.

7. Ibid.

8. J. B. Condliffe. *The Development of Australia.* Galt, Ontario, Canada: The Free Press of Glencoe, Collier Macmillan Ltd., 1964.

in the soils phase of the study. For countries where land constraints were otherwise formulated, only a general cropland restraint was specified and no minor crops were designated.

The procedure for projecting crop areas assumed that if projected total area reaches a boundary value, then no further total expansion can occur. In such cases, however, further adjustments in areas of individual crops were still permitted within the limits set by prespecified bounds on total cropland. These gains or losses for areas of individual crops do not reflect any specific economic considerations; rather, the adjustments are based on relative rates of change in individual area trends, a point explained subsequently.

CROP AREA PROJECTION

The total land area classified in the soil phase of the study as suitable for the general crops is denoted as *general cropland* (GC), and the total area designated suitable for a limited number of crops as *special cropland* (SC). The term *total projected area* (TPA) denotes the sum of the unconstrained area projections of all individual crops grown in a country and the term *special crop projected area* (SCPA) is the sum of the unconstrained area projections of crops designated as suitable for production on special cropland.

Total projected area and special crops area were first computed for 1985 and 2000 by adding individual area figures estimated by extrapolating area trends. Negative values were not admitted and any occurring were set upwards to zero. For each of the projection years, one of three mutually exclusive states was observed, and consistency with the above assumptions required one of three different computational procedures to arrive at final area estimates for individual crops. The three states were:

State 1: $TPA \leq GC + SC$ and $TPA - SCPA \leq GC$
State 2: $TPA - SCPA > GC$ and $SCPA \leq SC$
State 3: $TPA > GC + SC$ and $SCPA > SC$

Under state 1, the land constraints are entirely inoperative, and the final projected area for individual crops is simply the unconstrained extrapolations. Under state 2, the special crops projected area can be accommodated on the special cropland, but crops not designated as special crops cannot be accommodated on the general cropland. Thus, final area estimates for individual crops not in the latter category are computed by applying the scale factor $GC/(TPA - SCA)$ to their unconstrained area estimates. Final area estimates for the individual special crops are, again, the unconstrained extrapolations.

State 3 is most restrictive since the projected area of special crops exceeds the special cropland, and projected total crop area exceeds the sum of general cropland and special cropland. According to the assumptions used, the special crops compete on both kinds of cropland, CC and SC, while other crops compete only on the general cropland, GC. Final area estimates for crops not designated as special crops are obtained by multiplying their unconstrained area estimates by the scale factor GC/(TPA — SC), and final area estimate for the special crops is obtained by scaling down the unconstrained values with the factor SC/SCPA + [GC/(TPA — SC)] [1 — (SC/SCPA)]. In other words, the special crops are allocated to the special cropland (up to its capacity) and the residual then is allowed to "compete" with general crops for the general cropland.

This procedure allows special cropland to be zero and the area for general crops to be bounded, while area for special crops is unbounded. The same decision rules, and computations are applied for future years, under low and high levels of land constraints (where two are present), and for all countries whose area trends are estimated.

YIELD PROJECTIONS

Projections of yields are extrapolations of trends to 1985 and 2000 (negative quantities not being admitted.) Future production was then estimated as the product of projected area times projected yield. To conform with definitions used in the demand projections, projections of sugarcane and sugarbeet production were converted into sugar on a raw value basis. Similarly, projections of groundnut production were transformed to a shelled basis. When production projections were completed for individual crops, each country's food and feed data were aggregated into six production categories: cereals, sugar (raw value), starchy roots, pulses, vegetables and fruit, and oil crops. These categories were defined in the same manner as those in the demand projections, except that the three types of livestock products were omitted. Production projections for industrial and beverage crops are dropped at this point.

The demand projections include all commodities used by each country within each of the defined food classes. The production projections, as described thus far, differ conceptually since not all crops are included. The estimated coverage varies substantially, both by country and type of commodity. Generally, the cereals and sugar are well covered. Starchy roots and oil crops are reasonably well covered; pulses, somewhat less; and vegetables and fruit least well covered.

To make the production and demand projections conceptually comparable, corrections were necessary for undercoverage. Estimated 1960 production was first computed by evaluating all area, yield, and production trends for 1960. The resulting production estimates then were aggregated into the six food categories described above and compared to the corresponding production estimates derived from the USDA 1959–61 food balance sheets. The difference, if any, was interpreted as a measure of crop omissions from the time trend projections. An approximate adjustment then was made in future production estimates by assuming that the undercoverage percentage remained constant over time.

COMPARISONS OF PROJECTED PRODUCTION AND DEMAND

Comparisons of estimated future production and demand are as follows: total food and industrial demand for the commodities is computed for nine product classes. Feed demand then is derived from livestock demand and the total demand for the six food and feed crop classes is compared to corresponding production projections. Excess demand, or demand less production, is computed for each of the six product classes. Food and feed crop production is compared with demand estimates (including both direct and indirect demands). This procedure was applied to 96 countries for each time period, each combination of population and income, and two alternative upper bounds on cropland expansion.

The resulting comparisons were then aggregated three ways: first, aggregates were computed and summarized geographically for 21 regions covering the world. Second, the same data were aggregated according to income levels of countries. Third, the estimates for all countries were aggregated to form a 96-nation total.

PART TWO

Production and Demand Prospects

CHAPTER 5

Population and Income Projections

THE ANALYSIS in earlier chapters emphasized that the so-called world food problem of the 1960s was really only one facet of a more general dilemma. The broader problem is one of rapid population growth coupled with lagging economic development in many of the nations of Latin America, Asia, and Africa. We do not claim a full investigation of the future demographic and economic development prospects for these nations. However, assessment of future agricultural supply and demand conditions must recognize these important basic factors. The purpose of this chapter is to present and discuss the basis for the population and income projections used in comparing future agricultural productions and demands for the 96 countries.

The methods used to construct these estimates were discussed in Chapter 3. Three alternative population projections and two alternative total income trends were determined for each country. These, in turn, were used to estimate future food demand under nine alternative population and per capita income assumptions.

The future growth of population in the developing countries is by no means fully predictable. However, it is even more difficult to make precise projections of rates of economic development and per capita incomes. The two trends of total income estimated for each country are regarded as illustrative of the possible future paths which this variable may take, but neither trend is necessarily regarded as "most likely." Trends in total real personal consumption expenditures, the income variable most commonly used, were not especially vigorous during the 1950s in most of the developing countries. Linear time trends frequently fit the historical data quite well. On a per capita basis, real personal consumption expenditures were almost constant during the 1950s in many countries. When these linear trends in total real income were extrapolated to 1985 and 2000 and combined with future population estimates, the resulting projections of per capita real income for developing countries were frequently little above 1960 levels. In several cases, the estimates for the future were lower than those for 1960. Of course, this decline reflects the expectation of a more rapid population growth in the future than in the 1950s, particularly under the high variant population projections.

The high variant trend in total real income was obtained, in most cases, by fitting an exponential trend to historical data. The resulting estimates of future per capita real income more nearly approximate those appearing in development plans in low income countries.

While the difference between the two sets of income projections is somewhat arbitrary, each of the resulting estimates for the future can be associated with a particular concept of the role which population growth plays in bringing about a given level of income and relative demand pressure on food supplies. On the one hand, if we accept the argument that labor is redundant in the producing sectors of many developing countries, then there is no necessary reason why expansion of labor supplies through population growth should lead to comparable growth in employment and income. Under this argument an extrapolated linear trend based on past total income growth may be accepted without regard for the fact that its properties (constant annual increments to total income) do not match those of the population trends (in many cases, increasing annual increments to total population).

On the other hand, the general properties of an exponential function representing total income growth more nearly match the properties of most population trends. If the labor redundancy argument is denied and constant returns to scale in production are assumed, an argument exists favoring total income growth which at least matches the growth in labor. We can argue also from a demand perspective. To hypothesize or project the existence of a certain population by a specified date requires, for consistency, the hypothesis of production of goods and services at least sufficient to meet the subsistence needs of this population. Hence, a lower bound on income earned from production is implied in relation to population.

This framework is greatly oversimplified and does not deal with the real substance of economic development, namely, the creation of additional, modern productive capacity through technological progress and investment. But as noted previously, these matters are beyond the scope of the investigation which is directed to alternative estimates of total income trends reflecting past rates of growth and illustrating some possibilities for the future.

For present purposes we give only the results for the high variant total income trend and constant per capita income. Similarly, the production-demand comparisons presented in Chapters 7 to 9 do not include comparisons made under the low variant income projections. The two alternatives presented, constant per capita income and high variant income trends, illustrate the range of possible outcomes sufficiently well (without adding unduly to the data presentation).

POPULATION AND INCOME PROJECTIONS FOR THE AMERICAS

Tables 5.1 and 5.2 present population and per capita income projections for the 25 countries in the Americas. The relatively large variation in individual countries' population and per capita incomes for 2000 under low, medium, and high population growth assumptions partially illustrates the uncertainty associated with long-term demand projections. However, a dispersion this large is probably not unreasonable for a time horizon 30 years in the future.

Data given in Table 5.1 show a marked difference between the population growth prospects of the 4 countries of the Temperate Zone and those of the Tropical and Subtropical zones. Projections for the United States, Canada, Argentina, and Uruguay for 2000 vary from 45 to 105 percent above 1960 levels under the medium growth assumption. Only Jamaica, among the remaining 21 countries, is expected to fall short of doubling its 1960 population under the medium projections. Medium growth rate projections for Brazil and Mexico, the two largest tropical countries, are 200 and 265 percent, respectively, above 1960 levels by 2000. Venezuela's projected population increases are greatest in relative terms. Its medium projection for 1985 is nearly 2.5 times the 1960 population and the 2000 projection is approximately 4 times the 1960 level.

The aggregate effects of these growth differentials may be seen by noting that the medium variant projections imply the addition of 93 million people by 1985 in the four Temperate Zone countries while the remainder of the hemisphere adds 202 million. From 1960 to 2000, the corresponding additions are 170 million and 391 million. In 1960 the four Temperate Zone countries had about 54 percent of the hemisphere's population. Should the medium variant projections be realized, they would have only 45 percent by 1985 and 41 percent by 2000.

The population projections illustrate one dimension of future food demand in the Americas. A second dimension is implied in the per capita income projections presented in Table 5.2. Per capita income growth rates for the four Temperate Zone countries are roughly similar at about 2.1 to 2.6 percent per year under the medium assumption. Of course, the United States and Canada presently have the highest income levels in the hemisphere and their absolute increases in income are expected to be higher than for other countries. The rest of the hemisphere is characterized by great variability. For example, a continuation of Mexico's recent rate of economic development is estimated to result in a tripling of per capita incomes by 2000 under the medium population assumption, even though population increases over 3.5 times during the 40-year projection period. The projections for Brazil, the largest Latin American country, illustrate a situation found in several countries

TABLE 5.1. Projected Population for the Americas under Three Population Growth Assumptions (thousands)

Country	1960	Low Projections		Medium Projections		High Projections	
		1985	2000	1985	2000	1985	2000
United States	180,676	234,774	263,970	254,403	316,376	267,938	336,022
Canada	17,909	25,981	30,187	28,360	37,444	29,890	39,885
Mexico	34,988	75,635	111,452	80,811	127,703	83,122	139,332
British Honduras	90	186	276	199	316	204	345
Costa Rica	1,171	2,557	3,632	2,727	4,128	2,801	4,473
Cuba	6,797	9,767	11,305	10,749	13,626	11,207	14,410
Dominican Republic	3,030	6,345	8,573	7,200	11,646	7,597	12,629
El Salvador	2,442	4,813	6,702	5,127	7,583	5,261	8,189
Guatemala	3,765	7,230	10,108	7,702	11,444	7,907	12,372
Haiti	4,140	6,919	8,674	7,739	11,134	8,120	11,932
Honduras	1,838	3,884	5,592	4,144	6,374	4,259	6,926
Jamaica	1,607	1,987	2,180	2,162	2,516	2,244	2,635
Nicaragua	1,403	2,966	4,269	3,165	4,867	3,252	5,288
Panama	1,079	2,139	3,037	2,280	3,451	2,342	3,741
Trinidad and Tobago	844	1,438	1,754	1,599	2,204	1,675	2,352
Brazil	70,459	130,086	168,251	145,412	211,480	152,336	225,800
Argentina	20,956	28,279	32,239	29,956	35,407	31,105	38,609
Uruguay	2,491	3,003	3,290	3,172	3,564	3,277	3,811
Bolivia	3,696	6,279	8,011	7,004	9,990	7,332	10,649
Chili	7,627	12,391	15,156	13,198	17,051	13,819	19,217
Colombia	15,468	29,465	39,420	33,104	50,522	34,741	54,158
Ecuador	4,355	8,527	11,543	9,597	14,893	10,078	15,987
Paraguay	1,720	3,041	3,913	3,251	4,476	3,423	5,161
Peru	10,199	18,484	23,964	20,622	30,210	21,609	32,275
Venezuela	7,394	16,061	22,478	18,165	29,549	19,107	31,842

TABLE 5.2. Projected Indices of Real Per Capita Income for the Americas under Three Population Growth Assumptions (1960 = 100)

Country	Low Population		Medium Population		High Population	
	1985	2000	1985	2000	1985	2000
United States	183	274	169	228	160	215
Canada	212	357	194	288	184	270
Mexico	220	379	206	331	200	304
British Honduras	154	147	144	128	141	117
Costa Rica	139	191	131	168	127	155
Cuba	172	256	156	212	150	201
Dominican Republic	209	375	184	276	174	254
El Salvador	209	219	196	193	191	179
Guatemala	148	198	139	175	135	162
Haiti	89	90	80	70	76	66
Honduras	122	149	114	131	111	120
Jamaica	283	369	260	319	251	305
Nicaragua	195	197	183	173	178	159
Panama	188	291	176	256	171	236
Trinidad and Tobago	205	241	185	191	176	179
Brazil	190	209	170	167	162	156
Argentina	213	260	201	237	194	217
Uruguay	238	303	226	280	219	261
Bolivia	148	201	132	161	127	151
Chile	158	227	148	202	142	179
Colombia	178	276	158	215	151	201
Ecuador	164	245	146	190	139	177
Paraguay	234	425	218	372	208	322
Peru	193	314	172	249	165	233
Venezuela	219	230	193	175	184	163

when expected population growth rates are high. Under the medium population projection, the projected index of real per capita income for 2000 is only a modest 67 percent above the 1960 level. However, this represents a slight decline from 1985. This decline is more pronounced under the high population assumption. Whether such a turn of events will actually be realized can only be conjectured, given the means at hand for analysis and the length of the projection period. However, this result is consistent with our goal of evaluating the consequences of a continuation of trends in the major determinants of demand and supply. By no means is it an inconceivable outcome. The most pessimistic outlook for future per capita income is found in the projections for Haiti. The projections presented in Table 5.2 imply a decline in per capita income below the 1960 level even for the low population projection.

Finally, while the estimate for some Latin American countries suggests good progress toward improved living standards in the next three decades, the typical situation is one where past rates of income growth, if maintained, would little more than match the anticipated growth in population. Given the population growth rates typical of those in Latin America during recent decades, the increases in economic activity required to sustain rising living standards are very large indeed.

POPULATION AND INCOME PROJECTIONS FOR EUROPE, THE USSR, AND OCEANIA

Population projections for this group of countries contrast sharply with those for countries of the Western Hemisphere. For the entire group shown in Table 5.3, the medium variant 1985 projection is only 24 percent, or 157 million, above the 1960 total. By 2000 the increase over 1960 is 253 million or 39 percent. Even the high variant projections for this group result in much smaller percentage increases than in the low variant projections for the Americas.

Population prospects for all European countries are relatively homogeneous and none is expected to experience great increases. At the upper extreme, high variant projections for Poland result in only a 65 percent increase between 1960 and 2000 and for the Netherlands only 59 percent. In some European countries the demographic situation is believed to have reached a point of near-stability. For example, Ireland's medium variant population projections for 1985 and 2000 are slightly below the 1960 population, and the low variant projections for East Germany show less than a 1 percent increase by 2000.

The USSR's 214 million population in 1960 was about one-third of the group total, and growth rate prospects for that country, shown in Table 5.3, are the highest found in Western Europe. The medium variant projected population for 2000 is 65 percent greater than in

TABLE 5.3. Projected Population for Europe, USSR, and Oceania under Three Population Growth Assumptions (thousands)

Country	1960	Low Projections		Medium Projections		High Projections	
		1985	2000	1985	2000	1985	2000
Austria	7,081	7,134	7,219	7,330	7,488	7,525	7,754
Belgium-Luxembourg	9,467	10,351	10,735	10,693	11,422	11,033	12,104
Denmark	4,581	4,951	4,927	5,164	5,423	5,340	5,908
Finland	4,430	4,996	4,966	5,239	5,568	5,430	6,157
France	45,684	53,021	55,717	54,939	60,078	56,833	64,445
Ireland	2,834	2,728	2,727	2,788	2,801	2,862	2,889
Netherlands	11,480	14,111	15,152	14,691	16,698	15,262	18,264
Norway	3,581	4,044	4,019	4,246	4,527	4,403	5,024
Sweden	7,480	7,977	7,993	8,296	8,659	8,571	9,364
Switzerland	5,362	6,202	6,461	6,413	6,905	6,622	7,348
United Kingdom	52,508	54,564	54,407	56,387	58,055	58,122	61,739
West Germany	55,423	59,763	61,568	61,650	65,057	63,521	68,525
Greece	8,327	8,996	9,217	9,326	9,859	9,655	10,495
Italy	49,642	53,406	54,722	55,364	58,527	57,320	62,306
Portugal	8,826	9,202	9,367	9,518	9,912	9,834	10,452
Spain	30,303	34,212	35,314	35,555	38,217	36,894	41,111
Bulgaria	7,867	9,256	9,761	9,591	10,462	9,925	11,166
Czechoslovakia	13,654	15,719	16,466	16,268	17,551	16,816	18,640
East Germany	17,241	17,211	17,354	17,689	17,930	18,166	18,504
Hungary	9,984	10,549	10,827	10,876	11,350	11,204	11,873
Poland	29,703	38,362	41,504	39,928	45,395	41,489	49,355
Rumania	18,403	22,313	23,787	23,164	25,713	24,012	27,661
Yugoslavia	18,402	21,660	22,439	22,537	24,425	23,412	26,411
USSR	214,400	282,363	316,499	296,332	353,099	320,025	402,799
Australia	10,315	14,655	17,111	15,373	19,176	16,024	20,528
New Zealand	2,372	3,749	4,574	3,956	5,252	4,130	5,671

1960. In this respect, the Soviet projections are more similar to those for the United States and Canada than to other European countries.

Demographic situations and outlooks for Australia and New Zealand are quite different from others in the group. Medium variant population projections for 1985 are 52 percent above 1960, and projected 2000 population is about double that of the base period.

Of course, the Europe, USSR, and Oceania group includes a major share of the world's economically developed countries. None was placed in the low per capita income category in terms of 1960 rankings. The projected indices of real per capita income used in this study and presented in Table 5.4 suggest continued rising living standards under all alternative population assumptions. Of the 12 Northern European countries, all but Ireland were placed in the high per capita income class for 1960. In the most extreme case, Austria's per capita income level in 2000 is over eight times the 1960 level under all population assumptions. Projected per capita income indices for the four Southern European countries shown in Table 5.4, especially those for Greece, compare favorably with those for Northern Europe. However, base level incomes were generally lower for Southern European countries, all being placed in the medium per capita income class in the base year, 1960.

East Germany and the USSR were in the high per capita income class in the base year, but the other six countries of Eastern Europe were in the medium income category. Table 5.4 shows a continuing rise in income levels for these countries, with projected indices of per capita income ranging from 304 to 451 in 2000 under the medium population assumption.

Finally, Australia and New Zealand have a unique growth ranking within the group, both with respect to population and per capita income. Assuming medium population growth rates, they show indices of per capita income of only 217 for Australia and 161 for New Zealand in 2000.

POPULATION AND INCOME PROJECTIONS FOR AFRICA AND ASIA

Like the Americas, the Africa-Asia group includes countries having quite diverse population and income prospects. The group is dominated by developing countries having low incomes, rapid population growth, and only modest income growth trends. But interspersed in close geographic proximity are countries such as Israel and Japan with modern, growing economies and, in the case of Japan, a low population growth rate.

A few aggregate comparisons may be useful before considering individual country projections. The 45 Asian and African countries had a total 1960 population of nearly 1.2 billion. Under the medium assumption, 1985 population would be about 76 percent or 890 million

TABLE 5.4. Projected Indices of Real Per Capita Income for Europe, USSR, and Oceania under Three Population Growth Assumptions (1960 = 100)

Country	Low Population		Medium Population		High Population	
	1985	2000	1985	2000	1985	2000
Austria	390	876	380	844	370	815
Belgium-Luxembourg	212	338	205	318	199	300
Denmark	321	681	308	618	298	568
Finland	219	378	209	337	201	305
France	291	574	281	532	271	496
Ireland	321	348	216	339	211	329
Netherlands	235	414	226	376	218	344
Norway	193	310	184	275	177	248
Sweden	192	297	185	272	179	252
Switzerland	250	453	242	424	234	399
United Kingdom	207	329	200	308	194	290
West Germany	317	643	307	609	298	578
Greece	425	1,035	410	968	396	909
Italy	281	532	271	497	261	467
Portugal	245	422	237	398	229	378
Spain	215	355	207	328	199	305
Bulgaria	351	484	338	451	327	423
Czechoslovakia	243	429	235	403	227	379
East Germany	268	366	261	354	254	343
Hungary	331	461	321	440	312	420
Poland	240	437	230	399	222	367
Rumania	340	464	328	429	316	399
Yugoslavia	350	492	337	452	324	418
USSR	266	339	253	304	234	266
Australia	168	244	161	217	154	203
New Zealand	140	185	133	161	127	149

greater than 1960, and by 2000 the 45-country total would increase over 1.6 billion, or 140 percent, above 1960. Prospective changes such as these spawned the term "population explosion" and cause continuing concern about world food problems.

Table 5.5 shows that projected population growth for the seven North African countries (Algeria through the UAR) is quite large. Even under the low variant, Ethiopia's population increase exceeds 95 percent by 2000. The remaining six countries more than double their 1960 populations over the same period. Base period incomes were consistently low. Estimates of future income gains in Table 5.6, except those for Libya and the UAR, suggest small to modest growth in the future. The Morocco projections suggest actual declines under high population growth.

Substantial variability prevails among the population growth rates of the 11 countries of West Central Africa (Angola through Togo in Table 5.5). For the entire group, population increases by 166 percent between 1960 and 2000 under the low population assumption and by 239 percent under the high assumption. The extremes include increases of 57 percent under Angola's low projection and 315 percent under Ghana's high projection. Table 5.6 shows moderate income gains for some of the 11 countries, but income projections for countries with the largest populations (Nigeria, Congo, and Ghana) are little above the low levels of 1960.

Population growth is universally high among the seven countries of East Africa (Kenya through Uganda in Tables 5.5 and 5.6), and per capita income growth over the 40-year projection period is less than 2.5 percent per year in all but Tanganyika.

South Africa was the only African country placed in the medium per capita income class for 1960. Again, however, its per capita income projections are dominated by large increases in projected population. The result, even under the low population assumption (Table 5.6), is an income growth rate of only about 1.5 percent per year.

Among the eight countries of West Asia (Cyprus through Turkey in Tables 5.5 and 5.6), Israel and Cyprus stand out in three ways: both had high 1960 income levels, both are expected to realize less population growth than others in the group, and only Iran compares with them in future income growth prospects. Per capita income projections for Jordan, Syria, and Turkey reflect growth rates of about 1 percent annually through 2000 under all population assumptions.

Projections in Table 5.5 for populous India indicate large future population growth. Even the low variant projection for 2000, 838.5 million, dwarfs the estimates of any other country included in the study. Projections of per capita income for 2000 (Table 5.6) indicate an improvement of only 46 to 71 percent above the country's low 1960 level. Population in Pakistan and Ceylon is expected to grow at somewhat more rapid rates than in India. To the year 2000 under the medium

TABLE 5.5. Projected Population for Africa and Asia under Three Population Growth Assumptions (thousands)

Country	1960	Low Projections		Medium Projections		High Projections	
		1985	2000	1985	2000	1985	2000
Algeria	11,020	20,809	27,817	22,326	32,214	22,749	34,379
Ethiopia	20,000	29,819	38,969	31,324	42,765	33,159	48,367
Libya	1,195	1,923	2,398	2,052	2,714	2,083	2,856
Morocco	11,626	24,272	33,803	26,133	39,680	26,676	42,675
Sudan	11,770	20,251	26,061	21,656	29,797	22,029	31,567
Tunisia	4,168	6,710	8,383	7,157	9,488	7,270	9,991
United Arab Republic	25,952	49,983	67,162	53,651	77,911	54,681	83,229
Angola	4,642	6,124	7,282	6,457	8,016	6,768	8,799
Cameroun	4,097	5,576	6,897	5,896	7,686	6,197	8,544
Congo, Rwanda, Burundi	14,139	23,284	32,827	24,844	38,044	26,344	44,015
Ghana	6,777	13,916	21,848	14,646	24,351	15,627	28,180
Guinea	3,072	5,547	8,301	5,828	9,183	6,204	10,520
Ivory Coast	3,230	5,537	8,055	5,810	8,870	6,175	10,099
Liberia	980	1,276	1,512	1,328	1,610	1,397	1,750
Nigeria	50,000	102,706	158,032	108,007	175,588	115,127	202,337
Senegal	3,110	4,834	6,677	5,062	7,292	5,367	8,208
Sierra Leone	2,450	3,952	5,303	4,134	5,766	4,376	6,447
Togo	1,440	2,562	3,864	2,692	4,278	2,867	4,909
Kenya	8,115	14,505	21,263	15,344	23,909	16,356	27,939
Malagasy Republic	5,393	7,867	10,458	8,273	11,519	8,768	13,093
Malawi, Rhodesia, Zambia	10,350	20,363	30,836	21,582	34,911	23,049	41,168
Tanganyika	9,239	14,731	20,061	15,514	22,210	16,465	25,422
Uganda	6,677	10,331	13,699	10,861	15,080	11,509	17,129
Republic of South Africa	15,822	30,106	41,547	30,911	46,335	32,211	52,585
Cyprus	573	644	685	666	717	674	736

TABLE 5.5 (cont.)

Country	1960	Low Projections		Medium Projections		High Projections	
		1985	2000	1985	2000	1985	2000
Iran	20,182	33,123	40,668	35,090	43,765	36,722	47,393
Iraq	7,000	15,404	22,758	16,230	25,803	16,736	28,711
Israel	2,114	3,252	3,924	3,388	4,241	3,457	4,498
Jordan	1,695	3,730	5,470	3,928	6,192	4,049	6,878
Lebanon	1,793	3,338	4,477	3,499	4,964	3,591	5,403
Syria	4,682	10,322	15,293	10,888	17,347	11,228	19,309
Turkey	27,818	52,171	69,844	54,672	77,397	56,100	84,187
India	432,750	678,905	838,500	717,955	908,000	753,079	981,000
Ceylon	9,986	18,515	23,587	19,657	25,509	20,623	27,854
Pakistan	92,578	159,825	208,300	169,758	226,500	178,658	245,800
Japan	93,210	110,324	115,330	114,615	122,400	122,219	138,730
Burma	22,325	37,448	49,365	38,704	53,696	39,661	57,065
Cambodia	5,600	10,747	15,114	11,138	16,671	11,438	17,896
China (Taiwan)	10,162	17,600	21,258	18,912	24,781	19,675	26,917
Federation of Malaya	6,909	14,006	20,096	14,528	22,260	14,929	23,966
Indonesia	92,250	165,264	224,612	171,031	245,943	175,437	262,613
Philippines	27,407	63,253	97,976	65,818	110,275	67,796	120,073
South Korea	24,665	46,159	62,016	49,197	67,418	53,623	79,151
South Vietnam	14,100	22,352	27,405	23,028	29,335	23,539	30,816
Thailand	26,438	51,219	68,909	52,983	75,289	54,330	80,266

TABLE 5.6. Projected Indices of Real Per Capita Income for Africa and Asia under Three Population Growth Assumptions (1960 = 100)

Country	Low Population		Medium Population		High Population	
	1985	2000	1985	2000	1985	2000
Algeria	171	259	160	224	157	210
Ethiopia	151	154	144	140	136	124
Libya	334	399	313	352	308	335
Morocco	107	124	99	106	97	98
Sudan	149	158	139	138	137	131
Tunisia	186	287	174	254	171	241
United Arab Republic	257	498	239	429	235	402
Angola	265	319	252	290	240	204
Cameroun	165	178	156	160	149	144
Congo, Rwanda, Burundi	156	151	146	130	138	112
Ghana	114	122	109	109	102	94
Guinea	194	185	184	167	173	146
Ivory Coast	241	241	229	218	216	192
Liberia	269	324	258	304	246	280
Nigeria	111	118	105	106	99	92
Senegal	225	233	215	213	203	189
Sierra Leone	217	231	207	212	196	190
Togo	126	112	120	101	113	88
Kenya	116	123	110	110	103	94
Malagasy Republic	240	258	228	234	215	206
Malawi, Rhodesia, Zambia	132	155	125	137	117	116
Tanganyika	241	397	229	358	216	313
Uganda	267	293	254	266	289	234
Republic of South Africa	145	193	141	173	136	153
Cyprus	316	637	306	608	302	592

TABLE 5.6 (cont.)

Country	Low Population 1985	Low Population 2000	Medium Population 1985	Medium Population 2000	High Population 1985	High Population 2000
Iran	233	423	220	393	210	363
Iraq	174	262	165	231	160	208
Israel	309	377	296	349	290	329
Jordan	159	155	151	137	147	123
Lebanon	198	324	189	292	184	268
Syria	134	173	127	153	123	137
Turkey	153	159	146	144	143	132
India	135	171	128	158	122	146
Ceylon	116	145	109	133	103	121
Pakistan	187	281	176	268	168	247
Japan	401	566	386	533	362	470
Burma	188	283	182	261	177	245
Cambodia	243	437	235	396	229	369
China (Taiwan)	293	625	272	536	262	494
Federation of Malaya	150	204	145	184	141	171
Indonesia	127	152	123	138	120	130
Philippines	138	179	133	159	129	146
South Korea	275	548	258	504	237	430
South Vietnam	144	192	139	179	136	171
Thailand	202	341	196	312	191	293

variant, population is projected to increase by about 2.5 times. Per capita income growth in Ceylon is expected to be less rapid than in India. But regardless of its more rapid population growth rate, Pakistan's per capita income growth is estimated to exceed India's by a considerable margin.

The population and income outlook for Japan is unique in all of Asia. The medium variant projection (Table 5.5) suggests a 2000 population only 30 percent above the 1960 level. Per capita income was second only to Israel in 1960, but a continuation of recent income trends would result in about a fivefold increase by 2000 (Table 5.6).

Population growth rates, again, are generally high among the nine countries of Other East Asia (Burma through Thailand in Table 5.5) countries. South Vietnam, with the region's lowest growth rate, shows a 94 percent population increase by 2000 under the low assumption. The other extreme is the Philippines with a 341 percent population increase estimated under the high variant. Considerable variability among per capita income is projected in Table 5.6. Under the medium population assumption, incomes in 2000 increase by over 400 percent for Taiwan and South Korea, but only 38 percent for Indonesia.

CHAPTER 6

Crop Production Trends

THE PRIMARY PURPOSE in this chapter is a summarization of the basic trends in world crop production as they existed in the 1960s. The analysis included over 3,000 sets of data on either area or yield for 71 crops in 96 countries. We present results in only a summarized form, although a more complete presentation is available from the authors.

STRUCTURE OF THE CHAPTER

In this chapter and in later production-demand comparisons for 1895 and 2000, it is convenient to organize the presentation geographically. We first consider crop production trends for all countries in North and South America; next we consider the countries of Europe, including the USSR and Oceania; lastly we turn to Asia and Africa. We focus particularly on reporting measured changes in total planted area and average yields for the individual crops under study. One group of crops, the cereals, is singled out for particular examination. This group is selected because of its importance in both agricultural production and consumption, and also to lend a more specific context to the presentation.

Tables 6.1 through 6.6 present summary data of production trends. Estimates in Tables 6.1, 6.3, and 6.5 relate to all crops for which specific analysis of area and yield trends was possible. The number of crops included in estimates of the (a) total area, and (b) average yield for each country are shown in the first column.

Production trends for several crops were analyzed separately for countries where area and yield data were not available. Direct projections of production first were estimated for such crops. These then were used in later analyses of future production-demand comparisons. However, Tables 6.1, 6.3, and 6.5 do not reflect area and yield trends for these crops. The totals and averages frequently include estimates for nonfood crops such as cotton, hops, coffee, tea, etc., even though demand estimates and production-demand comparisons were not included in the subsequent analysis.

The values given for the 1960 crop area in Tables 6.1, 6.3, and 6.5

are not totals of the actual 1960 observations for individual crops but are figures obtained by evaluating the time trend function for each area to determine its "expected value" at 1960. The results then are summed to the 1960 expected values for all crops in each country. Total crop area trend is obtained by first differentiating the area trend for each crop with respect to time and evaluating the result at the value of t corresponding to 1960. The derivatives of all crop area trends for a country are then summed and divided by the total 1960 crop area to determine the yearly percentage change in total crop area. From the discussion of functional forms in Chapter 4, it is apparent that the percentage change rate is not constant throughout the projection period. However, the data indicate the predicted base period trends in cropland.

Average yield trends are computed in a similar fashion. For each crop, the function representing its yield trend is differentiated with respect to time and the result evaluated at a value of t corresponding to 1960. The derivative is then divided by the corresponding 1960 yield to estimate the percentage rate of change for the particular crop in 1960. Finally a weighted average of these values is constructed for each country, using weights proportional to 1960 areas of the individual crops.

Estimates in Tables 6.2, 6.4, and 6.6 relate specifically to the cereal crops. They are computed in a fashion similar to that described above. The cereals group includes wheat, rye, barley, oats, corn, rice, sorghum, and millet. Area and production estimates given in these tables are sums for all cereal crops grown in the country, and total production is divided by total area to obtain yield estimates. Yearly percentage changes in area and production are computed by the same method used to compute the trend in total crop area in Tables 6.1, 6.3, and 6.5, and the percentage trend in cereals yield is calculated as the difference between the production and area percentage changes.

PRODUCTION TRENDS IN THE AMERICAS

Data in Table 6.1 indicate considerable agricultural growth in the Americas as a whole. The trend in production, measured by the sum of percentage changes in areas and yield, exceeds 2 percent per year in 12 of 25 countries.

However, North and South America have more diversity in agricultural growth rates than the other major geographic areas. The tremendous capacity and increasing productivity of the U.S. agricultural sector are well known. The 2.9 percent average annual yield increase for 22 crops in the United States stands sharply above the corresponding estimates for nearly all other countries of the region. Neighboring Mexico is a country not yet among the world's wealthy, but experiencing very rapid agricultural and economic development. Farther south, particularly in Central America and the Caribbean, evidence of agricultural improvement is small. (The exception for Trinidad and Tobago can be ignored since data were available for yield trend analysis on only a single crop, sugarcane.) Yield trends for other countries of the region were strikingly low. In 20 of the 23 Latin American countries, estimated average increases in crop yields were 1 percent per year or less; in 15, trend was less than 0.5 percent per year.

The common method of achieving greater production in Latin America was through increases in cropland area. One-third of the Latin American countries had annual area increases for included crops of 2 percent per year or more. For 13 major crops in Venezuela the increase was 4.1 percent per year. Brazil, the largest Latin American country, had a 2.1 percent annual expansion in area for the 23 crops analyzed. In Mexico, for 22 crops included, not only was planted area expanded by 2.5 percent per year but also average yields increased by 2.4 percent annually.

Latin America corresponds remarkably to Malthus's view of the relation between population growth and food supplies. Food supply expansion has closely paralleled the expansion in planted area. This situation, coupled with the large projected population increases discussed in Chapter 5, is justification for concern over future food balances.

ANALYSIS BY COUNTRIES

We turn now to a more particular consideration of the trends estimated for selected countries. Table 6.2 shows that cereal crops, mostly wheat and corn, accounted for about three-fourths of the total crop area in the analysis of U.S. area trends. Because of the history of government policies affecting planted areas of wheat and corn, it was deemed inappropriate to either attach meaning to past trends in these variables or to base projections on them. Consequently, these two trends were established entirely on a priori grounds for purposes of projections. Only in these two instances were trends in variables affecting production established in this fashion. The assumed trends were specified after giving consideration to government policies of 1960 affecting wheat and corn area, current levels of idle cropland, the decline in oats area, and estimates of future crop demand and land needs. The net result (Table 6.2) is an assumption of 0.1 percent per year increase in cereals area for the United States. The corresponding yield trend estimate reflects not only the increasing yield trends of individual crops but also a "commodity mix effect," resulting from substitution of wheat and corn, two crops having relatively strong yield trends, for oats, a crop showing relatively less yield improvement. Among other U.S. crops studied, annual increase in soybeans area was 5 percent. Increasing yields were quite generally evident, though as Tables 6.1 and 6.2 show, yields of the cereals

TABLE 6.1. Aggregate Crop Area and Yield Trends in the Americas, 1960

Country	Number of Crops Included	Total Crop Area	Total Crop Area Change	Average Annual Yield Change
		(1,000 ha)	*(% per year)*	*(%)*
United States	22	85,103	0.7	2.9
Canada	17	21,229	0.1	1.2
Mexico	22	11,915	2.5	2.4
British Honduras	3	12	2.8	0.0
Costa Rica	3	98	1.6	0.0
Cuba	7	1,537	0	0.2
Dominican Republic	4	146	1.1	1.3
El Salvador	8	420	—0.1	1.0
Guatemala	9	1,086	1.7	0.5
Haiti	3	43	0	0.0
Honduras	12	560	3.3	0.0
Jamaica	6	44	—0.2	0.2
Nicaragua	7	373	2.6	0.2
Panama	4	204	1.9	0.0
Trinidad and Tobago	1	33	0	3.9
Brazil	23	24,184	2.1	0.3
Argentina	26	13,875	0.4	0.1
Uruguay	14	1,261	0.2	0.0
Bolivia	3	117	2.7	0.0
Chile	16	1,432	0.9	0.5
Colombia	9	1,549	0.7	0.7
Ecuador	15	743	2.4	1.0
Paraguay	11	321	0.9	0.1
Peru	17	1,344	1.3	0.2
Venezuela	13	708	4.1	0.4

were advancing more rapidly than the average of all U.S. crops. Trends reported in Tables 6.1 and 6.2 do not reflect results for several fruit crops for which only modest production growth was observed.

Cereals dominate production even more in Canada than in the United States. Tables 6.1 and 6.2 show that cereals accounted for over 92 percent of 1960 crop area for the 17 crops studied, and over 12 million of the 19.7 million hectare cereals area was wheat. As in the United States, increases in wheat and corn area approximately offset declines in oats, and yield trends were higher among the cereals than in other crops. Yield trends in Canada provide an illustration of a pattern frequently observed in our analysis of world agricultural production. Canada is, of course, an economically developed country with progressive agricultural practices. Yet the yield trend for all crops (Table 6.1) and the yield level and trend for cereals are not particularly outstanding by world standards. A comparison of results for Canada with those for Europe (Table 6.4) emphasizes this point. Throughout the world, there seems to be small yield improvement progress in countries having a major proportion of their agriculture concentrated in arid, rain-fed areas. The

TABLE 6.2. 1960 Cereals Area, Yield, Production, and Trends in the Americas

Country	1960 Area	Annual Area Change	1960 Yield	Annual Yield Change	1960 Production	Annual Production Change
	(1,000 ha)	(%)	(100 kg/ha)	(%)	(1,000 metric tons)	(%)
United States	64,373	0.1	23.7	3.8	152,660	3.9
Canada	19,667	0.0	14.4	1.4	28,410	1.4
Mexico	7,335	2.7	10.0	2.4	7,325	5.1
British Honduras	7	0.0	10.0	0.0	7	0.0
Costa Rica[a]	51	3.1	12.0	0.0	184	1.4
Cuba	272	0.0	13.9	0.6	378	0.6
Dominican Republic[a]	57	2.1	23.9	3.4	238	3.9
El Salvador	285	0.0	10.7	0.2	305	0.2
Guatemala	678	1.6	7.7	0.1	520	1.7
Haiti[b]	...	...	...	...	124	0.0
Honduras	453	3.0	8.1	0.0	368	3.0
Jamaica	9	0.0	12.2	0.0	11	0.0
Nicaragua	217	0.9	9.7	0.3	211	1.2
Panama	183	2.1	10.7	0.2	195	2.3
Brazil	10,731	2.8	13.2	0.4	14,217	3.2
Argentina	10,346	0.4	18.7	0.2	14,153	0.6
Uruguay	932	0.0	8.4	0.2	781	0.2
Bolivia[c]	26	5.0	15.6	0.0	413	0.5
Chile	1,120	0.7	14.2	0.8	1,590	1.5
Colombia	1,141	0.7	12.9	1.3	1,467	2.0
Ecuador	509	2.5	10.6	0.0	540	2.5
Paraguay	115	0.0	12.9	0.3	148	0.3
Peru	660	1.2	14.6	0.6	964	1.8
Venezuela	432	4.6	12.0	0.4	517	5.0

[a] Production estimates are for corn and rice. Area and yield estimates are for rice only.
[b] Only production trends were estimated.
[c] Production estimates are for wheat, barley, corn, and rice. Area and yield estimates are for rice only.

importance of wheat production in the western provinces accounts for this result in Canada, but other examples are observed later.

Over half the 11.9 million acres of the 22-crop total in Mexico was in corn. The upward trend in this crop's area was the major factor leading to the 2.7 percent annual increase in cereals area. The corn yield increase was estimated at 1.7 percent annually in 1960, though the yield level of 8.8 hundred kilograms per hectare was still relatively low. Wheat was a major contributor to the 2.4 percent annual increase in yields for all cereals reported in Table 6.2. The estimated yearly yield increase for wheat was 6.7 percent. Production increases were found among most of the crops studied. Both the area and yield increases for Mexico denote an agricultural sector of rapidly expanding output.

Crop coverage was much less complete in our analysis of trends for the 12 Central American and Caribbean countries. Cereals are less important and sugar and fruit are important as export crops throughout the region. Noteworthy yield increases were observed for rice, tobacco, and potatoes in Cuba and rice in the Dominican Republic. Most production gains in other countries were modest and those of significance tend to be associated with area expansion.

Next to the United States, in the Americas, Brazil has both the largest population and the greatest area under agricultural crops. Like most tropical countries of the hemisphere, area expansion has been the primary source of increased output. Table 6.1 shows an average yield trend of only 0.3 percent per year. For the crops analyzed, nearly 40 percent of total area was in cereals, and most of this was in corn and rice. A 0.6 percent annual increase for rice was the only positive yield trend among the cereals. Root crops, a dietary staple for many Brazilians, also showed distinct upward area trends in 1960. Yields of potatoes, sweet potatoes, yams, and cassava increased slightly more rapidly than yields of cereals.

Agriculture of Argentina and Uruguay is dominated by Temperate Zone crops, but estimates given in Tables 6.1 and 6.2 show that neither country made great progress in agricultural production as of 1960. Neither area nor yield of any major crop was increasing in Uruguay, while only corn and groundnuts area, and millet and sugarcane yields were increasing significantly in Argentina. Production of several Argentine fruit crops was tending upward, however. The increase in yields for all cereals shown for Uruguay (Table 6.2) arises from an increasing trend for rice area. Rice yields were considerably higher than yields of other cereals in Uruguay. Thus the commodity mix effect results in a positive total cereals yield trend, even though no positive yield trends were identified in our analysis for the five individual cereal crops grown in the country. While lack of progress in yield improvement is certainly not uncommon, the situation in Argentina and Uruguay is particularly interesting. In their natural resource endowments and climatological features, these countries are not unlike the world's leaders in achieving yield gains. These two, however, were making little or no progress in 1960.

Finally we consider the remaining seven countries in South America. Table 6.1 shows that rates of total crop area expansion ranged from 0.7 to 4.1 percent per year, while average annual yield increases varied from zero to only 1 percent in 1960. The analysis of cereals trends reported in Table 6.2 shows similar results. Bolivia's agricultural production data are extremely sketchy, but rice, potatoes, and sugar production were increasing through area expansion. No trend was evident for corn, the major cereal. Among other significant trends were the moderate wheat, rye, and corn yield increases in Chile. Noteworthy positive yield trends were reported for several significant crops in Colombia. However, no positive yield trend was evident for corn, the crop having the largest planted area. The 1 percent average annual yield increase shown for all crops studied in Ecuador reflects the substantial gains measured in sugarcane and groundnuts yield. As Table 6.2 shows, no yield progress was evident among the cereals. Analysis of trends for 11 crops in Paraguay revealed practically no progress in yields as of 1960, and only a moderate increase in total crop area. With the exception of a 2.4 percent per year increase in rice yields for Venezuela, measured yield increases in that country and Peru were limited to minor crops. The striking increase in Venezuela's total crop area reflects mostly the cereals area trend of 4.6 percent per year as shown in Table 6.2.

PRODUCTION TRENDS IN EUROPE, THE USSR, AND OCEANIA

This grouping includes a majority of the world's economically developed countries. With few exceptions, their agriculture is dominated by Temperate Zone crops, land productivity is high, and continued progress in yield improvement is clearly evident. Table 6.3 and 6.4 present summary results from our analysis of crop area and yield trends in these countries. Average annual crop yield changes estimated for 1960 (Table 6.3) exceeded 1 percent in 22 of the 26 countries studied. In 7, average yield increases in excess of 2 percent per year were estimated. Table 6.3 shows a pattern of cropland trends unlike that of Latin America. Over half the countries in this grouping exhibited negative trends in total crop area as of 1960. However, there were some instances of positive trends in total crop area. Among them, France and West Germany, where the crops studied totaled 16.5 million hectares in 1960, had an estimated 0.2 percent annual increase in cropland. Rumania had a 0.3 percent annual increase measured over 22 crops totaling 9.3 million hectares in 1960. The USSR, with a 144.4 million hectare total, was growing at 0.4 percent annually, and Australia, with 8.3 million hectares, was increasing at 3.4 percent.

TABLE 6.3. Aggregate Crop Area and Yield Trends in Europe, the USSR, and Oceania, 1960

Country	Number of Crops Included	Total Crop Area	Total Crop Area Change	Average Annual Yield Change
		(1,000 ha)	*(% per year)*	*(%)*
Austria	13	1,097	—0.4	2.6
Belgium and Luxembourg	13	743	—0.9	1.1
Denmark	10	1,351	0.8	0.9
Finland	9	1,119	0.4	0.8
France	22	10,666	0.2	2.9
Ireland	6	553	—2.3	1.8
Netherlands	13	754	—0.2	1.2
Norway	5	271	1.6	0.1
Sweden	12	1,494	0.3	1.6
Switzerland	10	224	—0.1	2.3
United Kingdom	12	3,618	—0.9	2.1
West Germany	17	5,846	0.2	1.6
Greece	22	2,239	0.2	2.4
Italy	24	8,636	—1.0	2.0
Portugal	10	2,548	0.0	0.5
Spain	25	8,757	—0.6	1.2
Bulgaria	23	3,401	—0.3	1.5
Czechoslovakia	20	3,618	—0.8	1.8
East Germany	17	3,215	—0.6	1.2
Hungary	21	4,031	—0.2	2.3
Poland	16	12,565	—0.3	1.7
Rumania	22	9,322	0.3	2.0
Yugoslavia	26	6,268	—0.1	1.8
USSR	16	141,399	0.4	1.3
Australia	21	8,255	3.4	1.4
New Zealand	12	156	1.1	1.0

Table 6.4 shows that cereals yields were generally high throughout Europe. The Soviet Union, with a 1960 average yield of only 860 kilograms per hectare on 113 million hectares, ranked decidedly lower than other countries of Europe or Oceania. This low average yield again reflects the large area of wheat grown on arid lands. A similar comment applies to the average cereals yield reported for Australia.

A comparison of yield data for cereals in Table 6.4 with yields of all crops in Table 6.3 shows that not only were cereals the most important class of crops studied but also that rates of increase in their yields were usually significantly greater than for the average of all crops. Switzerland was the only case where the reverse was true. The two rates were equal in Yugoslavia, the USSR, and Australia. These comparisons illustrate that insights can be gained from detailed commodity analysis of world production trends. While cereals deserve special attention, important developments can easily be missed if analysis is focused too narrowly on this group.

TABLE 6.4. 1960 Cereals Area, Yield, Production, and Trends in Europe, the USSR, and Oceania

Country	1960 Area	Annual Area Change	1960 Yield	Annual Yield Change	1960 Production	Annual Production Change
	(1,000 ha)	(%)	(100 kg/ha)	(%)	(1,000 metric tons)	(%)
Austria	873	-0.4	23.7	3.0	2,072	2.6
Belgium and Luxembourg	559	-0.8	34.7	1.5	1,942	0.7
Denmark	1,194	1.2	35.8	1.1	4,269	2.3
Finland	1,002	0.6	16.8	0.9	1,687	1.5
France	8,982	0.5	24.3	3.6	21,791	4.1
Ireland	428	-2.3	30.0	2.6	1,284	0.3
Netherlands	468	0.1	35.8	1.9	1,677	2.0
Norway	218	2.2	23.3	0.2	509	2.4
Sweden	1,251	0.6	22.7	1.8	2,842	2.4
Switzerland	160	0.3	31.7	2.0	507	2.3
United Kingdom	3,074	-0.7	31.9	2.8	9,793	2.1
West Germany	4,465	0.6	29.2	2.0	13,021	2.6
Greece	1,702	0.0	14.3	2.7	2,438	2.7
Italy	6,724	-1.1	20.6	2.1	13,844	1.0
Portugal	1,959	-0.1	8.2	0.9	1,608	0.8
Spain	7,246	-0.7	11.6	1.5	8,412	0.8
Bulgaria	2,571	-0.7	14.4	2.0	3,712	1.3
Czechoslovakia	2,667	-0.8	22.0	2.6	5,858	1.8
East Germany	2,023	-0.4	23.7	2.0	4,794	1.6
Hungary	3,413	-0.3	18.8	2.9	6,415	2.6
Poland	8,944	-0.9	15.9	2.4	14,247	1.5
Rumania	7,047	-0.6	14.1	2.2	9,949	1.6
Yugoslavia	5,406	-0.3	17.1	1.8	9,263	1.5
USSR	112,913	0.0	8.6	1.3	96,636	1.3
Australia	7,908	3.5	11.8	1.4	9,327	4.9
New Zealand	107	1.6	29.1	1.5	311	3.1

COUNTRY ANALYSIS

We turn now to some of the highlights of production trends in individual countries and areas within Europe, the USSR, and Oceania. Among the countries of Northern Europe, the first 12 listed in Tables 6.3 and 6.4, agricultural production trends show uniformities relatively easy to describe. Total crop area changes were, as was indicated earlier, only modestly positive or were negative. Outside the cereals group, potatoes were a significant crop in nearly every country, but the area trend was negative in every case. In all countries, area trends for rye and oats were either constant or declining, the latter being commonest.[1]

Barley is the major European feed grain. Its area was increasing in every country, often at a rapid rate. The reason is clear: rising incomes have increased demand for livestock products, and in turn, livestock feed demand has increased. Noteworthy increases in wheat area were indicated in Austria, the Netherlands, Switzerland, and West Germany; and France's corn area showed a significantly rising trend.

Agricultural production patterns were somewhat less consistent among the four countries of Southern Europe (Greece, Italy, Portugal, and Spain). Wheat and barley areas were increasing in Greece, but declines in areas of six other cereals resulted in the zero growth rate for total cereals area shown in Table 6.4. Yield increases were well distributed among crops, indicating broadly based progress in agricultural development. For the 24 crops whose area trends were analyzed in Italy, the net effect was a 1.0 percent annual decline in cropped area (Table 6.3). Most yield trends were distinctly positive as indicated by the 2.0 percent average yield increase reported in Table 6.3. However, the combined effect of area and yield trends is a mixed pattern of positive and negative production trends for individual crops. Portugal's agriculture, with no trend in total crop area and only 0.5 percent annual average increase in yields, was nearly stagnant. In Spain, five of the six cereal crops were declining in area. Only corn, a relatively minor crop, showed an increased area. Yield increases, most of them modest, prevailed for 16 of the crops studied.

Production of fruit crops is important in all countries of Southern Europe and generally showed strong positive trends in production. The trends were particularly notable in Italy and Spain. In Spain, for example, 11 of the 12 fruit crops studied showed positive trends ranging from 2.3 to 9.4 percent annually. Apparently some of the decline in total crop area for Italy and Spain (Table 6.3) results from the shift toward fruit production.

Analysis of trends for the seven Eastern European countries showed

---

1. It should be noted, however, that extrapolated area trends for these crops will reach zero before 2000 in many cases, and that the nature of the projection procedure is such that the trend in total cropland may recover and even become positive as the crops with negative area trends "drop out."

considerable diversity. Total 1960 crop area was falling at small to moderate rates in every country except Rumania. Each country produced substantial amounts of rye and oats, but areas of these crops were either constant or falling. Unlike the countries of Western Europe, there was no uniform tendency toward an increasing area of feed grains. In Czechoslovakia, however, barley area was increasing 1.0 percent annually and corn was increasing 2.0 percent. Small upward trends prevailed in Bulgaria and East Germany. Feed grains area was constant or falling in the remaining four countries. Wheat area was constant or falling in all countries. The total impact of these tendencies is a negative trend in total cereals area in each of the seven Eastern European countries. However, yield trends were generally upward, with average cereals yields increasing from 1.8 to 2.9 percent per year.

Outside the cereals group the situation was mixed. Potato area was declining in Czechoslovakia and East Germany but increasing in Hungary, Poland, Rumania, and Yugoslavia. Positive yield trends were generally evident in Bulgaria, Hungary, Poland, Rumania, and Yugoslavia. However, sunflower seed was the only significant noncereal crop showing a positive yield trend in Czechoslovakia, and no increasing yield trends were identified among the noncereal crops in Hungary. Production for several fruit crops in Bulgaria and Yugoslavia was sharply rising, much like that of countries in Southern Europe.

Data for 16 crops in the Soviet Union indicated that total area was rising 0.4 percent per year and average yields were increasing 1.3 percent (Table 6.3). Wheat accounted for about half the 113 million hectare total cereals area shown in Table 6.4. No wheat area trend was evident, but an upward yield trend of 1.4 percent per year was estimated. Decreases in rye and oats area were matched by increases in corn and barley. Outside the cereals group, substantial positive area trends were estimated for sugarbeets, dry peas, soybeans, sunflower seed, and cotton. Dry peas and sunflower seed showed the most noteworthy yield gains.

Rapidly rising wheat area was the primary cause of the 3.4 percent estimated annual increase in Australia's total cropland as of 1960. Unlike the trend in most countries, the estimated trend for oats area was positive. Yield trends were positive for five of the eight cereals grown, leading to the 1.4 percent annual total cereals yield increase shown in Table 6.4. Sugarcane, the only crop of any significance outside the cereals group, had a 2.9 percent yearly yield increase. New Zealand had an increasing area for only wheat, barley, and corn. Similarly, most of the yield increases were found in the cereals group.

PRODUCTION TRENDS IN AFRICA AND ASIA

Tables 6.5 and 6.6 summarize results for Africa and Asia. With the few exceptions noted earlier, the countries of this group are underde-

TABLE 6.5. Aggregate Crop Area and Yield Trends in Asia and Africa, 1960

Country	Number of Crops Included	Total Crop Area	Total Crop Area Change	Average Annual Yield Change
		(1,000 ha)	*(% per year)*	*(%)*
United Arab Republic	17	3,086	0.7	0.9
Cyprus	13	161	0.6	1.0
Iran	6	5,017	2.7	0.3
Iraq	13	2,910	1.3	0.1
Israel	18	189	1.7	2.5
Jordan	11	394	0.3	0.4
Lebanon	14	105	0.3	0.3
Syria	18	2,039	2.3	0.6
Turkey	25	13,058	1.9	0.6
India	23	133,051	0.7	1.1
Ceylon	12	840	0.8	2.3
Pakistan	19	21,720	0.8	0.9
Japan	27	6,469	—0.9	1.8
Burma	11	6,542	1.7	1.4
Cambodia	9	2,437	1.1	0.0
China (Taiwan)	22	1,429	0.1	2.5
Federation of Malaya	4	394	0.7	2.6
Indonesia	11	13,056	2.1	0.8
Philippines	14	5,749	2.3	0.4
South Korea	19	2,803	1.1	0.7
South Vietnam	12	2,533	3.8	0.1
Thailand	11	6,671	2.5	1.5

veloped. In general, area and yield data for African countries were not sufficiently reliable or complete to warrant continuation with our preliminary attempts at estimation. Thus the United Arab Republic is the only African country for which data are reported in Table 6.5 based on area and yield trends.

Trends in production (not based on separate yield and area estimates) of most major crops in Africa were estimated and results for the cereals are presented in Table 6.6. Cereals trends have somewhat less significance in many African countries because they dominate production and consumption to a lesser degree than in other world regions.

The estimates in Table 6.5 are somewhat similar to the results reported for Latin American countries and contrast sharply to those for the United States, Canada, and Europe. Expansion of crop area was a major factor in increasing production as of 1960. Japan was the only country with a negative trend in total crop area. Countries in this group showed somewhat more progress in yield improvement than most Latin American countries, but the trends estimated were not comparable to those of Europe and the United States. Cereals occupied two-thirds or more of the cropland for crops used in analyses of Asian trends, and area and yield increases for cereals were similar to those for all crops.

TABLE 6.6. 1960 Cereals Area, Yield, Production, and Trends in Asia and Africa

Country	1960 Area (1,000 ha)	Annual Area Change (%)	1960 Yield (100 kg/ha)	Annual Yield Change (%)	1960 Production (1,000 metric tons)	Annual Production Change (%)
Algeria[a]					2,115	0.0
Ethiopia[a]					2,683	1.9
Libya[a]					150	0.2
Morocco[a]					2,718	0.0
Sudan[a]					1,546	3.0
Tunisia[a]					650	0.0
United Arab Republic	1,974	0.7	27.3	1.3	5,390	2.0
Angola[a]					510	0.3
Cameroun[a]					402	1.0
Kinshasa, Congo, Rwanda, Burundi[a]					824	0.0
Ghana[a]					451	2.5
Guinea[a]					571	1.6
Ivory Coast[a]					364	7.2
Liberia[a]					179	0.0
Nigeria[a]					4,499	2.6
Senegal[a]					489	2.6
Sierra Leone[a]					473	4.1
Togo[a]					184	0.4
Kenya[a]					1,467	0.2
Malagasy Republic[a]					1,254	2.6
Malawi, Rhodesia, Zambia[a]					1,877	3.9
Tanganyika[a]					1,732	1.1
Uganda[a]					806	2.3
Republic of South Africa[a]					6,253	0.7
Cyprus	140	0.4	9.7	1.4	136	1.8

[a] Only production trends were estimated.

TABLE 6.6 (cont.)

Country	1960 Area	Annual Area Change	1960 Yield	Annual Yield Change	1960 Production	Annual Production Change
	(1,000 ha)	(%)	(100 kg/ha)	(%)	(1,000 metric tons)	(%)
Iran	4,554	2.9	10.1	0.2	4,590	3.1
Iraq	2,802	1.2	7.0	−0.3	1,970	0.9
Israel	143	2.2	12.9	2.9	184	5.1
Jordan	342	0.0	6.5	0.0	223	0.0
Lebanon	81	0.1	9.1	0.0	74	0.1
Syria	1,613	2.1	7.1	−0.1	1,149	2.0
Turkey	11,321	2.0	10.9	0.3	12,340	2.3
India	89,723	0.6	9.0	1.7	80,506	2.3
Ceylon	512	1.2	17.0	2.6	868	3.8
Pakistan	16,511	0.8	12.0	1.3	19,789	2.1
Japan	4,831	−1.0	39.7	2.5	19,155	1.5
Burma	5,069	1.6	15.5	1.3	7,832	2.9
Cambodia	2,344	1.2	11.2	0.0	2,614	1.2
China (Taiwan)	809	0.1	32.8	2.7	2,651	2.8
Federation of Malaya	375	0.8	23.9	2.7	898	3.5
Indonesia	9,856	2.0	15.6	0.9	15,378	2.9
Philippines	4,857	2.3	10.0	0.1	4,866	2.4
South Korea	2,276	1.3	19.6	0.9	4,471	2.2
South Vietnam	2,359	3.8	20.7	0.0	4,875	3.8
Thailand	6,166	2.3	14.2	1.6	8,752	3.9

87

Trends in cereals production in Africa were mixed, varying from zero to over 3 percent annual increases in moderately important cereal-producing countries.

COUNTRY TRENDS

We turn now to some of the more important production trends of individual countries. Contrasting results prevailed among the seven countries of North Africa; Ethiopia, Sudan, and the UAR had generally positive production trends. All but 2 of the 15 Ethiopian crops had positive trends, although millet and sorghum, the most important cereals, were increasing less rapidly than the average of all cereals. Eleven of 14 production trends analyzed for Sudan were increasing at annual rates of 2.1 percent or more (the remaining 3 were minor crops). Similar results were obtained for the UAR. On the other hand, only scattered positive production trends (none among leading crops) prevailed for Algeria, Libya, Morocco, and Tunisia.

Agricultural production in the 11 countries of West Central Africa centers around tropical crops, particularly sweet potatoes, yams, and cassava. The Congo, second most populous country in the region, was making the least progress in agricultural production by 1960. The effects of the civil war were very damaging to agriculture through the postwar period. Again, in Nigeria, the recent civil war makes the assessment of future production trends tenuous. Production trends in Ghana and Ivory Coast were more favorable than in other countries of the region and exceeded those for Angola and Nigeria, which also were predominantly positive. Production gains in Cameroun and Togo were generally small for significant crops. Agricultural production progress in the remaining countries was even less impressive.

Overall agricultural productivity trends among the five countries of East Africa were found to be little different from those in the West Central region. Output was by no means stagnant, but production increases were not large relative to population growth rates. Furthermore, the most notably positive production trends frequently were for export crops rather than subsistence crops. Among production trends analyzed for South Africa, sizable growth rates were found for many crops. However, no trend was evident in corn, the major subsistence crop.

Israel and Cyprus, two of the smaller countries of West Asia, were achieving significantly greater gains in agricultural productivity than other countries of the group. Increasing areas of barley and potatoes account for the rising areas of cereals and of all crops in Cyprus which had a 2.1 percent annual increase in barley yield. Area increases of 4.4 and 4.7 percent, respectively, were estimated for sorghum and wheat in Israel. These area increases were paralleled by 7.9 and 2.9 percent annual yield increases. Significantly positive yield trends were estimated for 8 of 14 noncereal crops, and very striking production increases were evident for each of the 10 Israeli fruit crops considered.

Moderately to rapidly rising trends in total cropland were measured in all other countries of the region. Iran's 2.7 percent increase (Table 6.5) reflects rising trends for barley, rice, and particularly wheat, the country's principal cereal. Four of the 7 crop yield trends analyzed were increasing, but production increases for the major cereals were solely a reflection of increasing area. Wheat area expansion was also the major determinant of Iraq's 1.3 percent yearly increase in total cropland. Increases in area and yield were limited to minor crops in both Jordan and Lebanon. However, 11 positive production trends were found among the 18 fruit crops. Expanding wheat and barley areas were major sources of the 2.3 and 1.9 percent annual total cropland increases, respectively, for Syria and Turkey. Among Syria's major crops, only cotton appeared to be increasing in yield. Our estimates showed Turkey's overall yield trends to be similar to those in Syria.

With the analysis referring to the period before the green revolution, crop area was increasing at about 0.7 percent per year in India and average yields for the 23 crops studied were rising at about 1.1 percent annually. Wheat, rice, and groundnuts accounted for a 750,000-hectare increase annually, but positive area trends were evident for many other crops. Yield trends for several major crops were slightly positive. The largest yield increase for any crop was 2.4 percent per year.

Total crop area was increasing by 0.8 percent per year in both Ceylon and Pakistan. Rice area was the most important contributor in both cases, but wheat, sugarcane, and cotton areas also showed significant increases in Pakistan. The trend in Ceylon tea yield was upward at 2.9 percent per year while rice yield was increasing 2.6 percent per year. Positive yield trends in Pakistan were limited to rice, cotton, linseed, and tobacco. However, none of these was large.

Japan's agricultural sector is commonly regarded as the most advanced in Asia. Estimates in Tables 6.5 and 6.6 substantiate this evaluation. Average cereal yields were estimated at 3,970 kilograms per hectare in 1960, the highest in the world. Also, the annual increase was 2.5 percent. The average annual increase for all Japanese crop yields was estimated to be 1.8 percent. Area for all crops and cereals was falling at about 1 percent per year, except for rice which was increasing slightly.

Trends in total crop area were positive in each of the nine countries in the final grouping, Other East Asia, where rice is the dominant crop. The rising rice area was the important contributor to area expansion of cereals cropland generally in every country except Taiwan. Over the entire group of Asian countries, cereals yields in Taiwan are second only to those of Japan.

Cereals yields were comparable in Burma and Thailand, where increases in both area and yields of most crops were significant. Table

6.5 shows that 1960 total crop area was increasing 1.7 percent annually in Burma and 2.5 percent in Thailand, while average yields were growing at 1.4 and 1.5 percent, respectively. Similar yield trends prevailed in Indonesia, the Philippines and South Korea. As a group, these three countries were making less rapid yield progress than Burma and Thailand. Rice area was the most significant upward trend for Cambodia and South Vietnam. Virtually no yield progress was measured for Cambodia and the only positive yield trends identified for South Vietnam were among minor crops. Few crops were included in the analysis of Malayan agricultural trends. Rice only is included in Table 6.6 and had an 0.8 percent annual increase in area and a 2.7 percent increase in yield.

This completes the summary of world agricultural production trends in 1960. Chapters which follow analyze continuation of these trends in terms of the balance between future food production and food demand in 1985 and 2000.

CHAPTER 7

Projected Agricultural Production-Demand Comparisons for the Americas

THE TIME TRENDS presented and discussed in preceding chapters provide the basic building blocks for comparing future production and demand. Production is projected subject to constraints on land expansion where possible, and demand projections are derived from projected population and income by methods described in Chapter 3.

Chapters 7, 8, and 9 examine the resulting comparisons of production and demand potentials for the 21 regions defined in Table 3.1. Our assessment of these results is intended to suggest where and when possible serious imbalances may occur, and to analyze the manner in which the factors under study simultaneously determine the outcome as they evolve through time.

For each region we begin by considering the production-demand balance as it existed in 1960. These estimates are based on USDA's food balances, except that they are adjusted as indicated in Chapter 3. Conceptually, then, the entries titled "domestic disappearance" include all demand for primary agricultural products. As we discuss the various future projections under different population and income assumptions, it is convenient to designate one set of projections, those made under assumptions of medium population and constant per capita income, as a "benchmark" set. This designation is somewhat arbitrary, and it carries no connotation that the associated projections are necessarily most likely to be realized or that they are in any sense "preferred." However, it will be seen that population growth is generally a much more important determinant of future food demand than is per capita income. It is frequently useful to focus on a set of projections which illustrate the effects of intermediate level population growth on future supply-demand balances while holding per capita income fixed at the base period level. The benchmark projections provide such a set. Results under other assumptions about the determinants of food demand (low or high population and increasing per capita income) can then be related to the benchmark projections.

In this chapter we consider the projections for the seven regions of the Western Hemisphere, beginning with the United States.

TABLE 7.1.   1960 Production-Demand Comparisons for the United States (1,000 metric tons)

Commodity	Domestic Disappearance			Production	Excess Demand
	Food and Industrial	Feed	Total		
Cereals	24,377	98,536	122,913	157,495	−34,582
Raw sugar	8,644	27	8,671	2,744	5,927
Root crops	10,604	1,519	12,123	12,796	−673
Pulses	1,115	2	1,117	1,042	75
Fruit, veg.	36,058	...	36,058	35,425	633
Oil crops	2,485	8,655	11,140	15,601	−4,462
Meat	16,871	...	16,871	16,868	3
Milk	49,012	1,225	50,237	55,955	−5,718
Eggs	3,755	...	3,755	3,842	−87

COMPARISONS FOR THE UNITED STATES

The adjusted 1960 food balance for the United States is shown in Table 7.1. Of course, total food demand contained a sizable component of livestock products. Because of this, over 80 percent of domestic disappearance of cereals and oil crops is represented by feed demand. Notwithstanding the large feed demand for these two commodity classes, production of both was well in excess of domestic disappearance. Production of sugar was only about 30 percent of domestic disappearance, but this was the only commodity group supplied primarily from imports.

United States income elasticities as of 1960 were among the world's lowest. The highest estimate, 0.21, was for meat. Among the negative elasticities estimated were −0.2 for cereals and milk and −0.1 for starchy roots, pulses, and eggs. Other elasticities used were 0.15 for fruit and vegetables, 0.1 for oil crops, and zero for sugar. It can be inferred that most future demand increases will come from population growth. Income increases will affect demand for cereals positively through feed demand since meat production is the largest determinant of feed demand, but a counterbalancing effect can be foreseen through reduced direct per capita consumption of cereals at higher income levels.

The U.S. production projections used in all production-demand comparisons are given in Table 7.2. Crops for which trends have been

TABLE 7.2.   1985 and 2000 Production Projections for the United States (1,000 metric tons)

Commodity	1985 Production	2000 Production
Cereals	331,533	460,448
Raw sugar	6,409	9,004
Root crops	18,147	21,006
Pulses	1,431	1,662
Fruit, veg.	37,681	35,725
Oil crops	50,805	78,308

estimated account for a very high percentage of total production in each product class except fruit and vegetables. There the estimated coverage is 60 percent.

Projected production increases are especially large for cereals and oil crops. However, we reemphasize a point made in Chapter 6 with respect to the U.S. cereals projections, namely, that positive area trends for wheat and corn were specified on a priori grounds. These trends, together with the yield trends for U.S. cereals, represent the potential cereals production capacity which could be mobilized if the nation were called upon to adopt a policy of expanding cereals acreage rather than restricting it as in the present and recent past. The total projected area of U.S. crops included in this analysis is 30.6 million hectares greater than in 1960 by the year 2000. Of this about 20 million represents increases in wheat and corn area.

With medium population growth and constant per capita income, production of cereals and oil crops would exceed domestic demand by a very large margin in 1985 (Table 7.4). The 158 million ton surplus of cereals production projected for 1985 amounts to about half of estimated domestic production, while only about 30 percent of the estimated 50 million tons production of oil crops would be required domestically. Fruit and vegetable production does not keep pace with domestic demand, even with per capita income assumed constant. Projected sugar production increases about keep pace with demand increases in 1985, leaving the country a net importer at about 1960 levels. Of course, no changes in demand for sugar are anticipated under other income assumptions because of the zero income elasticity.

All income effects are relatively minor as would be expected. With income increasing at historical rates and medium population growth, 1985 per capita income would be 69 percent above benchmark levels. However, the two corresponding estimates for meat demand shown in Table 7.4 differ by only 11 percent. Feed demands rise 6 percent above benchmark levels as the result of demand differences for meat, milk, and eggs. Changes in direct and indirect demand for cereals result in a net 3.3 percent increase above the benchmark level, while for oil crops, the total increase is 5.6 percent. It will be recalled that the income elasticity of direct demand for oil crops is positive, and that it is negative for cereals. Fruit and vegetables show an 8 percent increase in demand and a 30 percent increase in the projected deficit when comparing the benchmark projections for 1985 to those with increased per capita income.

Comparing 1985 demand under low versus high population assumptions, but with constant per capita income in both cases (Tables 7.3 and 7.5), demand for cereals is found to be 14 percent or 23 million tons greater. Of course, demand for all product classes differs by the same percent in this comparison since per capita income is the same. It is

TABLE 7.3. 1985 and 2000 Production-Demand Comparisons for the United States, Assuming Low Population Growth Rates (1,000 metric tons)

1985 Comparisons, Assuming Constant Per Capita Income

Commodity	Domestic Disappearance Food and Industrial	Feed	Total	Excess Demand
Cereals	31,676	128,039	159,715	−171,817
Raw sugar	11,233	35	11,268	4,859
Root crops	13,779	1,974	15,753	−2,393
Pulses	1,449	3	1,452	21
Fruit, veg.	46,854	...	46,854	9,173
Oil crops	3,229	11,246	14,475	−36,329
Meat	21,923		21,923	...
Milk	63,687	1,592	65,279	...
Eggs	4,880	...	4,880	...

2000 Comparisons, Assuming Constant Per Capita Income

Commodity	Domestic Disappearance Food and Industrial	Feed	Total	Excess Demand
Cereals	35,615	143,962	179,577	−280,871
Raw sugar	12,630	39	12,669	3,665
Root crops	15,493	2,219	17,712	−3,293
Pulses	1,630	3	1,633	−28
Fruit, veg.	52,681	...	52,681	16,957
Oil crops	3,631	12,645	16,276	−62,031
Meat	24,649		24,649	...
Milk	71,607	1,790	73,397	...
Eggs	5,487	...	5,487	...

1985 Comparisons, Assuming Historical Income Growth Rates

Commodity	Domestic Disappearance Food and Industrial	Feed	Total	Excess Demand
Cereals	29,061	137,039	166,100	−165,432
Raw sugar	11,233	37	11,270	4,861
Root crops	13,168	2,112	15,280	−2,865
Pulses	1,385	3	1,388	−42
Fruit, veg.	51,099	...	51,099	13,418
Oil crops	3,379	12,037	15,416	−35,389
Meat	24,704		24,704	...
Milk	58,166	1,704	59,870	...
Eggs	4,664	...	4,664	...

2000 Comparisons, Assuming Historical Income Growth Rates

Commodity	Domestic Disappearance Food and Industrial	Feed	Total	Excess Demand
Cereals	31,570	161,644	193,214	−267,233
Raw sugar	12,630	44	12,674	3,669
Root crops	14,540	2,492	17,032	−3,973
Pulses	1,529	4	1,533	−128
Fruit, veg.	60,633	...	60,633	24,909
Oil crops	3,868	14,198	18,066	−60,240
Meat	29,858		29,858	...
Milk	63,074	2,010	65,084	...
Eggs	5,149	...	5,149	...

TABLE 7.4. 1985 and 2000 Production-Demand Comparisons for the United States, Assuming Medium Population Growth Rates (1,000 metric tons)

1985 Comparisons, Assuming Constant Per Capita Income

Domestic Disappearance

Commodity	Food and Industrial	Feed	Total	Excess Demand
Cereals	34,324	138,744	173,068	−158,464
Raw sugar	12,172	38	12,210	5,801
Root crops	14,931	2,139	17,070	−1,076
Pulses	1,571	3	1,574	142
Fruit, veg.	50,772	...	50,772	13,090
Oil crops	3,499	12,187	15,686	−35,119
Meat	23,756	...	23,756	...
Milk	69,012	1,725	70,737	...
Eggs	5,288	...	5,288	...

1985 Comparisons, Assuming Historical Income Growth Rates

Domestic Disappearance

Commodity	Food and Industrial	Feed	Total	Excess Demand
Cereals	31,766	147,102	178,868	−152,665
Raw sugar	12,172	40	12,212	5,803
Root crops	14,335	2,268	16,603	−1,544
Pulses	1,508	3	1,511	80
Fruit, veg.	54,760	...	54,760	17,079
Oil crops	3,465	12,921	16,565	−34,239
Meat	26,368	...	26,386	...
Milk	63,608	1,829	65,437	...
Eggs	5,077	...	5,077	...

2000 Comparisons, Assuming Constant Per Capita Income

Domestic Disappearance

Commodity	Food and Industrial	Feed	Total	Excess Demand
Cereals	42,685	172,543	215,228	−245,219
Raw sugar	15,137	47	15,184	6,180
Root crops	18,568	2,660	21,228	223
Pulses	1,953	4	1,957	295
Fruit, veg.	63,140	...	63,140	27,415
Oil crops	4,351	15,155	19,506	−58,800
Meat	29,543	...	29,543	...
Milk	85,823	2,145	87,968	...
Eggs	6,576	...	6,576	...

2000 Comparisons, Assuming Historical Income Growth Rates

Domestic Disappearance

Commodity	Food and Industrial	Feed	Total	Excess Demand
Cereals	38,363	189,587	227,950	−232,497
Raw sugar	15,137	52	15,189	6,184
Root crops	17,554	2,923	20,477	−528
Pulses	1,846	4	1,850	189
Fruit, veg.	70,956	...	70,956	35,231
Oil crops	4,603	16,652	21,255	−57,051
Meat	34,663	...	34,663	...
Milk	76,701	2,357	79,058	...
Eggs	6,217	...	6,217	...

TABLE 7.5. 1985 and 2000 Production-Demand Comparisons for the United States, Assuming High Population Growth Rates (1,000 metric tons)

1985 Comparisons, Assuming Constant Per Capita Income

Commodity	Domestic Disappearance			Excess Demand
	Food and Industrial	Feed	Total	
Cereals	36,150	146,126	182,276	−149,256
Raw sugar	12,819	40	12,859	6,451
Root crops	15,725	2,253	17,978	−168
Pulses	1,654	3	1,657	226
Fruit, veg.	53,473	...	53,473	15,792
Oil crops	3,685	12,835	16,520	−34,284
Meat	25,020	...	25,020	...
Milk	72,684	1,817	74,501	...
Eggs	5,569	...	5,569	...

1985 Comparisons, Assuming Historical Income Growth Rates

Commodity	Domestic Disappearance			Excess Demand
	Food and Industrial	Feed	Total	
Cereals	33,657	153,993	187,650	−143,882
Raw sugar	12,819	42	12,861	6,453
Root crops	15,145	2,374	17,519	−627
Pulses	1,593	3	1,596	165
Fruit, veg.	57,258	...	57,258	19,576
Oil crops	3,826	13,526	17,352	−33,452
Meat	27,499	...	27,499	...
Milk	67,416	1,915	69,331	...
Eggs	5,364	...	5,364	...

2000 Comparisons, Assuming Constant Per Capita Income

Commodity	Domestic Disappearance			Excess Demand
	Food and Industrial	Feed	Total	
Cereals	45,336	183,257	228,593	−231,854
Raw sugar	16,077	50	16,127	7,123
Root crops	19,721	2,825	22,546	1,541
Pulses	2,074	4	2,078	416
Fruit, veg.	67,061	...	67,061	31,336
Oil crops	4,622	16,096	20,718	−57,589
Meat	31,377	...	31,377	...
Milk	91,153	2,279	93,432	...
Eggs	6,984	...	6,984	...

2000 Comparisons, Assuming Historical Income Growth Rates

Commodity	Domestic Disappearance			Excess Demand
	Food and Industrial	Feed	Total	
Cereals	40,957	199,918	240,875	−219,572
Raw sugar	16,077	55	16,132	7,127
Root crops	18,695	3,082	21,777	771
Pulses	1,966	4	1,971	309
Fruit, veg.	74,756	...	74,756	39,031
Oil crops	4,875	17,560	22,435	−55,871
Meat	36,418	...	36,418	...
Milk	81,909	2,486	84,395	...
Eggs	6,621	...	6,621	...

seen that the various demand projections differ more under the extreme population assumptions than under alternative income levels, even though the variability in income levels far exceeds the variability in population.

The estimates given in Tables 7.3 to 7.5 illustrate the effects of continued operation of the forces just described for another 15 years to 2000. Vigorous production trends, coupled with moderate population growth and severe damping of the effects of increased per capita income, result in extremely large production surpluses for cereals and oil crops under all demand assumptions. Extremes of cereals surpluses in amounts of 220 and 280 million metric tons are shown in Table 7.5 under the constant per capita income comparisons and in Table 7.3 under the increasing income comparisons, respectively. Corresponding extremes for oil crops are 56 and 62 million tons. Sugar, pulses, and root crops production grow at about the same rate as demand, while vegetables and fruit show increasing production deficits.

As a set, the U.S. projections indicate that agricultural capacity will be more than equal to any foreseeable domestic needs. Continued reliance on sugar imports from abroad is indicated, but it seems unlikely that the U.S. sugar market will offer major opportunities for expanded exports from developing countries because of modest population growth and inelasticity of demand with respect to income. Projected demand for fruit and vegetables outpaces production, but in view of the general pattern of excess production capacity shown in Tables 7.3 to 7.5, it would seem unwise to assume that production patterns would not adjust to meet at least part of this demand. The estimated production and excess supply figures shown for cereals must be interpreted in the light of the comments made earlier about area trends for wheat and corn. Actual production of cereals in future needs will be determined by exports and possibly by government policy decisions, should it be determined that government financing of noncommercial foreign demand is necessary.

In summary, the projections indicate that production capacity of U.S. agriculture will far exceed any of the possible patterns of domestic demand which have been postulated for the future, and that a sizable reserve will exist to satisfy export demands which may materialize.

COMPARISONS FOR CANADA

Estimates presented in Table 7.6 show that Canada produced about 50 percent more cereals (primarily wheat) than was required domestically in 1960. Like the United States, about 80 percent of domestic cereals demand was for livestock feed. Sugar, pulses, fruit and vegetables, and oil crops were all consumed in proportions well above domestic production. However, cereals are much more important to Canadian

TABLE 7.6.   1960 Production-Demand Comparisons for Canada (1,000 metric tons)

Commodity	Domestic Disappearance			Production	Excess Demand
	Food and Industrial	Feed	Total		
Cereals	2,941	11,474	14,415	21,774	—7,359
Raw sugar	967	. . .	967	208	759
Root crops	1,808	105	1,913	1,906	7
Pulses	89	7	96	59	37
Fruit, veg.	3,117	. . .	3,117	1,861	1,256
Oil crops	190	474	664	377	287
Meat	1,454	. . .	1,454	1,454	. . .
Milk	7,233	1,114	8,347	8,533	—184
Eggs	302	. . .	302	306	—4

agriculture than any of these crops, and in any event, the absolute deficit levels among the lesser commodity groups were not large.

Base period income elasticities for Canada are similar to those for the United States. Elasticities for vegetables and fruit, oil crops, and meat—0.25, 0.2, and 0.26, respectively—are the only positive values estimated. Cereals, pulses, and eggs are estimated to have income elasticities of about —0.1, milk's elasticity is —0.2, and the elasticity estimated for starchy roots is —0.3. Sugar again shows no response to per capita income changes. From these values it is clear that population growth will be the major determinant of future Canadian domestic demand as was true for the United States.

Projected Canadian production of the six commodity groups under study are presented in Table 7.7. Most of the increases shown are reflections of yield improvement. Total cropland expansion proceeds very slowly. The constraint set on cropland expansion exceeds the 1960 total by only 3.6 percent, but estimated area time trends remain unbounded until after 1985.

Crops for which trends are estimated account for nearly all of estimated 1960 production in each product class except fruit and vegetables. Only 55 percent coverage is estimated for this class.

Projections to 1985 under assumptions of medium population growth and constant per capita income, the benchmark assumptions, are presented in Table 7.9. Cereals production advances well above de-

TABLE 7.7.   1985 and 2000 Production Projections for Canada (1,000 metric tons)

Commodity	1985 Production	2000 Production
Cereals	37,674	42,944
Raw sugar	268	296
Root crops	3,256	4,023
Pulses	67	67
Fruit, veg.	1,994	2,062
Oil crops	912	1,237

TABLE 7.8. 1985 and 2000 Production-Demand Comparisons for Canada, Assuming Low Population Growth Rates (1,000 metric tons)

1985 Comparisons, Assuming Constant Per Capita Income

Commodity	Domestic Disappearance			Excess Demand
	Food and Industrial	Feed	Total	
Cereals	4,267	16,648	20,915	−16,758
Raw sugar	1,403	…	1,403	1,135
Root crops	2,623	152	2,775	−481
Pulses	129	11	140	73
Fruit, veg.	4,522	…	4,522	2,528
Oil crops	275	688	963	52
Meat	2,109	…	2,109	…
Milk	10,494	1,616	12,110	…
Eggs	438	…	438	…

1985 Comparisons, Assuming Historical Income Growth Rates

Commodity	Domestic Disappearance			Excess Demand
	Food and Industrial	Feed	Total	
Cereals	4,026	18,360	22,386	−15,287
Raw sugar	1,403	…	1,403	1,135
Root crops	2,239	167	2,406	−849
Pulses	123	12	135	68
Fruit, veg.	5,370	…	5,370	3,376
Oil crops	306	759	1,065	153
Meat	2,521	…	2,521	…
Milk	9,443	1,783	11,226	…
Eggs	415	…	415	…

2000 Comparisons, Assuming Constant Per Capita Income

Commodity	Domestic Disappearance			Excess Demand
	Food and Industrial	Feed	Total	
Cereals	4,957	19,343	24,300	−18,643
Raw sugar	1,630	…	1,630	1,334
Root crops	3,047	176	3,223	−799
Pulses	150	13	163	96
Fruit, veg.	5,254	…	5,254	3,193
Oil crops	320	800	1,120	−117
Meat	2,451	…	2,451	…
Milk	12,192	1,878	14,070	…
Eggs	509	…	509	…

2000 Comparisons, Assuming Historical Income Growth Rates

Commodity	Domestic Disappearance			Excess Demand
	Food and Industrial	Feed	Total	
Cereals	4,580	22,900	27,480	−15,463
Raw sugar	1,630	…	1,630	1,334
Root crops	2,455	209	2,664	−1,358
Pulses	140	15	155	88
Fruit, veg.	6,927	…	6,927	4,865
Oil crops	369	947	1,316	79
Meat	3,262	…	3,262	…
Milk	10,557	2,223	12,780	…
Eggs	473	…	473	…

TABLE 7.9. 1985 and 2000 Production-Demand Comparisons for Canada, Assuming Medium Population Growth Rates (1,000 metric tons)

Commodity	1985 Comparisons, Assuming Constant Per Capita Income				1985 Comparisons, Assuming Historical Income Growth Rates			
	Domestic Disappearance			Excess Demand	Domestic Disappearance			Excess Demand
	Food and Industrial	Feed	Total		Food and Industrial	Feed	Total	
Cereals	4,657	18,172	22,829	−14,843	4,416	19,804	24,220	−13,453
Raw sugar	1,531	...	1,531	1,263	1,531	...	1,531	1,263
Root crops	2,863	166	3,029	−227	2,476	181	2,657	−599
Pulses	141	12	153	86	135	13	148	81
Fruit, veg.	4,936	...	4,936	2,942	5,754	...	5,754	3,759
Oil crops	300	751	1,051	140	331	819	1,150	238
Meat	2,302	...	2,302	...	2,699	...	2,699	...
Milk	11,455	1,764	13,219	...	10,397	1,923	12,320	...
Eggs	478	...	478	...	455	...	455	...

Commodity	2000 Comparisons, Assuming Constant Per Capita Income				2000 Comparisons, Assuming Historical Income Growth Rates			
	Domestic Disappearance			Excess Demand	Domestic Disappearance			Excess Demand
	Food and Industrial	Feed	Total		Food and Industrial	Feed	Total	
Cereals	6,149	23,993	30,142	−12,801	5,723	27,590	33,313	−9,630
Raw sugar	2,022	...	2,022	1,726	2,022	...	2,022	1,726
Root crops	3,780	219	3,999	−24	3,108	251	3,359	−663
Pulses	186	16	202	135	175	18	193	126
Fruit, veg.	6,517	...	6,517	4,456	8,241	...	8,241	6,179
Oil crops	397	992	1,389	152	452	1,141	1,593	356
Meat	3,040	...	3,040	...	3,876	...	3,876	...
Milk	15,124	2,330	17,454	...	13,273	3,679	15,952	...
Eggs	631	...	631	...	591	...	591	...

mand by 1985, and the production surplus is projected to double the 1960 level. However, in the succeeding 15 years, demand increases 32 percent while projected production rises about 15 percent. Thus the rising cereals production surplus projected for the 25-year period, 1960–85, levels off and then falls slightly between 1985 and 2000 as domestic demand catches up. With medium population growth and continued income increases, Canada's cereals surplus in 2000 is estimated to be only about 2.3 million tons greater than in 1960. In the same 15-year period, production deficits for raw sugar and fruit and vegetables increase.

Assuming a high income variant and medium population, the per capita income estimate for 1985 is 94 percent above the 1960 level. Comparing the associated demand estimates, also shown in Table 7.9, meat demand increases 17 percent above the level estimated in the benchmark projections while demands for milk and eggs fall 7 and 5 percent, respectively. The net effect on feed demand is a 9 percent increase. Total demand for cereals is only 6 percent higher, since off-setting changes occur in demand for food and industrial uses. Demand for root crops lies 12 percent lower, reflecting the influence of a large negative income elasticity. Demands for fruit and vegetables are 16 percent higher, and the production deficit increases 0.8 million tons.

Generally, the projections suggest that by the year 2000 Canadian production may not outdistance demand to nearly the extent found in the analysis of U.S. projections. We point out again that this comparison must be qualified because of the nature of the area trends for U.S. wheat and corn. Cereals production continues to dominate Canadian agriculture, and a sizable production surplus is still in evidence under all assumptions as to future trends in factors affecting demand. Assumptions of low population and constant per capita income result in a production surplus of 18.6 million tons of cereals in 2000, as shown in Table 7.8. At the other extreme, projections given in Table 7.10 for high population and increasing income assumptions yield only a 7.7 million ton surplus. This later value is only slightly above the 1960 level. Vegetables and fruit is the only product class for which large deficits are projected in 2000. Deficits ranging up to 6.6 million tons are estimated.

To summarize, the projections for Canada indicate that production will keep pace with domestic demand for the most part. Maintenance of Canada's traditional role as a grain exporter seems likely, though rapid population and income growth at home could limit export expansion.

COMPARISONS FOR MEXICO

Table 7.11 shows that Mexico was not a major importer of any class of food in 1960, but that moderate exports of sugar and vegetables and

TABLE 7.10. 1985 and 2000 Production-Demand Comparisons for Canada, Assuming High Population Growth Rates (1,000 metric tons)

1985 Comparisons, Assuming Constant Per Capita Income

Commodity	Domestic Disappearance			Excess Demand
	Food and Industrial	Feed	Total	
Cereals	4,909	19,152	24,061	−13,612
Raw sugar	1,614	...	1,614	1,346
Root crops	3,017	175	3,192	−63
Pulses	149	12	161	94
Fruit, veg.	5,203	...	5,203	3,208
Oil crops	317	792	1,109	197
Meat	2,427	...	2,427	...
Milk	12,073	1,860	13,933	...
Eggs	504	...	504	...

1985 Comparisons, Assuming Historical Income Growth Rates

Commodity	Domestic Disappearance			Excess Demand
	Food and Industrial	Feed	Total	
Cereals	4,668	20,725	25,393	−12,279
Raw sugar	1,614	...	1,614	1,346
Root crops	2,631	189	2,820	−435
Pulses	142	13	155	89
Fruit, veg.	5,996	...	5,996	4,001
Oil crops	347	857	1,204	292
Meat	2,811	...	2,811	...
Milk	11,019	2,012	13,031	...
Eggs	481	...	481	...

2000 Comparisons, Assuming Constant Per Capita Income

Commodity	Domestic Disappearance			Excess Demand
	Food and Industrial	Feed	Total	
Cereals	6,550	25,557	32,107	−10,836
Raw sugar	2,154	...	2,154	1,858
Root crops	4,026	233	4,259	236
Pulses	199	17	216	148
Fruit, veg.	6,942	...	6,942	4,881
Oil crops	423	1,057	1,480	242
Meat	3,238	...	3,238	...
Milk	16,109	2,481	18,590	...
Eggs	672	...	672	...

2000 Comparisons, Assuming Historical Income Growth Rates

Commodity	Domestic Disappearance			Excess Demand
	Food and Industrial	Feed	Total	
Cereals	6,111	29,138	35,249	−7,694
Raw sugar	2,154	...	2,154	1,858
Root crops	3,333	266	3,599	−424
Pulses	186	19	205	138
Fruit, veg.	8,668	...	8,668	6,607
Oil crops	479	1,205	1,684	447
Meat	4,076	...	4,076	...
Milk	14,202	2,829	17,031	...
Eggs	631	...	631	...

fruit were sustained. Feed demand accounted for about 13 percent of total domestic demand for cereals, and about half of oil crops demand was for livestock feed.

Because of lower income levels and more restricted diets, estimates of 1960 income elasticities are quite different for Mexico than for Canada and the United States. Values near 0.2 are estimated for cereals, starchy roots, and pulses; 0.3 for milk; 0.4 for sugar; 0.5 for oil crops; 0.6 for meat and for vegetables and fruit; and 0.7 for eggs. Thus it is assured that future income as well as population will affect demand significantly, even for the staple foods, and that improved nutrition will result if these demands are realized.

Commodity coverage in our trend analysis of Mexican production is relatively good. Root crops and pulses are the only commodity classes for which time trend estimates on significant quantities of production are lacking.

The projections for Mexico are especially interesting because they reflect conditions in a country where economic development is progressing rapidly and substantial changes are being realized in both food production and demand. Comparing production estimates in Tables 7.11 and 7.12 it is seen that cereals production is projected to be over 3.5 times larger in 2000 than in 1960. Production increases for other crops range from about 2 to over 4 times 1960 levels.

Both low and high variant constraints on cropland expansion are established for Mexico, but the trend in total area is not bounded before 2000 under the low constraint even though projected area of crops increases 89 percent in the 40-year projection period. Estimates based on data from the special soils study show that the projected crop area for 2000 would still leave unexploited about 30 percent of the low land variant potential.

Consider first the production-demand comparisons for 1985 under medium population growth and constant per capita income assumptions. Table 7.14 shows that production increases generally exceed the

TABLE 7.11. 1960 Production-Demand Comparisons for Mexico (1,000 metric tons)

Commodity	Food and Industrial	Feed	Total	Production	Excess Demand
Cereals	6,186	904	7,090	6,993	97
Raw sugar	1,194	...	1,194	1,597	−403
Root crops	412	19	431	426	5
Pulses	787	...	787	790	−3
Fruit, veg.	3,262	13	3,275	3,607	−332
Oil crops	203	204	407	411	−4
Meat	691	...	691	721	−30
Milk	2,915	51	2,966	2,803	163
Eggs	166	...	166	166	...

TABLE 7.12. 1985 and 2000 Production Projections for Mexico (1,000 metric tons)

Commodity	1985 Production	2000 Production
Cereals	18,022	25,631
Raw sugar	3,842	4,845
Root crops	1,078	1,628
Pulses	2,495	4,000
Fruit, veg.	5,972	7,352
Oil crops	1,252	1,728

rise in demand. By 1985 only fruit and vegetables show a less favorable production-demand comparison than in 1959–61, and surplus produc tion is indicated in all other product classes.

However, noteworthy deviations from benchmark results are obtained with different population and income assumptions. If continued income growth is assumed, 1985 per capita income would more than double. Comparing the two sets of 1985 projections given in Table 7.14, food and industrial demand for cereals increases 11 percent, feed demand increases 31 percent, and total domestic demand for cereals increases 13 percent or 2.2 million tons over benchmark levels. Thus, rather than a 1.6 million ton surplus, a 0.5 million ton deficit is projected.

A fairly general principle can be illustrated by comparing cereals demand for Mexico with that for the United States and for Canada. In low income countries, large relative increases in feed demand occur when incomes rise, but since feed demand is initially small, the effect on total cereals demand is also small. The total change in cereals demand related to increasing incomes comes primarily from expanded direct demand. In high income countries, direct demand for cereals often falls with higher incomes, though demand for feed usually will still advance. And since livestock feed makes up a large share of total demand in such countries, the total may well expand still further.

With medium population growth, Mexican demand for meat and for fruit and vegetables is 43 percent higher in 1985 under an assumption of growing income than with a constant per capita income assumption. Qualitatively similar comparisons result in other product classes.

Varying the population assumptions while holding per capita income constant results in less striking contrasts. The 1985 demands under high population assumptions shown in Table 7.15 are only about 10 percent greater than low population demands shown in Table 7.13.

By 2000, the forces discussed above create quite divergent results, depending on population and income assumptions. For the benchmark assumptions the cereals and fruit and vegetables categories shown in Table 7.14 indicate a deficit, while other product classes show modest surpluses. However, under assumptions of high population and continued income growth, population is 4.0 times the 1960 level and per

TABLE 7.13. 1985 and 2000 Production-Demand Comparisons for Mexico, Assuming Low Population Growth Rates (1,000 metric tons)

| | 1985 Comparisons, Assuming Constant Per Capita Income | | | | 1985 Comparisons, Assuming Historical Income Growth Rates | | | |
| | Domestic Disappearance | | | | Domestic Disappearance | | | |
Commodity	Food and Industrial	Feed	Total	Excess Demand	Food and Industrial	Feed	Total	Excess Demand
Cereals	13,372	1,955	15,327	−2,694	14,902	2,614	17,516	−505
Raw sugar	2,582	...	2,582	−1,259	3,211	...	3,211	−630
Root crops	890	40	930	−147	1,030	54	1,084	5
Pulses	1,701	...	1,701	−794	1,897	...	1,897	−598
Fruit, veg.	7,051	29	7,080	1,107	10,379	38	10,417	4,445
Oil crops	440	440	880	−372	613	589	1,202	−50
Meat	1,494	...	1,494	...	2,199	...	2,199	...
Milk	6,301	111	6,412	...	7,420	148	7,568	...
Eggs	359	...	359	...	556	...	556	...

| | 2000 Comparisons, Assuming Constant Per Capita Income | | | | 2000 Comparisons, Assuming Historical Income Growth Rates | | | |
| | Domestic Disappearance | | | | Domestic Disappearance | | | |
Commodity	Food and Industrial	Feed	Total	Excess Demand	Food and Industrial	Feed	Total	Excess Demand
Cereals	19,705	2,880	22,585	−3,044	22,752	4,454	27,206	1,576
Raw sugar	3,805	...	3,805	−1,039	5,108	...	5,108	263
Root crops	1,312	59	1,371	−256	1,662	91	1,753	125
Pulses	2,506	...	2,506	−1,493	2,904	...	2,904	−1,095
Fruit, veg.	10,390	42	10,432	3,080	18,703	65	18,768	11,416
Oil crops	648	649	1,297	−430	1,080	1,003	2,083	355
Meat	2,201	...	2,201	...	3,962	...	3,962	...
Milk	9,286	163	9,449	...	11,581	253	11,834	...
Eggs	529	...	529	...	1,022	...	1,022	...

TABLE 7.14. 1985 and 2000 Production-Demand Comparisons for Mexico, Assuming Medium Population Growth Rates (1,000 metric tons)

1985 Comparisons, Assuming Constant Per Capita Income

Commodity	Domestic Disappearance			Excess Demand
	Food and Industrial	Feed	Total	
Cereals	14,288	2,088	16,376	−1,645
Raw sugar	2,759	...	2,759	−1,082
Root crops	951	43	994	−84
Pulses	1,817	...	1,817	−677
Fruit, veg.	7,533	30	7,563	1,592
Oil crops	470	470	940	−311
Meat	1,596	...	1,596	...
Milk	6,733	119	6,852	...
Eggs	383	...	383	...

1985 Comparisons, Assuming Historical Income Growth Rates

Commodity	Domestic Disappearance			Excess Demand
	Food and Industrial	Feed	Total	
Cereals	15,828	2,737	18,565	−543
Raw sugar	3,388	...	3,388	−453
Root crops	1,088	56	1,144	66
Pulses	2,014	...	2,014	−481
Fruit, veg.	10,790	40	10,830	4,858
Oil crops	639	616	1,255	3
Meat	2,286	...	2,286	...
Milk	7,854	155	8,009	...
Eggs	577	...	577	...

2000 Comparisons, Assuming Constant Per Capita Income

Commodity	Domestic Disappearance			Excess Demand
	Food and Industrial	Feed	Total	
Cereals	22,578	3,300	25,878	248
Raw sugar	4,360	...	4,360	−484
Root crops	1,503	68	1,571	−57
Pulses	2,872	...	2,872	−1,128
Fruit, veg.	11,905	48	11,953	4,601
Oil crops	743	743	1,486	−241
Meat	2,522	...	2,522	...
Milk	10,639	187	10,826	...
Eggs	606	...	606	...

2000 Comparisons, Assuming Historical Income Growth Rates

Commodity	Domestic Disappearance			Excess Demand
	Food and Industrial	Feed	Total	
Cereals	25,888	4,937	30,825	5,194
Raw sugar	5,764	...	5,764	919
Root crops	1,863	101	1,964	336
Pulses	3,302	...	3,302	−697
Fruit, veg.	20,458	72	20,530	13,178
Oil crops	1,187	1,112	2,299	571
Meat	4,334	...	4,334	...
Milk	13,118	280	13,398	...
Eggs	1,114	...	1,114	...

TABLE 7.15. 1985 and 2000 Production-Demand Comparisons for Mexico, Assuming High Population Growth Rates (1,000 metric tons)

1985 Comparisons, Assuming Constant Per Capita Income

Commodity	Domestic Disappearance			Excess Demand
	Food and Industrial	Feed	Total	
Cereals	14,696	2,148	16,844	−1,177
Raw sugar	2,838	...	2,838	−1,003
Root crops	978	44	1,022	−55
Pulses	1,869	...	1,869	−625
Fruit, veg.	7,749	31	7,780	1,808
Oil crops	483	484	967	−284
Meat	1,642	...	1,642	...
Milk	6,925	122	7,047	...
Eggs	394	...	394	...

1985 Comparisons, Assuming Historical Income Growth Rates

Commodity	Domestic Disappearance			Excess Demand
	Food and Industrial	Feed	Total	
Cereals	16,238	2,791	19,029	1,007
Raw sugar	3,465	...	3,465	−375
Root crops	1,114	57	1,171	92
Pulses	2,066	...	2,066	−429
Fruit, veg.	10,968	41	11,009	5,036
Oil crops	651	629	1,280	27
Meat	2,324	...	2,324	...
Milk	8,045	158	8,203	...
Eggs	585	...	585	...

2000 Comparisons, Assuming Constant Per Capita Income

Commodity	Domestic Disappearance			Excess Demand
	Food and Industrial	Feed	Total	
Cereals	24,634	3,601	28,235	2,604
Raw sugar	4,757	...	4,757	−87
Root crops	1,640	74	1,714	85
Pulses	3,133	...	3,133	−866
Fruit, veg.	12,989	53	13,042	5,689
Oil crops	810	811	1,621	−106
Meat	2,752	...	2,752	...
Milk	11,608	204	11,812	...
Eggs	661	...	661	...

2000 Comparisons, Assuming Historical Income Growth Rates

Commodity	Domestic Disappearance			Excess Demand
	Food and Industrial	Feed	Total	
Cereals	28,103	5,268	33,371	7,740
Raw sugar	6,220	108	6,220	1,375
Root crops	2,004	108	2,112	484
Pulses	3,583	...	3,583	−417
Fruit, veg.	21,642	77	21,719	14,367
Oil crops	1,260	1,186	2,446	718
Meat	4,585	...	4,585	...
Milk	14,195	299	14,494	...
Eggs	1,175	...	1,175	...

capita income triples. Total domestic demand for cereals, shown in Table 7.15, increases to 4.7 times the 1960 level and a 7.7 million ton deficit results, i.e., 23 percent of estimated total demand. Feed demand accounts for about 16 percent of total cereals demand under these assumptions. Total 2000 cereals demand shown in Table 7.13 for low population and continued income growth is reduced 18 percent as compared to the corresponding estimate in Table 7.15 where high population is assumed. This is the result of assuming that the same total income in 2000 is distributed among 20 percent fewer people.

In summary, Mexico's agricultural output is projected to rise very rapidly, indeed. Area and yield increases both contribute significantly. That possible future deficits might occur is less surprising than that the projected deficits are not larger in view of the extremely large projected demand figures. Reductions in population growth could cause very substantial export surpluses in future years, and certainly no crises in food supplies are evident in the projections to 1985. However, continued growth of Mexican population at present very high rates could outdistance the present vigorous growth rates for agricultural output by a time near the turn of the century.

COMPARISONS FOR CENTRAL AMERICA
AND THE CARIBBEAN

Table 7.16 shows the 1960 food balance for the 12 countries comprising the Central America and Caribbean region. In 1960 the region's population was about one-third greater than that of Canada. Its production-demand comparisons contrast sharply, however. The region imported about one-third of its cereals in 1960; it was a major exporter of sugar; and lesser quantities of fruit and vegetables were exported. Feed demand was only 8 percent of total cereals demand.

Income elasticities for the 12 countries of this region are variable, but in general they are relatively high. Unweighted averages of esti-

TABLE 7.16. 1960 Production-Demand Comparisons for Central America and the Caribbean (1,000 metric tons)

Commodity	Domestic Disappearance			Production	Excess Demand
	Food and Industrial	Feed	Total		
Cereals	3,522	297	3,819	2,566	1,253
Raw sugar	972	...	972	8,264	−7,292
Root crops	2,032	50	2,082	2,034	48
Pulses	334	...	334	339	−5
Fruit, veg.	4,520	251	4,771	6,405	−1,634
Oil crops	110	85	195	133	62
Meat	607	...	607	588	19
Milk	2,092	153	2,245	1,989	256
Eggs	95	...	95	95	0

mated elasticities for cereals, sugar, starchy roots, pulses, and vegetables and fruit all fall in the range 0.18 to 0.35. Average elasticities for oil crops and the three livestock product classes range from 0.65 to 0.87. The lowest elasticity, 0.18, is for starchy roots, and the 0.87 value estimated for eggs is the highest.

Among the 12 countries, crops for which time trends are estimated account for nearly all production of cereals and sugar. Coverage for other crops is variable, but root crops are covered least well throughout the region.

Crop area is unbounded under the land constraint in all 12 countries. The aggregate of all crop area trends estimated for the region projects a total crop area in 2000 which is 42 percent above the 1960 level. However, area trends could not be estimated for a number of important crops. Upper bounds on cropland expansion were derived from the special soils analysis for 6 of the 12 countries in the region. Of the 6, only Haiti's 1960 crop area was near the estimated potential. Estimated expansion potentials for the other 5 were far above any conceivable levels which might be reached by 2000. Three of the countries omitted from the soils analysis had no crops for which estimated area trends were positive. Upper bounds on cropland expansion for the remaining 3 countries were routinely set at levels which would leave the area trends unbounded. The above crop area expansion, together with yield trends estimated for the region, results in the production projections shown in Table 7.17. Very modest growth is indicated relative to 1960.

Considering results for the benchmark projections to 1985, Table 7.17 shows that demand rises substantially faster than production through the intermediate projection period for all product classes except sugar. Projected cereals production for 1985 is about half the level estimated for domestic demand. The corresponding fraction for root crops and pulses is about two-thirds, and for oil crops it is about two-fifths. A deficit of about 1.8 million tons of fruit and vegetables is projected, while in 1960 the region exported about 1.6 million tons annually.

Assuming high income levels and medium population, 1985 de-

TABLE 7.17. 1985 and 2000 Production Projections for Central America and the Caribbean (1,000 metric tons)

Commodity	1985 Production	2000 Production
Cereals	3,683	4,245
Raw sugar	9,251	9,724
Root crops	2,341	2,448
Pulses	471	520
Fruit, veg.	7,781	8,554
Oil crops	137	150

mands for meat, milk, and eggs are, respectively, 32, 21, and 48 percent higher than under the benchmark variant for the same year. Feed demand is a small proportion of cereals demand, and the corresponding comparison for total cereals demand shows only a 12 percent increase. Production deficits under these assumptions are higher, of course (Table 7.19). Deficits of 4.6, 1.6, and 2.8 million tons are estimated for cereals, root crops, and fruit and vegetables, respectively.

Comparing demand projections to 1985 in Tables 7.18 and 7.20, it is seen that high population assumptions result in food demands which are 12 percent higher than those under low population assumptions when per capita income is held constant.

By 2000, very sizable deficits are projected for cereals, root crops, and fruit and vegetables when medium population growth and constant per capita income are assumed. Table 7.19 shows that cereals production, though 65 percent above 1960 levels, accounts for only 40 percent of projected domestic demand, and a 6.5 million ton deficit is projected. Sugar continues to be produced in surplus at near 1960 levels, but root crops and fruit and vegetables production are little over half of estimated domestic demand.

Other assumptions on population and income result in differences in demand qualitatively like those discussed for the 1985 estimates. High population and continuing income growth result in about 9 million ton annual deficits for cereals and fruit and vegetables in 2000 as shown in Table 7.20. The most favorable balance occurs under low population and constant per capita income assumptions, but Table 7.18 still indicates a 4.8 million ton deficit of cereals.

To summarize, projections for this region indicate that agricultural production is rising much less rapidly than demand throughout the projection period. Expanding crop areas are more important than yields in their effects on production. Sugar production continues to keep pace, and a continuing export surplus is projected. Evidence suggests that more rapid crop area expansion could be sustained, but the desirability of this has not been analyzed.

Although the demand projections are high relative to production, they reflect only modest increases in per capita income. Given the income elasticities of food demand typical of the region, any major breakthroughs in economic development trends will need to be accompanied by far greater than historical growth in agricultural productivity to avoid severe strains on food markets.

COMPARISONS FOR BRAZIL

Table 7.21 shows that in 1960 Brazil was surplus in every crop product class except cereals. Cereals demand exceeded production by 1.4 million tons. Feed demand for cereals was 38 percent of total cereals demand, and about 20 percent of root crops demand was for feed.

TABLE 7.18. 1985 and 2000 Production-Demand Comparisons for Central America and the Caribbean, Assuming Low Population Growth Rates (1,000 metric tons)

1985 Comparisons

	Assuming Constant Per Capita Income				Assuming Historical Income Growth Rates			
	Domestic Disappearance				Domestic Disappearance			
Commodity	Food and Industrial	Feed	Total	Excess Demand	Food and Industrial	Feed	Total	Excess Demand
Cereals	6,289	556	6,845	3,162	7,083	739	7,822	4,140
Raw sugar	1,679	...	1,679	−7,570	1,857	...	1,857	−7,392
Root crops	3,346	79	3,425	1,084	3,539	103	3,642	1,301
Pulses	595	...	595	125	665	...	665	195
Fruit, veg.	8,292	456	8,748	967	9,274	632	9,906	2,125
Oil crops	178	136	314	177	233	190	423	286
Meat	1,041	...	1,041	...	1,441	...	1,441	...
Milk	3,718	290	4,008	...	4,658	365	5,023	...
Eggs	170	...	170	...	271	...	271	...

2000 Comparisons

	Assuming Constant Per Capita Income				Assuming Historical Income Growth Rates			
	Domestic Disappearance				Domestic Disappearance			
Commodity	Food and Industrial	Feed	Total	Excess Demand	Food and Industrial	Feed	Total	Excess Demand
Cereals	8,322	752	9,074	4,829	10,017	1,175	11,192	6,947
Raw sugar	2,178	...	2,178	−7,545	2,558	...	2,558	−7,165
Root crops	4,176	97	4,273	1,826	4,551	155	4,706	2,258
Pulses	780	...	780	260	935	...	935	416
Fruit, veg.	11,022	613	11,635	3,081	12,969	1,005	13,974	5,420
Oil crops	220	165	385	236	357	276	633	483
Meat	1,344	...	1,344	...	2,222	...	2,222	...
Milk	4,894	400	5,294	...	6,895	523	7,418	...
Eggs	224	...	224	...	471	...	471	...

TABLE 7.19. 1985 and 2000 Production-Demand Comparisons for Central America and the Caribbean, Assuming Medium Population Growth Rates (1,000 metric tons)

1985 Comparisons, Assuming Constant Per Capita Income

Commodity	Food and Industrial	Feed	Total	Excess Demand
Cereals	6,826	601	7,427	3,745
Raw sugar	1,828	...	1,828	−7,421
Root crops	3,709	87	3,796	1,455
Pulses	652		652	182
Fruit, veg.	9,084	494	9,578	1,796
Oil crops	197	151	348	211
Meat	1,134		1,134	...
Milk	4,039	311	4,350	...
Eggs	185		185	...

1985 Comparisons, Assuming Historical Income Growth Rates

Commodity	Food and Industrial	Feed	Total	Excess Demand
Cereals	7,531	765	8,296	4,613
Raw sugar	1,988	...	1,988	−7,261
Root crops	3,834	108	3,942	1,601
Pulses	711		711	240
Fruit, veg.	9,942	650	10,592	2,811
Oil crops	239	200	439	303
Meat	1,492		1,492	...
Milk	4,896	382	5,278	...
Eggs	273		273	...

2000 Comparisons, Assuming Constant Per Capita Income

Commodity	Food and Industrial	Feed	Total	Excess Demand
Cereals	9,837	880	10,717	6,473
Raw sugar	2,591	...	2,591	−7,182
Root crops	5,225	120	5,345	2,897
Pulses	947		947	427
Fruit, veg.	13,427	721	14,148	5,594
Oil crops	273	205	478	327
Meat	1,601		1,601	...
Milk	5,790	458	6,248	...
Eggs	267		267	...

2000 Comparisons, Assuming Historical Income Growth Rates

Commodity	Food and Industrial	Feed	Total	Excess Demand
Cereals	11,380	1,261	12,641	8,396
Raw sugar	2,943	...	2,943	−6,780
Root crops	5,437	171	5,608	3,161
Pulses	1,080		1,080	560
Fruit, veg.	15,126	1,065	16,191	7,636
Oil crops	380	311	691	541
Meat	2,390		2,390	...
Milk	7,676	572	8,248	...
Eggs	478		478	...

TABLE 7.20. 1985 and 2000 Production-Demand Comparisons for Central America and the Caribbean, Assuming High Population Growth Rates (1,000 metric tons)

1985 Comparisons

Commodity	1985 Comparisons, Assuming Constant Per Capita Income — Domestic Disappearance				1985 Comparisons, Assuming Historical Income Growth Rates — Domestic Disappearance			
	Food and Industrial	Feed	Total	Excess Demand	Food and Industrial	Feed	Total	Excess Demand
Cereals	7,070	621	7,691	4,008	7,728	776	8,504	4,822
Raw sugar	1,896	...	1,896	−7,353	2,047	...	2,047	−7,202
Root crops	3,877	91	3,968	1,628	3,964	110	4,074	1,733
Pulses	678	...	678	208	731	...	731	260
Fruit, veg.	9,445	511	9,956	2,174	10,239	658	10,897	3,116
Oil crops	206	157	363	226	242	205	447	311
Meat	1,177	...	1,177	...	1,514	...	1,514	...
Milk	4,184	321	4,505	...	5,000	389	5,389	...
Eggs	191	...	191	...	274	...	274	...

2000 Comparisons

Commodity	2000 Comparisons, Assuming Constant Per Capita Income — Domestic Disappearance				2000 Comparisons, Assuming Historical Income Growth Rates — Domestic Disappearance			
	Food and Industrial	Feed	Total	Excess Demand	Food and Industrial	Feed	Total	Excess Demand
Cereals	10,583	950	11,533	7,288	12,032	1,300	13,332	9,088
Raw sugar	2,782	...	2,782	−6,942	3,113	...	3,113	−6,610
Root crops	5,593	128	5,721	3,273	5,728	177	5,905	3,457
Pulses	1,019	...	1,019	499	1,138	...	1,138	619
Fruit, veg.	14,471	777	15,248	6,695	16,031	1,091	17,122	8,568
Oil crops	291	218	509	360	388	322	710	560
Meat	1,717	...	1,717	...	2,458	...	2,458	...
Milk	6,226	495	6,721	...	8,011	602	8,613	...
Eggs	287	...	287	...	482	...	482	...

TABLE 7.21. 1960 Production-Demand Comparisons for Brazil (1,000 metric tons)

Commodity	Domestic Disappearance			Production	Excess Demand
	Food and Industrial	Feed	Total		
Cereals	9,354	5,730	15,084	13,669	1,415
Raw sugar	2,955	. . .	2,955	3,757	—802
Root crops	15,658	4,251	19,909	20,152	—243
Pulses	1,774	. . .	1,774	1,955	—181
Fruit, veg.	8,114	112	8,226	8,537	—311
Oil crops	437	239	676	759	—83
Meat	2,032	. . .	2,032	2,089	—57
Milk	4,877	213	5,090	5,046	44
Eggs	281	. . .	281	281	. . .

The structure of Brazilian food demand in 1960 is quite similar to that in Central America. Estimated income elasticity of demand for sugar is zero, and for pulses, 0.1. Starchy roots and cereals are estimated to have elasticities of 0.2 and 0.28, respectively, while for fruit and vegetables the estimate is 0.4. Estimates for oil crops and the three livestock product classes are distributed in the range 0.5 to 0.61.

Crops for which time trends are estimated account for at least 87 percent of production in each of the six product classes.

Low and high variant constraints on cropland expansion are estimated for Brazil, but area trends are unbounded even under the low variant constraint. Total projected crop area in 2000 is 81 percent above the 1960 total, but according to estimates derived from the special soils study, the area projected for 2000 by extending time trends is only about 25 percent of the low variant potential. Like many tropical regions, Brazil's soil resources are judged to be of much lower productivity than soils in the temperate zones; however, the estimated quantity of unused but potentially cropable land is striking indeed.

Comparing production estimates for 1960 shown in Table 7.21 with projected production shown in Table 7.22, it is apparent that increasing yield trends were evident in Brazil as well as the rising area trends mentioned above. Projected production for 2000 is more than double the 1960 estimate in each class except pulses. Cereals and root crops, two of the main staple foods in Brazil, each increases about 2.5 times over the 40-year projection period.

TABLE 7.22. 1985 and 2000 Production Projections for Brazil (1,000 metric tons)

Commodity	1985 Production	2000 Production
Cereals	26,106	33,851
Raw sugar	7,959	10,750
Root crops	40,576	54,678
Pulses	2,865	3,293
Fruit, veg.	16,539	21,269
Oil crops	2,663	4,064

Turning now to results for the intermediate projection period, a somewhat mixed pattern is projected under the benchmark assumptions. Estimates presented in Table 7.24 show a 5 million ton cereals deficit. Both sugar and oil crops are produced in increasingly surplus quantities through 1985, while a trend toward slight deficits is observed for the other three product classes.

Under assumptions of medium population and continued income growth, per capita income is 70 percent higher than under benchmark assumptions for 1985, and demands for meat, milk, and eggs are found to be 31, 32, and 26 percent higher. Feed demand increases 31 percent. Total demand for cereals increases 19 percent, and demand for root crops rises 13 percent. The net effect in terms of production-demand comparisons is a 10.9 million ton deficit of cereals and a 5.9 million ton deficit of root crops. Only sugar, with an income elasticity of zero, retains the same position; all other demands rise in comparison to results for the benchmark assumptions.

Effects of alternative population assumptions are illustrated by comparing the constant per capita income estimates for 1985 shown in Tables 7.23 and 7.25. Demand for each commodity is 17 percent greater under the high population assumption. Assuming continued income growth and high population in 1985, demand for cereals is 37 percent higher than under low population and constant per capita income assumptions.

Under the benchmark assumptions, sugar and oil crops continue to be produced in surplus quantities in 2000. Table 7.24 shows that deficits are projected for the other four commodity classes. Compared to 1960, total cereals demand increases 3 times by 2000, while production rises 2½ times. While this is a very substantial production increase, it does not match the increase in demand, and it falls even further below demand under other assumptions.

With high population and increasing income, estimated cereals production for 2000 is only 60 percent of demand, and a 22.3 million ton deficit is projected as shown in Table 7.25. Assumptions of low population and high income result in the projections shown in Table 7.23. Cereals demand is 20 percent lower and the deficit is halved. Minimum cereals demand is estimated when low population and constant per capita income are assumed.

In summary, Brazil's agricultural output shows quite vigorous growth, primarily from cropland expansion. Evidence available suggests that the cropland trend can continue, or even accelerate if necessary. However, total demand has a potential for even more rapid growth. Demand projections and production-demand comparisons for 2000 show great variability, largely because of uncertain population trends. The low and high population estimates for that date differ by almost 60 million; a figure little below total 1960 population for the country. Income

TABLE 7.23. 1985 and 2000 Production-Demand Comparisons for Brazil, Assuming Low Population Growth Rates (1,000 metric tons)

1985 Comparisons, Assuming Constant Per Capita Income

Commodity	Domestic Disappearance			Excess Demand
	Food and Industrial	Feed	Total	
Cereals	17,270	10,580	27,850	1,744
Raw sugar	5,456	...	5,456	-2,502
Root crops	28,909	7,848	36,757	-3,818
Pulses	3,275	...	3,275	411
Fruit, veg.	14,981	207	15,188	-1,351
Oil crops	807	442	1,249	-1,413
Meat	3,752	...	3,752	...
Milk	9,004	393	9,397	...
Eggs	519	...	519	...

1985 Comparisons, Assuming Historical Income Growth Rates

Commodity	Domestic Disappearance			Excess Demand
	Food and Industrial	Feed	Total	
Cereals	19,555	14,547	34,102	7,996
Raw sugar	5,456	...	5,456	-2,502
Root crops	31,641	10,791	42,432	1,856
Pulses	3,434	...	3,434	569
Fruit, veg.	18,814	284	19,098	2,559
Oil crops	1,117	607	1,724	-937
Meat	5,143	...	5,143	...
Milk	12,517	541	13,058	...
Eggs	685	...	685	...

2000 Comparisons, Assuming Constant Per Capita Income

Commodity	Domestic Disappearance			Excess Demand
	Food and Industrial	Feed	Total	
Cereals	22,337	13,684	36,021	2,170
Raw sugar	7,056	...	7,056	-3,693
Root crops	37,391	10,150	47,541	-7,135
Pulses	4,236	...	4,236	943
Fruit, veg.	19,376	267	19,643	-1,624
Oil crops	1,044	571	1,615	-2,448
Meat	4,852	...	4,852	...
Milk	11,646	509	12,155	...
Eggs	671	...	671	...

2000 Comparisons, Assuming Historical Income Growth Rates

Commodity	Domestic Disappearance			Excess Demand
	Food and Industrial	Feed	Total	
Cereals	25,604	19,613	45,217	11,366
Raw sugar	7,056	...	7,056	-3,693
Root crops	41,298	14,548	55,846	1,168
Pulses	4,463	...	4,463	1,170
Fruit, veg.	25,104	383	25,487	4,219
Oil crops	1,507	819	2,326	-1,737
Meat	6,932	...	6,932	...
Milk	16,896	729	17,625	...
Eggs	919	...	919	...

TABLE 7.24. 1985 and 2000 Production-Demand Comparisons for Brazil, Assuming Medium Population Growth Rates (1,000 metric tons)

1985 Comparisons, Assuming Constant Per Capita Income

| Commodity | Domestic Disappearance | | | Excess Demand |
	Food and Industrial	Feed	Total	
Cereals	19,305	11,826	31,131	5,025
Raw sugar	6,098	...	6,098	−1,860
Root crops	32,315	8,772	41,087	512
Pulses	3,661	...	3,661	796
Fruit, veg.	16,746	231	16,977	438
Oil crops	902	494	1,396	−1,266
Meat	4,194	...	4,194	...
Milk	10,065	440	10,505	...
Eggs	580	...	580	...

1985 Comparisons, Assuming Historical Income Growth Rates

| Commodity | Domestic Disappearance | | | Excess Demand |
	Food and Industrial	Feed	Total	
Cereals	21,523	15,489	37,012	10,906
Raw sugar	6,098	...	6,098	−1,860
Root crops	34,968	11,489	46,457	5,880
Pulses	3,814	...	3,814	950
Fruit, veg.	20,284	302	20,586	4,048
Oil crops	1,188	647	1,835	−827
Meat	5,478	...	5,478	...
Milk	13,308	576	13,884	...
Eggs	733	...	733	...

2000 Comparisons, Assuming Constant Per Capita Income

| Commodity | Domestic Disappearance | | | Excess Demand |
	Food and Industrial	Feed	Total	
Cereals	28,076	17,199	45,275	11,424
Raw sugar	8,869	...	8,869	−1,880
Root crops	46,998	12,758	59,756	5,078
Pulses	5,324	...	5,324	2,081
Fruit, veg.	24,355	336	24,691	3,422
Oil crops	1,312	718	2,030	−2,033
Meat	6,099	...	6,099	...
Milk	14,638	640	15,278	...
Eggs	843	...	843	...

2000 Comparisons, Assuming Historical Income Growth Rates

| Commodity | Domestic Disappearance | | | Excess Demand |
	Food and Industrial	Feed	Total	
Cereals	31,218	22,346	53,564	19,713
Raw sugar	8,869	...	8,869	−1,880
Root crops	50,755	16,575	67,330	12,653
Pulses	5,541	...	5,541	2,249
Fruit, veg.	29,327	436	29,763	8,494
Oil crops	1,714	933	2,647	−1,416
Meat	7,904	...	7,904	...
Milk	19,195	831	20,026	...
Eggs	1,059	...	1,059	...

TABLE 7.25. 1985 and 2000 Production-Demand Comparisons for Brazil, Assuming High Population Growth Rates (1,000 metric tons)

	1985 Comparisons, Assuming Constant Per Capita Income				1985 Comparisons, Assuming Historical Income Growth Rates			
	Domestic Disappearance				Domestic Disappearance			
Commodity	Food and Industrial	Feed	Total	Excess Demand	Food and Industrial	Feed	Total	Excess Demand
Cereals	20,224	12,389	32,613	6,508	22,389	15,888	38,277	12,171
Raw sugar	6,389	...	6,389	−1,569	6,389	...	6,389	−1,569
Root crops	33,854	9,190	43,044	2,468	36,442	11,786	48,228	7,652
Pulses	3,835	...	3,835	971	3,985	...	3,985	1,120
Fruit, veg.	17,544	242	17,786	1,246	20,934	310	21,244	4,695
Oil crops	945	517	1,462	−1,199	1,218	663	1,881	−780
Meat	4,393	...	4,393	...	5,621	...	5,621	...
Milk	10,544	461	11,005	...	13,643	591	14,234	...
Eggs	608	...	608	...	754	...	754	...

	2000 Comparisons, Assuming Constant Per Capita Income				2000 Comparisons, Assuming Historical Income Growth Rates			
	Domestic Disappearance				Domestic Disappearance			
Commodity	Food and Industrial	Feed	Total	Excess Demand	Food and Industrial	Feed	Total	Excess Demand
Cereals	29,977	18,364	48,341	14,490	32,991	23,153	56,144	22,293
Raw sugar	9,470	...	9,470	−1,279	9,470	...	9,470	−1,279
Root crops	50,180	13,622	63,802	9,125	53,784	17,175	70,959	16,281
Pulses	5,685	...	5,685	2,392	5,893	...	5,893	2,600
Fruit, veg.	26,004	359	26,363	5,094	30,631	452	31,083	9,814
Oil crops	1,401	767	2,168	−1,895	1,775	967	2,742	−1,321
Meat	6,512	...	6,512	...	8,192	...	8,192	...
Milk	15,629	683	16,312	...	19,870	861	20,731	...
Eggs	901	...	901	...	1,101	...	1,101	...

TABLE 7.26. 1960 Production-Demand Comparisons for Argentina and Uruguay (1,000 metric tons)

Commodity	Domestic Disappearance			Production	Excess Demand
	Food and Industrial	Feed	Total		
Cereals	4,839	3,275	8,114	13,879	—5,765
Raw sugar	909	. . .	909	972	—63
Root crops	2,298	178	2,475	2,499	—23
Pulses	60	. . .	60	63	—3
Fruit, veg.	5,450	. . .	5,450	5,392	58
Oil crops	314	355	669	1,120	—451
Meat	2,297	. . .	2,297	2,850	—553
Milk	4,998	27	5,025	5,185	—160
Eggs	161	. . .	161	178	—17

trends are important in determining demand because of high income elasticities. However, projected per capita income growth for Brazil is, again, relatively limited. Even under the low population alternative, projected income for 2000 is only about double that of 1960. As was true of Central America and the Caribbean, more rapid progress in raising per capita incomes could intensify the pressure on food supplies unless corresponding progress is achieved in agriculture.

COMPARISONS FOR ARGENTINA AND URUGUAY

The 1960 food balances for Argentina and Uruguay show qualitative similarities to those of the North American Temperate Zone countries. Data in Table 7.26 show that the region was an exporter in every product class except fruit and vegetables in 1960. Cereals were produced in amounts about 70 percent greater than domestic demand, leaving about 5.8 million tons for export (either as cereals or livestock). Feed demand accounted for about 40 percent of total cereals demand and about half of total oil crops demand.

The pattern of income elasticities for this region is somewhat unique in that 1960 elasticities for livestock products are relatively small while those for crop products are more characteristic of low income countries. Elasticity estimates for meat and milk are less than 0.1, and the estimate for eggs is about 0.3. The estimated elasticity for cereals is 0.19. Root crops demand is estimated to have a near-zero income elasticity, while for pulses, oil crops, and vegetables and fruit, estimates range from about 0.4 to 0.6. Thus per capita income changes will primarily affect food and industrial demand and they will have little effect on feed demand.

At least 75 percent of the production in each product class is accounted for by crops included in this study. Neither country's crop area trends were bounded before 2000; area expansion projected by extending time trends amounts to only 14 percent over the 40-year period.

TABLE 7.27. 1985 and 2000 Production Projections for Argentina and Uruguay
             (1,000 metric tons)

Commodity	1985 Production	2000 Production
Cereals	16,816	17,815
Raw sugar	1,489	1,789
Root crops	3,602	4,244
Pulses	101	118
Fruit, veg.	8,306	9,886
Oil crops	1,528	1,756

However, the total area projected for 2000 is only about one-third of the low variant potential estimated from the special soils analysis. Of course, this is a much more significant unused potential than some, for it occurs in a region where soil productivity is higher than in many other areas.

In Chapter 6 we noted that historical trends in area and yield were both quite limited for these two countries. Table 7.27 shows the implications of these trends in terms of 1985 and 2000 food production. Projected 2000 production is less than twice that of 1960 for each of the six commodity groups.

Table 7.29 shows that no significant changes occur in the projected production-demand comparisons through 1985 when medium population growth and constant per capita income are assumed. Production and demand grow along essentially parallel time paths. Assuming historical income trends, per capita incomes in 1985 are about twice as high as under benchmark assumptions. Total cereals demand lies 6.7 percent higher, and the production surplus is reduced from 5.3 to 4.6 million tons. Vegetables and fruit demand increases 35 percent, and the production-demand balance changes from a 0.6 million ton surplus to a 2.1 million ton deficit. Other changes are less significant.

Population differences under the various assumptions are not great in 1985, and projections based on low and high population assumptions, along with constant per capita income, show only small demand differences as evidenced in Tables 7.28 and 7.30.

Continuing for another 15 years under benchmark assumptions, production-demand comparisons for 2000 are very similar to those in 1985. Table 7.30 shows that deficits do occur under high population and increasing income assumptions but none is of crisis proportion. A 3.8 million ton deficit of fruit and vegetables is the most striking figure. Production surpluses are maintained for cereals and oil crops, but at reduced levels. Other deficits are minor. Under these assumptions, total demand for cereals increases 94 percent, root crops demand increases 75 percent, and demand for fruit and vegetables increases 152 percent above 1960 levels. These values appear small compared to corresponding demand increases in other Latin American countries, and the reason why production more easily keeps pace with demand is clear.

TABLE 7.28. 1985 and 2000 Production-Demand Comparisons for Argentina and Uruguay, Assuming Low Population Growth Rates (1,000 metric tons)

	1985 Comparisons, Assuming Constant Per Capita Income				1985 Comparisons, Assuming Historical Income Growth Rates			
	Domestic Disappearance			Excess Demand	Domestic Disappearance			Excess Demand
Commodity	Food and Industrial	Feed	Total		Food and Industrial	Feed	Total	
Cereals	6,465	4,384	10,849	−5,965	7,122	4,498	11,620	−5,195
Raw sugar	1,213	...	1,213	−275	1,486	...	1,486	−2
Root crops	3,069	240	3,309	−292	2,955	245	3,200	−401
Pulses	80	...	80	−20	117	...	117	16
Fruit, veg.	7,281	...	7,281	−1,025	10,042	...	10,042	1,737
Oil crops	420	473	893	−633	546	486	1,032	−494
Meat	3,052	...	3,052	...	3,076	...	3,076	...
Milk	6,648	32	6,680	...	6,984	34	7,018	...
Eggs	214	...	214	...	275	...	275	...

	2000 Comparisons, Assuming Constant Per Capita Income				2000 Comparisons, Assuming Historical Income Growth Rates			
	Domestic Disappearance			Excess Demand	Domestic Disappearance			Excess Demand
Commodity	Food and Industrial	Feed	Total		Food and Industrial	Feed	Total	
Cereals	7,347	4,985	12,332	−5,482	8,212	5,144	13,356	−4,458
Raw sugar	1,377	...	1,377	−411	1,744	...	1,744	−44
Root crops	3,487	273	3,769	−482	3,335	281	3,616	−627
Pulses	91	...	91	−26	143	...	143	26
Fruit, veg.	8,273	...	8,273	−1,613	12,239	...	12,239	2,353
Oil crops	478	538	1,016	−789	659	556	1,215	−540
Meat	3,462	...	3,462	...	3,493	...	3,493	...
Milk	7,543	35	7,578	...	8,009	38	8,047	...
Eggs	243	...	243	...	330	...	330	...

TABLE 7.29. 1985 and 2000 Production-Demand Comparisons for Argentina and Uruguay, Assuming Medium Population Growth Rates (1,000 metric tons)

| | 1985 Comparisons, Assuming Constant Per Capita Income | | | | 1985 Comparisons, Assuming Historical Income Growth Rates | | | |
| | Domestic Disappearance | | | | Domestic Disappearance | | | |
Commodity	Food and Industrial	Feed	Total	Excess Demand	Food and Industrial	Feed	Total	Excess Demand
Cereals	6,847	4,643	11,490	-5,325	7,507	4,756	12,263	-4,552
Raw sugar	1,284	...	1,284	-204	1,557	...	1,557	68
Root crops	3,251	254	3,505	-97	3,135	259	3,394	-206
Pulses	85	...	85	-15	121	...	121	20
Fruit, veg.	7,710	...	7,710	-595	10,415	...	10,415	2,109
Oil crops	445	501	946	-580	569	514	1,083	-444
Meat	3,232	...	3,232	...	3,256	...	3,256	...
Milk	7,040	34	7,074	...	7,373	36	7,409	...
Eggs	227	...	227	...	286	...	286	...

| | 2000 Comparisons, Assuming Constant Per Capita Income | | | | 2000 Comparisons, Assuming Historical Income Growth Rates | | | |
| | Domestic Disappearance | | | | Domestic Disappearance | | | |
Commodity	Food and Industrial	Feed	Total	Excess Demand	Food and Industrial	Feed	Total	Excess Demand
Cereals	8,060	5,470	13,530	-4,283	8,952	5,630	14,582	-3,232
Raw sugar	1,511	...	1,511	-278	1,886	...	1,886	97
Root crops	3,826	300	4,126	-117	3,668	308	3,976	-266
Pulses	99	...	99	-17	152	...	152	34
Fruit, veg.	9,076	...	9,076	-810	13,009	...	13,009	3,123
Oil crops	525	590	1,115	-641	704	608	1,312	-443
Meat	3,796	...	3,796	...	3,828	...	3,828	...
Milk	8,271	38	8,309	...	8,739	41	8,780	...
Eggs	226	...	266	...	353	...	353	...

TABLE 7.30. 1985 and 2000 Production-Demand Comparisons for Argentina and Uruguay, Assuming High Population Growth Rates (1,000 metric tons)

| | 1985 Comparisons, Assuming Constant Per Capita Income | | | | 1985 Comparisons, Assuming Historical Income Growth Rates | | | |
| | Domestic Disappearance | | | | Domestic Disappearance | | | |
Commodity	Food and Industrial	Feed	Total	Excess Demand	Food and Industrial	Feed	Total	Excess Demand
Cereals	7,107	4,820	11,927	−4,888	7,767	4,931	12,698	−4,118
Raw sugar	1,333	...	1,333	−155	1,605	...	1,605	116
Root crops	3,374	264	3,638	35	3,259	269	3,528	−73
Pulses	88	...	88	−12	123	...	123	22
Fruit, veg.	8,003	...	8,003	−303	10,662	...	10,662	2,356
Oil crops	462	520	982	−544	583	533	1,117	−410
Meat	3,354	...	3,354	...	3,378	...	3,378	...
Milk	7,306	35	7,341	...	7,635	37	7,672	...
Eggs	235	...	235	...	294	...	294	...

| | 2000 Comparisons, Assuming Constant Per Capita Income | | | | 2000 Comparisons, Assuming Historical Income Growth Rates | | | |
| | Domestic Disappearance | | | | Domestic Disappearance | | | |
Commodity	Food and Industrial	Feed	Total	Excess Demand	Food and Industrial	Feed	Total	Excess Demand
Cereals	8,775	5,958	14,733	−3,081	9,685	6,116	15,801	−2,013
Raw sugar	1,644	...	1,644	−144	2,024	...	2,024	235
Root crops	4,165	327	4,492	249	4,005	335	4,340	96
Pulses	108	...	108	−9	160	...	160	42
Fruit, veg.	9,881	...	9,881	−5	13,746	...	13,746	3,860
Oil crops	571	642	1,213	−542	748	660	1,408	−347
Meat	4,129	...	4,129	...	4,162	...	4,162	...
Milk	8,999	41	9,040	...	9,465	44	9,509	...
Eggs	289	...	289	...	375	...	375	...

123

TABLE 7.31. 1960 Production-Demand Comparisons for Other South America
(1,000 metric tons)

Commodity	Domestic Disappearance Food and Industrial	Feed	Total	Production	Excess Demand
Cereals	6,488	796	7,284	5,575	1,709
Raw sugar	1,865	...	1,865	2,210	—345
Root crops	5,646	786	6,432	6,514	—82
Pulses	514	...	514	570	—56
Fruit, veg.	9,796	279	10,075	11,405	—1,330
Oil crops	204	140	344	181	163
Meat	1,362	...	1,362	1,373	—11
Milk	4,550	314	4,864	4,154	710
Eggs	197	...	197	172	25

Summarization of projections for Argentina and Uruguay can be
done briefly. Small area and yield increases raise production somewhat
less rapidly than demand, and projected demand increases are only
moderate. Unused cropland potential in these countries represents a
substantial stock of idle agricultural capacity. Perhaps more important,
yield trends in this region are distinctly less vigorous than in regions
with similar climate and soils, and potential agricultural capacity in
this form may be even larger.

COMPARISONS FOR OTHER SOUTH AMERICA

Table 7.31 shows that 1960 domestic production of cereals in the
seven countries of this region was only about 76 percent of domestic
demand. Root crops and vegetables and fruit were important components of domestic demand, as were cereals. Feed demand was concentrated primarily in cereals and root crops, but in both cases, the feed
component was only slightly over 10 percent of total domestic demand.
Thus, even though livestock products demand, and hence feed demand,
may show sizable relative gains in response to per capita income increases, the income effect through demand for food and industrial uses
will be more important. Modest production surpluses are recorded for
all other crop product classes except oil crops.

The structure of agricultural demand for this region as a whole is
similar to that for Brazil and Central America. The unweighted average 1960 income elasticity for starchy roots, 0.2, is the lowest among the
nine produce classes. Estimated elasticities for cereals, sugar, pulses,
and vegetables and fruit lie in the range 0.33 to 0.41. Oil crops and
the three livestock product groups have estimated income elasticities
ranging from 0.71 to 0.81. Again, per capita income as well as population will be a major determinant of food demand to the extent that
income growth may be realized.

Except for fruit and vegetables, a high proportion of production in each product class is accounted for by crops for which area and yield or production trends have been estimated. However, only about half of all fruit and vegetable production is included. Almost no production data were available for pulses, oil crops, and fruit and vegetables in Bolivia, but production in that country is a small fraction of the regional totals.

As in every other Latin American country, crop area expansion is unbounded through 2000 in each of the seven countries of the region. The total of all area trends included rises 53 percent above the 1960 level by 2000, but the total projected for 2000 amounts to only 20 percent of the low variant cropland potential estimated from the special soils study. Production projections shown in Table 7.32 show the major staples increasing at moderate rates.

Table 7.34 shows that production in each of the three major product classes rises less rapidly than demand through 1985 under benchmark assumptions. Total demand in 1985 increases over 1960 levels by about 100 percent for cereals, root crops, and fruit and vegetables. Projected cereals and fruit and vegetables deficits are about 6 million tons each, and the estimated deficit for root crops is about 3.8 million tons.

Assumptions of continued historical income trends result in demand increases for cereals, root crops, and fruit and vegetables of 15, 12, and 20 percent, respectively, relative to 1985 demands under the benchmark assumptions. In tonnage, the deficits are 8.2, 5.4, and 10.2 million tons, respectively, for cereals, root crops, and fruit and vegetables.

Alternative population assumptions for 1985 result in about the same demand variability as observed under 1985 income alternatives. Demand for all product classes is about 16 percent higher in 1985 with medium population and trend level income than with medium population and constant per capita income.

By 2000, sizable deficits are projected under all demand variants and in every product class except sugar. Minimum and maximum cereals deficits for 2000 shown in Tables 7.33 and 7.35 are 6.8 and 17.4 million tons. Other product classes show similar contrasts in the projections to 2000.

TABLE 7.32. 1985 and 2000 Production Projections for Other South America (1,000 metric tons)

Commodity	1985 Production	2000 Production
Cereals	8,739	10,697
Raw sugar	4,122	5,246
Root crops	9,309	10,568
Pulses	711	788
Fruit, veg.	15,074	17,046
Oil crops	532	726

TABLE 7.33. 1985 and 2000 Production-Demand Comparisons for Other South America, Assuming Low Population Growth Rates (1,000 metric tons)

Commodity	1985 Comparisons, Assuming Constant Per Capita Income				1985 Comparisons, Assuming Historical Income Growth Rates			
	Domestic Disappearance			Excess Demand	Domestic Disappearance			Excess Demand
	Food and Industrial	Feed	Total		Food and Industrial	Feed	Total	
Cereals	11,905	1,439	13,344	4,605	13,691	2,025	15,716	6,977
Raw sugar	3,517	...	3,517	−604	4,018	...	4,018	−103
Root crops	10,366	1,437	11,803	2,495	11,401	2,126	13,527	4,218
Pulses	968	...	968	257	1,162	...	1,162	451
Fruit, veg.	18,383	534	18,917	3,843	22,785	775	23,560	8,487
Oil crops	400	274	674	142	599	387	986	455
Meat	2,526	...	2,526	...	3,678	...	3,678	...
Milk	8,543	581	9,124	...	12,084	818	12,903	...
Eggs	372	...	372	...	586	...	586	...

Commodity	2000 Comparisons, Assuming Constant Per Capita Income				2000 Comparisons, Assuming Historical Income Growth Rates			
	Domestic Disappearance			Excess Demand	Domestic Disappearance			Excess Demand
	Food and Industrial	Feed	Total		Food and Industrial	Feed	Total	
Cereals	15,600	1,868	17,468	6,770	19,150	3,121	22,271	11,573
Raw sugar	4,670	...	4,670	−576	5,690	...	5,690	444
Root crops	13,598	1,882	15,480	4,912	15,488	3,454	18,942	8,374
Pulses	1,283	...	1,283	496	1,703	...	1,703	915
Fruit, veg.	24,347	715	25,062	8,016	33,546	1,332	34,878	17,832
Oil crops	539	368	907	182	972	582	1,554	829
Meat	3,327	...	3,327	...	5,890	...	5,890	...
Milk	11,309	764	12,073	...	18,948	1,315	20,263	...
Eggs	494	...	494	...	1,057	...	1,057	...

TABLE 7.34. 1985 and 2000 Production-Demand Comparisons for Other South America, Assuming Medium Population Growth Rates (1,000 metric tons)

1985 Comparisons

	1985 Comparisons, Assuming Constant Per Capita Income				1985 Comparisons, Assuming Historical Income Growth Rates			
	Domestic Disappearance				Domestic Disappearance			
Commodity	Food and Industrial	Feed	Total	Excess Demand	Food and Industrial	Feed	Total	Excess Demand
Cereals	13,211	1,581	14,792	6,053	14,876	2,115	16,991	8,252
Raw sugar	3,929	...	3,929	−192	4,399	...	4,399	277
Root crops	11,521	1,573	13,094	3,784	12,517	2,218	14,735	5,426
Pulses	1,079	...	1,079	368	1,255	...	1,255	544
Fruit, veg.	20,477	599	21,076	6,002	24,508	809	25,317	10,243
Oil crops	447	305	752	220	626	410	1,036	504
Meat	2,805	...	2,805	...	3,841	...	3,841	...
Milk	9,497	645	10,142	...	12,731	859	13,590	...
Eggs	414	...	414	...	603	...	603	...

2000 Comparisons

	2000 Comparisons, Assuming Constant Per Capita Income				2000 Comparisons, Assuming Historical Income Growth Rates			
	Domestic Disappearance				Domestic Disappearance			
Commodity	Food and Industrial	Feed	Total	Excess Demand	Food and Industrial	Feed	Total	Excess Demand
Cereals	19,451	2,281	21,732	11,034	23,008	3,451	26,459	15,761
Raw sugar	5,915	...	5,915	668	6,933	...	6,933	1,686
Root crops	16,991	2,274	19,265	8,697	19,010	3,804	22,814	12,246
Pulses	1,618	...	1,618	831	2,020	...	2,020	1,233
Fruit, veg.	30,661	916	31,577	14,531	39,531	1,477	41,008	23,962
Oil crops	685	467	1,152	426	1,064	664	1,728	1,002
Meat	4,160	...	4,160	...	6,522	...	6,522	...
Milk	14,193	957	15,150	...	21,518	1,483	23,001	...
Eggs	623	...	623	...	1,119	...	1,119	...

TABLE 7.35. 1985 and 2000 Production-Demand Comparisons for Other South America, Assuming High Population Growth Rates (1,000 metric tons)

	1985 Comparisons, Assuming Constant Per Capita Income				1985 Comparisons, Assuming Historical Income Growth Rates			
	Domestic Disappearance				Domestic Disappearance			
Commodity	Food and Industrial	Feed	Total	Excess Demand	Food and Industrial	Feed	Total	Excess Demand
Cereals	13,857	1,658	15,515	6,777	15,447	2,161	17,608	8,869
Raw sugar	4,123	...	4,123	1	4,572	...	4,572	450
Root crops	12,086	1,653	13,739	4,430	13,055	2,273	15,328	6,019
Pulses	1,132	...	1,132	421	1,299	...	1,299	588
Fruit, veg.	21,488	629	22,117	7,043	25,306	823	26,129	11,055
Oil crops	469	321	790	258	638	421	1,059	527
Meat	2,943	...	2,943	...	3,917	...	3,917	...
Milk	9,967	677	10,644	...	13,022	877	13,899	...
Eggs	435	...	435	...	611	...	611	...

	2000 Comparisons, Assuming Constant Per Capita Income				2000 Comparisons, Assuming Historical Income Growth Rates			
	Domestic Disappearance				Domestic Disappearance			
Commodity	Food and Industrial	Feed	Total	Excess Demand	Food and Industrial	Feed	Total	Excess Demand
Cereals	21,070	2,499	23,569	12,872	24,572	3,612	28,184	17,486
Raw sugar	6,375	...	6,375	1,129	7,371	...	7,371	2,125
Root crops	18,389	2,529	20,918	10,350	20,429	4,042	24,471	13,903
Pulses	1,749	...	1,749	962	2,138	...	2,138	1,351
Fruit, veg.	33,155	984	34,139	17,093	41,747	1,521	43,268	26,222
Oil crops	743	506	1,249	524	1,099	693	1,792	1,066
Meat	4,511	...	4,511	...	6,774	...	6,774	...
Milk	15,386	1,035	16,421	...	22,459	1,543	24,002	...
Eggs	674	...	674	...	1,142	...	1,142	...

In summary, though production levels increase over the 40-year projection period, projected demand grows much more rapidly, even by 1985. Expanding crop areas account for most of the production increases projected, while positive yield trends contribute in only a minor way. Unused land resources are estimated to exceed projected land use in 2000 by a wide margin. Both population and income are important determinants of future demand, and if both high population and high income trends are assumed, a continuation of recent production trends would result in output levels which are less than half the projected demands.

CHAPTER 8

Production-Demand Comparisons
for Europe, the USSR, and Oceania

IN THIS CHAPTER we continue with our consideration of the projected production-demand comparisons for 1985 and 2000. Here, however, we focus on a geographic group which includes most of the world's economically developed countries, and the results are much more uniform than those observed among the regions of the Western Hemisphere.

Before beginning, it may be well to note one implicit assumption which underlies all projections of this study, but which is probably most critical in projections for Europe. In estimating demand for livestock feed, no explicit attempt was made to estimate demand for forage crops or for grazing. Similarly, no production projections were made, nor were production-demand comparisons computed for these products. Implicitly, we assume that grazing and roughage crops will not be limiting factors, an assumption justified in most cases. Projected demand increases for livestock products in developing countries frequently are large in relative terms. However, the usual case is one where the demands placed upon the agricultural sector continue to be predominantly for production of food crops, and demand for feed is small relative to total agricultural demand. Too, these food crops frequently yield by-products which have feeding value as roughages, and not all livestock require roughage or grazing. Projections for some other developed countries such as the United States or Canada show large increases in demand for livestock products, but rarely does scarcity of land for producing the necessary roughage appear.

Caution should be exercised in interpretation of the projections for Europe. Projected population increases generally are not large, but the demand for livestock products grows substantially because of increasing per capita incomes. At the same time, agricultural land is not in abundance, as in North America. Thus, in some cases, the reported comparisons of production and demand may not provide a complete assessment of the ability of indigenous agriculture to meet the future demands which may be placed upon it. This possible shortcoming is tempered somewhat by the fact that root crops are used more extensively as a

130

livestock feed in Europe than in other areas, and to some extent these serve a function similar to that served by roughage crops in North American livestock operations. Although root crops production is generally falling or slowly increasing, it is included in the production-demand comparisons, and the omission of hay and pasture may therefore be somewhat less detrimental.

The present level and projected level of per capita incomes in Europe, plus expected lower population growth rates plus the potential surpluses (relative to domestic consumption) of cereals in such regions as North America, provide a supply source for European countries. Under current and prospective international economic and market conditions of the world, we would expect European countries to draw cereals for livestock production to them, through imports, rather than to allow them to flow to less developed countries as human food. While these are the conditions we would project to prevail, other things being equal, the following discussion does not automatically incorporate them into the analysis. Rather, we make only indigenous comparisons for European countries, namely, the trends of production relative to potential demand trends within the countries.

With the above qualification in mind, we begin our examination of the production-demand comparisons with those for Northern Europe.

COMPARISONS FOR NORTHERN EUROPE

Table 8.1 shows that Northern Europe was a major food deficit area in 1960. Deficits were recorded for every commodity group except milk and estimated cereals production was nearly 24 million tons less than domestic disappearance. Of course, this is a region where average incomes are high, and the structure of demand in 1960 reflects these high income and consumption levels. Estimated income elasticities for cereals and root crops are —0.3 and —0.37, respectively. Pulses, oil crops, and milk have estimated elasticities which are positive, but near zero; the

TABLE 8.1. 1960 Production-Demand Comparisons for Northern Europe (1,000 metric tons)

Commodity	Domestic Disappearance			Production	Excess Demand
	Food and Industrial	Feed	Total		
Cereals	32,691	55,285	87,976	64,057	23,919
Raw sugar	9,152	34	9,186	6,829	2,357
Root crops	31,760	32,955	64,715	63,550	1,165
Pulses	1,380	348	1,728	761	967
Fruit, veg.	48,971	1,181	50,152	37,073	13,079
Oil crops	2,508	5,883	8,391	359	8,032
Meat	13,806	. . .	13,806	13,151	655
Milk	58,581	30,987	89,568	89,833	—265
Eggs	2,759	. . .	2,759	2,689	70

TABLE 8.2. 1985 and 2000 Production Projections for Northern Europe (1,000 metric tons)

Commodity	1985 Production	2000 Production
Cereals	119,793	155,505
Raw sugar	9,645	10,471
Root crops	34,500	13,588
Pulses	440	463
Fruit, veg.	50,516	57,788
Oil crops	589	650

elasticity estimate for sugar is 0.18; and the highest estimates, 0.37 to 0.30, are associated with vegetables and fruit, meat, and eggs. Each figure reported above is a simple average of estimates for the 12 countries, but intercountry variability in elasticity estimates is small.

Crops for which time trends were estimated account for nearly all production for every product class except fruit and vegetables for which coverage is only about 40 percent. Table 8.2 shows projected cereals production for 2000 to be over 2.4 times the 1960 level, and significant increases are also estimated for sugar and fruit and vegetables. Projected root crops production declines steadily from 63.6 million tons in 1960 to 13.6 million tons in 2000. Total projected expansion in crop area by 2000 is only 4.7 percent for this region, and most of this occurs in France. Area trends for Norway and Sweden become bounded by 1970; Denmark, United Kingdom, and West Germany area trends are bounded between 1970 and 1985; and area trends in Finland, France, Netherlands, and Switzerland are bounded between 1985 and 2000. Total crop area estimates for Austria, Belgium-Luxembourg, and Ireland are lower in 2000 than in 1960.

Table 8.4 presents results for 1985 under benchmark assumptions. The significant production-demand comparisons are those for cereals and root crops. A dramatic shift is projected for cereals, with the 1960 deficit erased and a surplus of 20.4 million tons projected by 1985. (Of course, cereals expansion would substitute for root crops under market-correcting forces.)

Table 8.4 also shows the interesting projections which result for conditions of medium population and high income growth. Relative to the 1985 benchmark variant, feed components of demand for cereals and root crops increase 18.0 and 13.1 million tons, respectively, while respective food components fall 5.5 and 8.9 million tons. A surplus of 7.9 million tons of cereals is projected (keeping in mind the meaning attached to the term *surplus* as denoted elsewhere and emphasized in the outset of Chapter 10), along with a 42.9 million ton deficit of root crops. Production of fruit and vegetables and oil crops also is short of demand, with deficits somewhat larger than in 1960. The deficit of root crops reflects the absolute decline in production; the demand increase is moderate.

TABLE 8.3. 1985 and 2000 Production-Demand Comparisons for Northern Europe, Assuming Low Population Growth Rates (1,000 metric tons)

1985 Comparisons

	1985 Comparisons, Assuming Constant Per Capita Income — Domestic Disappearance				1985 Comparisons, Assuming Historical Income Growth Rates — Domestic Disappearance			
Commodity	Food and Industrial	Feed	Total	Excess Demand	Food and Industrial	Feed	Total	Excess Demand
Cereals	35,705	60,372	96,077	−23,715	30,249	78,351	108,600	−11,192
Raw sugar	9,965	38	10,003	358	11,166	51	11,217	1,572
Root crops	34,808	35,984	70,792	36,293	26,028	49,054	75,082	40,583
Pulses	1,513	383	1,896	1,455	1,577	480	2,057	1,616
Fruit, veg.	54,471	1,138	55,609	5,093	64,892	1,397	66,289	15,773
Oil crops	2,772	6,390	9,162	8,573	2,844	8,312	11,156	10,568
Meat	15,145	...	15,145	...	20,325	...	20,325	...
Milk	64,396	34,230	98,626	...	67,982	45,368	113,350	...
Eggs	3,003	...	3,003	...	3,752	...	3,752	...

2000 Comparisons

	2000 Comparisons, Assuming Constant Per Capita Income — Domestic Disappearance				2000 Comparisons, Assuming Historical Income Growth Rates — Domestic Disappearance			
Commodity	Food and Industrial	Feed	Total	Excess Demand	Food and Industrial	Feed	Total	Excess Demand
Cereals	36,616	61,814	98,430	−57,074	29,607	90,220	119,827	−35,677
Raw sugar	10,186	40	10,226	−214	11,798	59	11,857	1,386
Root crops	35,760	37,063	72,823	59,234	24,722	57,773	82,495	68,907
Pulses	1,552	392	1,944	1,481	1,637	545	2,182	1,719
Fruit, veg.	56,331	1,138	57,469	−318	71,858	1,535	73,393	15,605
Oil crops	2,857	6,527	9,384	8,734	2,953	9,561	12,514	11,863
Meat	15,558	...	15,558	...	23,999	...	23,999	...
Milk	66,105	35,253	101,358	...	70,932	53,125	124,057	...
Eggs	3,076	...	3,076	...	4,103	...	4,103	...

The high variant population projection for 1985 exceeds the low variant projection by only 6.8 percent, so demand variability shown in Tables 8.3 and 8.5 is not great. Under the benchmark assumptions for 2000, cereals production is projected at 2.4 times 1960 production, and a 50 million ton surplus is estimated. A 64 million ton deficit of root crops is projected as production falls to 21 percent of 1960 levels.

Under the assumptions of medium population and high income, the projected cereals surplus is 28 million tons in 2000 and the root crops deficit is 73.9 million (Table 8.4). Feed components of cereals and root crops demand are estimated to be about 75 percent of total demands. Results associated with other demand variants do not differ greatly.

Although the balance among individual product classes is uneven, the projections imply expansionary forces for European agriculture, which can cause production to grow more rapidly than demand. A given tonnage of root crops is produced with fewer resources than the same tonnage of cereals. Hence, the adjustment needs implied between future cereals and root crops production should cause no problems. Moreover, the assumption adopted that cereals and root crops will be fed to livestock in constant proportions through future years need not hold true, and market relationships should allow a ready shift of resources between the two crop categories. Rising yields are the dominant factor affecting agricultural output. However, barley area increases, largely at the expense of other crops, already have been important in raising cereals production. Projected population increases are small and income elasticities are relatively low, so future demand increases are only moderate.

The implications for long-term U.S. grain exports to this area are clear, though adjustments in livestock feeding practices, indicated above, and future internal agricultural policies in Europe may result in cereals surpluses less dramatic than those indicated in this study.

COMPARISONS FOR SOUTHERN EUROPE

Table 8.6 shows Southern Europe produced 5.4 million tons of vegetables and fruit in excess of domestic demand in 1960, but that domestic production was below domestic demand for all other product classes. Again, the cereals category where a 6.6 million ton deficit or 21 percent of domestic demand was recorded is most notable. Cereals were the major feed commodity, and feed demand accounted for 40 percent of total domestic disappearance in this product class.

Estimated income elasticities for root crops and pulses demand in Southern Europe are near zero, and the elasticity estimate for cereals is —0.21. Oil crops and vegetables and fruit demands have elasticities estimated at 0.3. Elasticity estimates for sugar and the three livestock product classes are in the range 0.6 to 0.76.

TABLE 8.4. 1985 and 2000 Production-Demand Comparisons for Northern Europe, Assuming Medium Population Growth Rates (1,000 metric tons)

1985 Comparisons, Assuming Constant Per Capita Income

Commodity	Domestic Disappearance			Excess Demand
	Food and Industrial	Feed	Total	
Cereals	36,938	62,480	99,418	-20,374
Raw sugar	10,312	40	10,352	707
Root crops	36,016	37,185	73,201	38,701
Pulses	1,565	396	1,961	1,521
Fruit, veg.	56,365	1,163	57,528	7,012
Oil crops	2,869	6,612	9,481	8,893
Meat	15,669	...	15,669	...
Milk	66,668	35,430	102,098	...
Eggs	3,106	...	3,106	...

1985 Comparisons, Assuming Historical Income Growth Rates

Commodity	Domestic Disappearance			Excess Demand
	Food and Industrial	Feed	Total	
Cereals	31,408	80,490	111,898	-7,895
Raw sugar	11,530	52	11,582	1,938
Root crops	27,098	50,311	77,409	42,909
Pulses	1,631	493	2,124	1,683
Fruit, veg.	66,810	1,421	68,231	17,715
Oil crops	2,943	8,537	11,480	10,891
Meat	20,852	...	20,852	...
Milk	70,297	46,590	116,887	...
Eggs	3,861	...	3,861	...

2000 Comparisons, Assuming Constant Per Capita Income

Commodity	Domestic Disappearance			Excess Demand
	Food and Industrial	Feed	Total	
Cereals	39,192	66,258	105,450	-50,054
Raw sugar	10,921	43	10,964	492
Root crops	38,294	39,463	77,757	64,169
Pulses	1,662	422	2,084	1,621
Fruit, veg.	60,366	1,169	61,535	3,747
Oil crops	3,066	6,991	10,057	9,407
Meat	16,662	...	16,662	...
Milk	70,961	37,792	108,753	...
Eggs	3,290	...	3,290	...

2000 Comparisons, Assuming Historical Income Growth Rates

Commodity	Domestic Disappearance			Excess Demand
	Food and Industrial	Feed	Total	
Cereals	31,827	95,526	127,353	-28,151
Raw sugar	12,613	63	12,676	2,209
Root crops	26,682	60,782	87,464	73,876
Pulses	1,753	579	2,332	1,869
Fruit, veg.	76,356	1,568	77,924	20,136
Oil crops	3,167	10,114	13,281	12,631
Meat	25,339	...	25,339	...
Milk	76,038	56,197	132,235	...
Eggs	4,363	...	4,363	...

TABLE 8.5. 1985 and 2000 Production-Demand Comparisons for Northern Europe, Assuming High Population Growth Rates (1,000 metric tons)

1985 Comparisons

	1985 Comparisons, Assuming Constant Per Capita Income — Domestic Disappearance				1985 Comparisons, Assuming Historical Income Growth Rates — Domestic Disappearance			
Commodity	Food and Industrial	Feed	Total	Excess Demand	Food and Industrial	Feed	Total	Excess Demand
Cereals	38,126	64,498	102,624	−17,168	32,532	82,518	115,050	−4,743
Raw sugar	10,645	41	10,686	1,042	11,878	53	11,932	2,287
Root crops	37,179	38,356	75,535	41,035	28,137	51,527	79,664	45,164
Pulses	1,616	409	2,025	1,584	1,682	506	2,188	1,748
Fruit, veg.	58,210	1,194	59,404	8,888	68,670	1,450	70,120	19,603
Oil crops	2,963	6,824	9,787	9,199	3,038	8,750	11,788	11,198
Meat	16,176	...	16,176	...	21,357	...	21,357	...
Milk	68,842	36,582	105,424	...	72,507	47,753	120,261	...
Eggs	3,206	...	3,206	...	3,966	...	3,966	...

2000 Comparisons

	2000 Comparisons, Assuming Constant Per Capita Income — Domestic Disappearance				2000 Comparisons, Assuming Historical Income Growth Rates — Domestic Disappearance			
Commodity	Food and Industrial	Feed	Total	Excess Demand	Food and Industrial	Feed	Total	Excess Demand
Cereals	41,770	70,700	112,470	−43,034	34,068	100,736	134,804	−20,700
Raw sugar	11,656	46	11,702	1,230	13,425	66	13,491	3,020
Root crops	40,826	41,857	82,683	69,095	28,667	63,733	92,400	78,812
Pulses	1,773	451	2,224	1,761	1,868	614	2,482	2,018
Fruit, veg.	64,402	1,206	65,608	7,820	80,804	1,608	82,412	24,623
Oil crops	3,275	7,454	10,729	10,079	3,381	10,658	14,039	13,389
Meat	17,766	...	17,766	...	26,654	...	26,654	...
Milk	75,816	40,324	116,140	...	81,126	59,204	140,330	...
Eggs	3,505	...	3,505	...	4,620	...	4,620	...

TABLE 8.6.   1960 Production-Demand Comparisons for Southern Europe
(1,000 metric tons)

Commodity	Domestic Disappearance			Production	Excess Demand
	Food and Industrial	Feed	Total		
Cereals	19,065	12,670	31,735	25,149	6,586
Raw sugar	2,052	. . .	2,052	1,724	328
Root crops	9,183	994	10,177	10,071	106
Pulses	1,051	813	1,864	1,739	125
Fruit, veg.	34,963	478	35,441	40,836	−5,395
Oil crops	5,779	460	6,239	5,467	772
Meat	2,360	. . .	2,360	2,070	290
Milk	9,086	7,220	16,306	15,188	1,118
Eggs	749	. . .	749	646	103

Except for fruit and vegetables where coverage is estimated to be about 70 percent, crops included in the study cover a high percentage of production in each product class. Estimated crop area trends for Greece are bounded by 1970, but trends in total area for the other three countries of the region are negative. For the region, the sum of all estimated areas in 2000 is 21 percent below the sum of the trend values in 1960. Estimates presented in Table 8.7 reveal increasing production in each commodity group. Cereals production grows much less rapidly than in Northern Europe. The most rapid growth is in fruit and vegetables production.

Estimates presented in Table 8.9 project production to rise more rapidly than demand through 1985 under benchmark assumptions. Steadily growing surpluses are projected for sugar and root and oil crops, while excess demand for pulses varies little from the 1960 level. By 1985 the cereals deficit is reduced to 3.8 million tons, and a 37.4 million ton surplus is projected for fruit and vegetables.

Under variants of high income and medium population, significant differences are observed (Table 8.9). Relative to the 1985 benchmark variant, the feed component of cereals demand increases 9.7 million tons as a result of increased per capita income and relatively high income elasticities for livestock products. However, the food and industrial component falls 4.3 million tons. The net result is an excess demand estimate of 9.2 million tons, 2.6 million tons higher than in 1960. De-

TABLE 8.7.   1985 and 2000 Production Projections for Southern Europe
(1,000 metric tons)

Commodity	1985 Production	2000 Production
Cereals	32,081	34,629
Raw sugar	2,984	3,815
Root crops	13,635	15,627
Pulses	1,746	1,740
Fruit, veg.	77,407	99,170
Oil crops	7,146	8,010

mands in other product classes increase accordingly, but a 28.7 million ton surplus of fruit and vegetables remains. (Again we refer the reader to Chapter 10 for meaning of the term *surplus* and the equilibrating forces expected to modify it on an individual crop basis.)

Comparing Tables 8.8 and 8.10, with constant per capita income, variations between low and high population assumptions result in 1985 demands 3.6 percent below and above benchmark levels, respectively. Even under high population and increasing per capita income, the total 1985 production-demand balance compares well with 1960. Cereals and oil crops deficits increase 3.9 and 1.4 million tons, respectively, from 1960; but the deficit for root crops decreases 1.3 million tons, and the surplus of fruit and vegetables increases 21.7 million tons.

Considering the benchmark projections in 2000, production-demand comparisons are similar to those in 1985. However, the surplus of fruit and vegetables rises to 56.7 million tons.

Tables 8.8 and 8.10 for the year 2000, with increasing per capita income, show that future population growth can introduce considerable uncertainty into the demand estimates and the production-demand comparisons. For both tables, the estimated cereals deficit in 2000 is above the 1960 level, but it is 5.6 million tons greater when high population (Table 8.8) is assumed rather than low population (Table 8.10). Of course, per capita incomes are not the same under these two sets of projections. If they were, the comparison would be more striking.

Under the most pressing demand assumptions, high population and increasing income, production-demand comparisons show a mixed pattern in 2000. Table 8.10 (high population) shows excess demand for cereals increasing to 8.7 million tons above 1960 levels. However, the production surplus for fruit and vegetables is 41 million tons.

Falling areas for a number of field crops, rising yields for these same crops, and very strongly positive production trends for fruit and vegetables are the dominant features reflected in the production projections for Southern Europe. Projections of slow population growth, along with moderate or high income growth, result in demand projections which are not generally excessive relative to production. Cereals may be an exception. The feed component of cereals demand is quite responsive to per capita income growth because of the relatively high income elasticity of demand for livestock products.

COMPARISONS FOR EASTERN EUROPE

Table 8.11 presents base period production-demand comparisons for Eastern Europe. In 1960, Eastern Europe's production of cereals was 4.5 million tons less than domestic demand. However, production in each other crop product class except oil crops exceeded demand by a small margin. Domestic disappearances of cereals and root crops were large. Slightly less than half the total demand for these products was

TABLE 8.8. 1985 and 2000 Production-Demand Comparisons for Southern Europe, Assuming Low Population Growth Rates (1,000 metric tons)

1985 Comparisons

	1985 Comparisons, Assuming Constant Per Capita Income				1985 Comparisons, Assuming Historical Income Growth Rates			
	Domestic Disappearance				Domestic Disappearance			
Commodity	Food and Industrial	Feed	Total	Excess Demand	Food and Industrial	Feed	Total	Excess Demand
Cereals	20,765	13,842	34,607	2,526	16,442	23,571	40,013	7,932
Raw sugar	2,232	...	2,232	−751	3,431	...	3,431	447
Root crops	10,084	1,078	11,162	−2,472	9,691	1,868	11,559	−2,076
Pulses	1,148	894	2,042	296	1,148	1,499	2,647	901
Fruit, veg.	38,046	516	38,562	−38,844	46,310	911	47,221	−30,184
Oil crops	6,319	497	6,816	−328	8,014	869	8,883	1,737
Meat	2,565	...	2,565	...	4,537	...	4,537	...
Milk	9,884	7,835	17,719	...	14,309	13,630	27,939	...
Eggs	815	...	815	...	1,314	...	1,314	...

2000 Comparisons

	2000 Comparisons, Assuming Constant Per Capita Income				2000 Comparisons, Assuming Historical Income Growth Rates			
	Domestic Disappearance				Domestic Disappearance			
Commodity	Food and Industrial	Feed	Total	Excess Demand	Food and Industrial	Feed	Total	Excess Demand
Cereals	21,314	14,213	35,527	898	14,111	30,227	44,338	9,709
Raw sugar	2,291	...	2,291	−1,523	4,290	...	4,290	475
Root crops	10,363	1,106	11,469	−4,157	9,548	2,411	11,959	−3,667
Pulses	1,179	919	2,098	358	1,179	1,930	3,109	1,369
Fruit, veg.	39,046	528	39,574	−59,555	50,216	1,156	51,372	−47,798
Oil crops	6,490	510	7,000	−1,010	9,109	1,124	10,233	2,223
Meat	2,632	...	2,632	...	5,857	...	5,857	...
Milk	10,143	8,036	18,179	...	17,150	17,596	34,746	...
Eggs	837	...	837	...	1,674	...	1,674	...

TABLE 8.9. 1985 and 2000 Production-Demand Comparisons for Southern Europe, Assuming Medium Population Growth Rates (1,000 metric tons)

| | 1985 Comparisons, Assuming Constant Per Capita Income | | | | 1985 Comparisons, Assuming Historical Income Growth Rates | | | |
| | Domestic Disappearance | | | | Domestic Disappearance | | | |
Commodity	Food and Industrial	Feed	Total	Excess Demand	Food and Industrial	Feed	Total	Excess Demand
Cereals	21,540	14,360	35,900	3,818	17,225	24,076	41,301	9,220
Raw sugar	2,316	...	2,316	−667	3,512	...	3,512	528
Root crops	10,464	1,118	11,582	−2,052	10,086	1,907	11,993	−1,640
Pulses	1,191	928	2,119	372	1,191	1,531	2,722	975
Fruit, veg.	39,462	535	39,997	−37,408	47,811	932	48,743	−28,662
Oil crops	6,556	515	7,071	−73	8,253	887	9,140	1,995
Meat	2,660	...	2,660	...	4,631	...	4,631	...
Milk	10,251	8,125	18,376	...	14,678	13,919	28,597	...
Eggs	846	...	846	...	1,343	...	1,343	...

| | 2000 Comparisons, Assuming Constant Per Capita Income | | | | 2000 Comparisons, Assuming Historical Income Growth Rates | | | |
| | Domestic Disappearance | | | | Domestic Disappearance | | | |
Commodity	Food and Industrial	Feed	Total	Excess Demand	Food and Industrial	Feed	Total	Excess Demand
Cereals	22,861	15,254	38,115	3,485	15,481	31,674	47,155	12,526
Raw sugar	2,457	...	2,457	−1,357	4,505	...	4,505	690
Root crops	11,137	1,185	12,322	−3,304	10,330	2,524	12,854	−2,773
Pulses	1,265	988	2,253	513	1,265	2,022	3,287	1,548
Fruit, veg.	41,871	565	42,436	−56,733	53,572	1,213	54,785	−44,385
Oil crops	6,967	545	7,512	−497	9,671	1,176	10,847	2,837
Meat	2,822	...	2,822	...	6,129	...	6,129	...
Milk	10,877	8,610	19,487	...	18,096	18,419	36,515	...
Eggs	897	...	897	...	1,755	...	1,755	...

TABLE 8.10. 1985 and 2000 Production-Demand Comparisons for Southern Europe, Assuming High Population Growth Rates (1,000 metric tons)

| | 1985 Comparisons, Assuming Constant Per Capita Income | | | | 1985 Comparisons, Assuming Historical Income Growth Rates | | | |
| | Domestic Disappearance | | | | Domestic Disappearance | | | |
Commodity	Food and Industrial	Feed	Total	Excess Demand	Food and Industrial	Feed	Total	Excess Demand
Cereals	22,312	14,877	37,189	5,108	18,013	24,567	42,580	10,498
Raw sugar	2,398	...	2,398	−585	3,591	...	3,591	607
Root crops	10,844	1,158	12,002	−1,633	10,482	1,946	12,428	−1,206
Pulses	1,234	961	2,195	449	1,234	1,561	2,795	1,048
Fruit, veg.	40,876	554	41,430	−35,976	49,295	952	50,247	−27,159
Oil crops	6,792	534	7,326	180	8,489	905	9,394	2,248
Meat	2,756	...	2,756	...	4,723	...	4,723	...
Milk	10,619	8,415	19,034	...	15,039	14,200	29,239	...
Eggs	876	...	876	...	1,372	...	1,372	...

| | 2000 Comparisons, Assuming Constant Per Capita Income | | | | 2000 Comparisons, Assuming Historical Income Growth Rates | | | |
| | Domestic Disappearance | | | | Domestic Disappearance | | | |
Commodity	Food and Industrial	Feed	Total	Excess Demand	Food and Industrial	Feed	Total	Excess Demand
Cereals	24,398	16,289	40,687	6,058	16,867	33,062	49,929	15,299
Raw sugar	2,621	...	2,621	−1,193	4,712	...	4,712	897
Root crops	11,907	1,263	13,170	−2,455	11,112	2,631	13,743	−1,883
Pulses	1,351	1,056	2,407	668	1,351	2,110	3,461	1,722
Fruit, veg.	44,679	602	45,281	−53,889	56,872	1,268	58,140	−41,030
Oil crops	7,441	581	8,022	12	10,221	1,225	11,446	3,436
Meat	3,010	...	3,010	...	6,391	...	6,391	...
Milk	11,606	9,180	20,786	...	19,012	19,210	38,222	...
Eggs	958	...	958	...	1,832	...	1,832	...

TABLE 8.11.   1960 Production-Demand Comparisons for Eastern Europe (1,000 metric tons)

| Commodity | Domestic Disappearance | | | Production | Excess Demand |
	Food and Industrial	Feed	Total		
Cereals	29,411	34,076	63,487	58,978	4,509
Raw sugar	3,019	...	3,019	4,295	—1,276
Root crops	25,267	36,591	61,858	65,564	—3,705
Pulses	622	116	738	763	—25
Fruit, veg.	18,834	611	19,445	20,179	—734
Oil crops	772	1,284	2,056	1,536	520
Meat	4,774	...	4,774	4,858	—84
Milk	19,658	11,374	31,032	30,793	239
Eggs	845	...	845	965	—120

for food and industrial use, and the remainder was for feed. The estimated average income elasticity in seven countries of Eastern Europe is —0.16 for cereals and root crops and 0.14 for pulses. The range is 0.35 to 0.44 for sugar, vegetables and fruit, milk, and eggs, while income elasticities for oil crops and meat are estimated at 0.51 and 0.55, respectively. Thus income will have counteracting effects on future demand for these commodities through the negative income elasticity of demand for direct consumption and through the positive income elasticity of demand for livestock products.

Again, the only significant omissions from the time trend estimates of production are found in the fruit and vegetables category.

The total crop area in 2000, estimated from area trends, is about 2 percent less than for 1960. Area trends for Rumania are bounded between 1970 and 1985; Bulgaria and Yugoslavia area expansion is bounded between 1985 and 2000; and trends for the other four countries in the region are unbounded throughout the projection period. Thus the production increases shown in Table 8.12 are primarily the result of increasing yields.

Turning now to the projected production-demand comparisons, Table 8.14 shows that under the benchmark variant excess demand falls steadily through 1985 for all product classes except root crops. By 1985 the 1960 cereals deficit is changed to a 4.3 million ton annual surplus, and the 0.7 million ton fruit and vegetables surplus increases to 11.3

TABLE 8.12.   1985 and 2000 Production Projections for Eastern Europe (1,000 metric tons)

Commodity	1985 Production	2000 Production
Cereals	81,537	94,836
Raw sugar	9,401	12,734
Root crops	72,012	75,390
Pulses	1,294	1,562
Fruit, veg.	34,865	42,681
Oil crops	3,416	4,460

million tons. However, in the same 25-year period, the 3.7 million ton surplus of root crops changes to a 5.2 million ton deficit. (It is hoped central planners would correct through resource shifts as is emphasized in the outset of Chapter 10.)

The effect of higher 1985 incomes on demand is pronounced. Table 8.14 shows that income growth at historical rates causes feed components of cereals and root crops demand to increase by 17.3 and 13.0 million tons, respectively. Food and industrial components decrease 5.8 and 8.0 million tons, respectively. The 4.3 million ton cereals surplus estimated for 1985 under the benchmark assumptions becomes a 7.1 million ton deficit, and the deficit for root crops increases from 5.2 to 10.3 million tons. As the population is moved from the low to the high variant while per capita income is constant, a 7 percent change in 1985 demand is shown in Tables 8.13 and 8.15.

Projections to 2000 under benchmark assumptions are qualitatively the same as in 1985. Production increases more than meet demand increases brought about by population growth, except in the case of root crops where a 10 million ton deficit is projected.

Demand projections for 2000 under the various assumptions of population are quite diverse (Tables 8.13 and 8.15). Assuming low population and high income, only root crops production falls short of projected demand in 2000. But under variants of high population and high income, deficits of 10.9 million tons for cereals and 22.7 million tons for root crops are projected. Sugar production is in surplus by 6.5 million tons, but estimated excess demands for other commodities are not greatly different from 1960 levels. Demand is sharply upward in movement to the high income variant.

In summary, projections of demand in most countries of Eastern Europe reflect slow to moderate population growth, though Poland, the largest country in the region, also shows the highest population growth rate. Demand for some important commodities is quite responsive to income growth, and while projected production values increase more rapidly than population, demands resulting from high population and income growth are in excess of production for important commodities.

COMPARISONS FOR THE USSR

The Soviet Union's base period food balance is characterized by large production figures but near self-sufficiency in every product class. Table 8.16 shows 1960 production surpluses for cereals, root crops, and oil crops. However, production and domestic demand differed by no more than 10 percent in any product classes. Cereals and root crops were again the major livestock feeds. The feed component was about 32 percent of total demand for each of these product classes.

Two negative income elasticities are estimated for 1960 demand:

TABLE 8.13. 1985 and 2000 Production-Demand Comparisons for Eastern Europe, Assuming Low Population Growth Rates (1,000 metric tons)

	1985 Comparisons, Assuming Constant Per Capita Income				1985 Comparisons, Assuming Historical Income Growth Rates			
	Domestic Disappearance				Domestic Disappearance			
Commodity	Food and Industrial	Feed	Total	Excess Demand	Food and Industrial	Feed	Total	Excess Demand
Cereals	34,767	39,684	74,451	−7,085	28,939	56,839	85,778	4,241
Raw sugar	3,512	...	3,512	−5,888	4,861	...	4,861	−4,539
Root crops	30,012	44,433	74,445	2,434	22,009	57,394	79,403	7,391
Pulses	728	134	862	−431	786	219	1,005	−288
Fruit, veg.	22,001	721	22,722	−12,142	31,929	1,131	33,060	−1,805
Oil crops	887	1,497	2,384	−1,031	1,314	2,277	3,591	175
Meat	5,533	...	5,533	...	7,902	...	7,902	...
Milk	23,139	13,319	36,458	...	27,639	17,864	45,503	...
Eggs	920	...	920	...	1,364	...	1,364	...

	2000 Comparisons, Assuming Constant Per Capita Income				2000 Comparisons, Assuming Historical Income Growth Rates			
	Domestic Disappearance				Domestic Disappearance			
Commodity	Food and Industrial	Feed	Total	Excess Demand	Food and Industrial	Feed	Total	Excess Demand
Cereals	36,661	41,694	78,355	−16,480	28,259	64,353	92,612	−2,223
Raw sugar	3,692	...	3,692	−9,041	5,611	...	5,611	−7,122
Root crops	31,752	47,327	79,079	3,689	18,514	65,952	84,466	9,076
Pulses	763	141	904	−657	853	253	1,106	−454
Fruit, veg.	23,125	751	23,876	−18,804	37,438	1,258	38,696	−3,984
Oil crops	930	1,575	2,505	−1,955	1,539	2,609	4,148	−312
Meat	5,812	...	5,812	...	8,939	...	8,939	...
Milk	24,399	14,051	38,450	...	30,149	20,309	50,458	...
Eggs	1,031	...	1,031	...	1,590	...	1,590	...

144

TABLE 8.14. 1985 and 2000 Production-Demand Comparisons for Eastern Europe, Assuming Medium Population Growth Rates (1,000 metric tons)

1985 Comparisons

	1985 Comparisons, Assuming Constant Per Capita Income — Domestic Disappearance				1985 Comparisons, Assuming Historical Income Growth Rates — Domestic Disappearance			
Commodity	Food and Industrial	Feed	Total	Excess Demand	Food and Industrial	Feed	Total	Excess Demand
Cereals	36,065	41,132	77,197	−4,338	30,221	58,456	88,677	7,140
Raw sugar	3,639	...	3,639	−5,761	4,994	...	4,994	−4,406
Root crops	31,129	46,123	77,252	5,240	23,153	59,167	82,320	10,309
Pulses	755	139	894	−399	813	225	1,038	−255
Fruit, veg.	22,811	749	23,560	−11,304	32,762	1,166	33,928	−936
Oil crops	919	1,551	2,470	−944	1,347	2,338	3,685	270
Meat	5,733	...	5,733	...	8,125	...	8,125	...
Milk	23,994	13,802	37,796	...	28,559	18,380	46,939	...
Eggs	1,016	...	1,016	...	1,400	...	1,400	...

2000 Comparisons

	2000 Comparisons, Assuming Constant Per Capita Income — Domestic Disappearance				2000 Comparisons, Assuming Historical Income Growth Rates — Domestic Disappearance			
Commodity	Food and Industrial	Feed	Total	Excess Demand	Food and Industrial	Feed	Total	Excess Demand
Cereals	39,492	44,756	84,248	−10,587	30,884	68,294	99,178	4,343
Raw sugar	3,960	...	3,960	−8,773	5,930	...	5,930	−6,802
Root crops	34,203	51,172	85,375	9,984	20,638	70,654	91,292	15,902
Pulses	822	151	973	−588	914	266	1,180	−381
Fruit, veg.	24,851	815	25,666	−17,014	39,497	1,348	40,845	−1,835
Oil crops	995	1,689	2,684	−1,775	1,620	2,757	4,377	−82
Meat	6,230	...	6,230	...	9,480	...	9,480	...
Milk	26,245	15,078	41,323	...	32,286	21,578	53,864	...
Eggs	1,105	...	1,105	...	1,677	...	1,677	...

TABLE 8.15. 1985 and 2000 Production-Demand Comparisons for Eastern Europe, Assuming High Population Growth Rates (1,000 metric tons)

1985 Comparisons, Assuming Constant Per Capita Income

Commodity	Domestic Disappearance			Excess Demand
	Food and Industrial	Feed	Total	
Cereals	37,361	42,577	79,938	−1,598
Raw sugar	3,766	...	3,766	−5,633
Root crops	32,243	47,808	80,051	8,039
Pulses	783	143	926	−367
Fruit, veg.	23,619	777	24,396	−10,468
Oil crops	951	1,605	2,556	−859
Meat	5,932	...	5,932	...
Milk	24,846	14,284	39,130	...
Eggs	1,051	...	1,051	...

1985 Comparisons, Assuming Historical Income Growth Rates

Commodity	Domestic Disappearance			Excess Demand
	Food and Industrial	Feed	Total	
Cereals	31,506	60,046	91,552	10,016
Raw sugar	5,125	...	5,125	−4,275
Root crops	24,307	60,910	85,217	13,205
Pulses	840	230	1,070	−223
Fruit, veg.	33,582	1,200	34,782	−82
Oil crops	1,380	2,397	3,777	362
Meat	8,344	...	8,344	...
Milk	29,470	18,887	48,357	...
Eggs	1,435	...	1,435	...

2000 Comparisons, Assuming Constant Per Capita Income

Commodity	Domestic Disappearance			Excess Demand
	Food and Industrial	Feed	Total	
Cereals	42,349	47,844	90,193	−4,641
Raw sugar	4,230	...	4,230	−8,503
Root crops	36,684	55,073	91,757	16,367
Pulses	882	160	1,042	−519
Fruit, veg.	26,591	880	27,471	−15,210
Oil crops	1,061	1,804	2,865	−1,594
Meat	6,651	...	6,651	...
Milk	28,111	16,118	44,229	...
Eggs	1,179	...	1,179	...

2000 Comparisons, Assuming Historical Income Growth Rates

Commodity	Domestic Disappearance			Excess Demand
	Food and Industrial	Feed	Total	
Cereals	33,566	72,181	105,747	10,911
Raw sugar	6,245	...	6,245	−6,488
Root crops	22,847	75,338	98,185	22,796
Pulses	975	278	1,253	−308
Fruit, veg.	41,520	1,437	42,957	275
Oil crops	1,700	2,902	4,602	142
Meat	10,015	...	10,015	...
Milk	34,423	22,839	57,262	...
Eggs	1,763	...	1,763	...

TABLE 8.16. 1960 Production-Demand Comparisons for the USSR (1,000 metric tons)

	Domestic Disappearance				
Commodity	Food and Industrial	Feed	Total	Production	Excess Demand
Cereals	63,446	30,087	93,533	97,901	−4,368
Raw sugar	7,006	...	7,006	6,286	720
Root crops	55,836	25,806	81,642	85,082	−3,440
Pulses	1,244	1,761	3,005	2,950	55
Fruit, veg.	21,884	...	21,884	20,838	1,046
Oil crops	1,222	2,678	3,900	4,169	−269
Meat	8,034	...	8,034	8,060	−26
Milk	33,986	18,225	52,211	52,667	−456
Eggs	1,509	...	1,509	1,498	11

−0.22 for cereals and −0.3 for root crops. Estimated elasticity for pulses is 0.1 and for oil crops, 0.8. Elasticities for sugar, vegetables and fruit, and the livestock product classes are estimated in the range 0.35 to 0.53.

Only about 10 percent of fruit and vegetables production is accounted for by crops included in this study, but coverage of the remaining product classes is essentially complete. Expansion of total crop area is bounded before 1970 at a level 2.5 percent greater than the total of all estimated areas in 1960. However, significant adjustments occur among individual crops. A large increase in barley area is projected. Cereals production increases only about 40 percent by 2000, but sugar, fruit and vegetables, oil crops, and pulses have much larger gains (Table 8.17). Root crops production remains essentially constant.

Significant changes in production-demand balances are projected by 1985 under the benchmark assumptions (Table 8.19). Cereals and root crops both move from surplus to deficit positions. The deficit is 4.2 million tons for cereals and 24.6 million tons for root crops. Large surpluses are projected for pulses and fruit and vegetables. Other projections in Table 8.19 show a negative effect of high income on total demand for cereals and root crops. Reduced demand for food and industrial uses more than offsets increased feed demand, though deficits are still projected for 1985. Surpluses of sugar, pulses, fruit and vegetables, and oil crops are all reduced, but each maintains an excess position.

TABLE 8.17. 1985 and 2000 Production Projections for the USSR (1,000 metric tons)

Commodity	1985 Production	2000 Production
Cereals	125,110	139,585
Raw sugar	15,071	17,976
Root crops	88,294	86,155
Pulses	26,548	37,848
Fruit, veg.	60,251	78,640
Oil crops	10,330	12,853

TABLE 8.18. 1985 and 2000 Production-Demand Comparisons for the USSR, Assuming Low Population Growth Rates (1,000 metric tons)

	1985 Comparisons, Assuming Constant Per Capita Income				1985 Comparisons, Assuming Historical Income Growth Rates			
	Domestic Disappearance				Domestic Disappearance			
Commodity	Food and Industrial	Feed	Total	Excess Demand	Food and Industrial	Feed	Total	Excess Demand
Cereals	83,558	39,636	123,194	-1,915	65,591	56,962	122,553	-2,556
Raw sugar	9,227	...	9,227	-5,843	13,736	...	13,736	-1,334
Root crops	73,536	33,996	107,532	19,238	51,973	48,858	100,831	12,537
Pulses	1,639	2,320	3,959	-22,588	1,799	3,335	5,134	-21,413
Fruit, veg.	28,821	...	28,821	-31,429	38,680	...	38,680	-21,569
Oil crops	1,609	3,528	5,137	-5,192	2,867	5,071	7,938	-2,391
Meat	10,581	...	10,581	...	15,338	...	15,338	...
Milk	44,759	24,009	68,768	...	62,294	34,504	96,798	...
Eggs	1,987	...	1,987	...	2,763	...	2,763	...

	2000 Comparisons, Assuming Constant Per Capita Income				2000 Comparisons, Assuming Historical Income Growth Rates			
	Domestic Disappearance				Domestic Disappearance			
Commodity	Food and Industrial	Feed	Total	Excess Demand	Food and Industrial	Feed	Total	Excess Demand
Cereals	93,660	44,427	138,087	-1,497	68,522	68,521	137,043	-2,541
Raw sugar	10,342	...	10,342	-7,633	16,651	...	16,651	-1,324
Root crops	82,426	38,106	120,532	34,377	52,259	58,773	111,032	24,876
Pulses	1,837	2,601	4,438	-33,409	2,061	4,011	6,072	-31,774
Fruit, veg.	32,305	...	32,305	-46,334	46,099	...	46,099	-32,540
Oil crops	1,803	3,955	5,758	-7,094	3,564	6,100	9,664	-3,189
Meat	11,860	...	11,860	...	18,515	...	18,515	...
Milk	50,170	26,912	77,082	...	72,890	41,506	114,396	...
Eggs	2,227	...	2,227	...	3,314	...	3,314	...

Effects of low and high population with per capita income constant are illustrated in Tables 8.18 and 8.20, respectively. Demand changes in 1985 are —4.7 and 8.0 percent, respectively, relative to results under benchmark assumptions.

The alternative projections to 2000 are quite diverse, depending on demand assumptions. Under benchmark assumptions (Table 8.19), a cereals deficit of 14.5 million tons is projected along with a root crops deficit of 48.3 million tons. Surpluses are estimated for the other four product classes, and those for fruit and vegetables and pulses, 42.6 and 32.9 million tons, respectively, are especially large.

The range between low and high population estimates for year 2000 is 86.3 million and is reflected in corresponding demand projections. Estimates in Table 8.18 for low population and high income show surpluses for all product classes except root crops where the deficit is 24.9 million tons. However, under high levels of both population and income, the 2000 projection for cereals in Table 8.20 indicates a deficit of 35.2 million tons. A 57.7 million ton deficit is projected for root crops. Changes among other commodities are less significant.

In summary, population growth projected for the Soviet Union is not extremely high by world standards. It is, nevertheless, substantial. Moreover, the production-demand comparisons imply that the future population magnitude will be an important determinant of actual future food balances within the USSR. Output expansion, based primarily on yield increases, is not presently increasing at a pace to match the most extreme of the several demand projections.

COMPARISONS FOR OCEANIA

Estimates in Table 8.21 for 1960 show the Oceania region was a surplus producer in every product class except oil crops. A surplus of 6 million tons of cereals, nearly two-thirds of production, is shown. High levels of demand for livestock products are estimated, and cereals and fruit and vegetables are the most important crop commodities in domestic demand. The feed component was about 45 percent of cereals total demand in 1960.

Estimated income elasticities for the two countries of Oceania are very low. The 1960 value for cereals is —0.21, and estimates for sugar, root crops, meat, milk, and eggs are near zero. Highest estimates are for oil crops, pulses, and fruit and vegetables, 0.10, 0.17, and 0.22, respectively. Thus population growth will be the major determinant of future domestic demand.

Time trend estimates are included for crops which account for essentially all production in each product class except fruit and vegetables and oil crops. About 80 percent coverage is estimated for these two product classes.

TABLE 8.19. 1985 and 2000 Production-Demand Comparisons for the USSR, Assuming Medium Population Growth Rates (1,000 metric tons)

1985 Comparisons, Assuming Constant Per Capita Income

Commodity	Domestic Disappearance			Excess Demand
	Food and Industrial	Feed	Total	
Cereals	87,692	41,596	129,288	4,179
Raw sugar	9,683	...	9,683	−5,387
Root crops	77,174	35,678	112,852	24,558
Pulses	1,720	2,435	4,155	−22,392
Fruit, veg.	30,247	...	30,247	−30,003
Oil crops	1,689	3,703	5,392	−4,938
Meat	11,104	...	11,104	...
Milk	46,973	25,197	72,170	...
Eggs	2,085	...	2,085	...

1985 Comparisons, Assuming Historical Income Growth Rates

Commodity	Domestic Disappearance			Excess Demand
	Food and Industrial	Feed	Total	
Cereals	69,767	58,903	128,670	3,561
Raw sugar	14,182	...	14,182	−888
Root crops	55,662	50,523	106,185	17,891
Pulses	1,880	3,448	5,328	−21,219
Fruit, veg.	40,083	...	40,083	−20,167
Oil crops	2,944	5,243	8,187	−2,142
Meat	15,850	...	15,850	...
Milk	64,734	35,680	100,414	...
Eggs	2,860	...	2,860	...

2000 Comparisons, Assuming Constant Per Capita Income

Commodity	Domestic Disappearance			Excess Demand
	Food and Industrial	Feed	Total	
Cereals	104,491	49,565	154,056	14,470
Raw sugar	11,538	...	11,538	−6,437
Root crops	91,957	42,513	134,470	48,315
Pulses	2,050	2,902	4,952	−32,896
Fruit, veg.	36,041	...	36,041	−42,599
Oil crops	2,012	4,412	6,424	−6,428
Meat	13,231	...	13,231	...
Milk	55,972	30,024	85,996	...
Eggs	2,484	...	2,484	...

2000 Comparisons, Assuming Historical Income Growth Rates

Commodity	Domestic Disappearance			Excess Demand
	Food and Industrial	Feed	Total	
Cereals	78,962	74,100	153,062	13,476
Raw sugar	17,945	...	17,945	−30
Root crops	61,321	63,557	124,878	38,723
Pulses	2,277	4,338	6,615	−31,232
Fruit, veg.	50,049	...	50,049	−28,590
Oil crops	3,800	6,596	10,396	−2,457
Meat	19,991	...	19,991	...
Milk	79,860	44,886	124,746	...
Eggs	3,588	...	3,588	...

TABLE 8.20. 1985 and 2000 Production-Demand Comparisons for the USSR, Assuming High Population Growth Rates (1,000 metric tons)

1985 Comparisons, Assuming Constant Per Capita Income

Domestic Disappearance

Commodity	Food and Industrial	Feed	Total	Excess Demand
Cereals	94,703	44,922	139,625	14,516
Raw sugar	10,458	...	10,458	−4,613
Root crops	83,344	38,531	121,875	33,581
Pulses	1,858	2,630	4,488	−22,060
Fruit, veg.	32,665	...	32,665	−27,585
Oil crops	1,824	3,999	5,823	−4,507
Meat	11,992	...	11,992	...
Milk	50,729	27,211	77,940	...
Eggs	2,252	...	2,252	...

1985 Comparisons, Assuming Historical Income Growth Rates

Domestic Disappearance

Commodity	Food and Industrial	Feed	Total	Excess Demand
Cereals	76,948	62,100	139,048	13,938
Raw sugar	14,914	...	14,914	−157
Root crops	62,036	53,265	115,301	27,007
Pulses	2,016	3,635	5,651	−20,896
Fruit, veg.	42,408	...	42,408	−17,842
Oil crops	3,067	5,528	8,595	−1,734
Meat	16,693	...	16,693	...
Milk	68,750	37,617	106,367	...
Eggs	3,019	...	3,019	...

2000 Comparisons, Assuming Constant Per Capita Income

Domestic Disappearance

Commodity	Food and Industrial	Feed	Total	Excess Demand
Cereals	119,198	56,541	175,739	36,154
Raw sugar	13,162	...	13,162	−4,813
Root crops	104,901	48,497	153,398	67,243
Pulses	2,338	3,310	5,648	−32,199
Fruit, veg.	41,114	...	41,114	−37,526
Oil crops	2,295	5,033	7,328	−5,524
Meat	15,094	...	15,094	...
Milk	63,850	34,249	98,099	...
Eggs	2,834	...	2,834	...

2000 Comparisons, Assuming Historical Income Growth Rates

Domestic Disappearance

Commodity	Food and Industrial	Feed	Total	Excess Demand
Cereals	93,529	81,293	174,822	35,237
Raw sugar	19,604	...	19,604	1,628
Root crops	74,096	69,727	143,823	57,668
Pulses	2,567	4,759	7,326	−30,521
Fruit, veg.	55,199	...	55,199	−23,440
Oil crops	4,093	7,236	11,329	−1,523
Meat	21,890	...	21,890	...
Milk	88,890	49,243	138,133	...
Eggs	3,944	...	3,944	...

TABLE 8.21.   1960 Production-Demand Comparisons for Oceania (1,000 metric tons)

| Commodity | Domestic Disappearance | | | Production | Excess Demand |
	Food and Industrial	Feed	Total		
Cereals	1,944	1,615	3,559	9,604	—6,045
Raw sugar	684	. . .	684	1,336	—652
Root crops	711	. . .	711	718	—7
Pulses	37	1	38	44	—6
Fruit, veg.	1,958	. . .	1,958	2,228	—270
Oil crops	51	20	71	20	51
Meat	1,570	. . .	1,570	2,385	—815
Milk	5,825	1,868	7,693	11,765	—4,072
Eggs	133	. . .	133	152	—19

Only high variant cropland expansion constraints have been specified for these two countries. Australia's cropland expansion is limited
between 1970 and 1985, but New Zealand's is unbounded through 2000.
Under these constraints, projected 2000 total crop area for the region
expands 45 percent above the 1960 level. Much of the increased potential cereals production indicated in Table 8.22 is due to increases
in planted area. Area of cereals in Australia was increasing at about 3.5
percent per year as of 1960.

Table 8.24 shows that under benchmark assumptions the cereals
surplus increases considerably, and a sugar surplus about equal to domestic disappearance emerges. Estimated 1985 cereals production is
nearly twice the 1960 level, and a 13.4 million ton surplus results. Small
deficits are projected for root crops, pulses, and oil crops, but all are
relatively insignificant.

Per capita incomes are over 50 percent greater under medium population and high income assumptions, but as expected, Table 8.24 shows
1985 demands little different from those found under benchmark assumptions. Oceania and the USSR are the only regions for which estimated total cereals demand is actually lower under a high income assumption than under a constant per capita income assumption. Comparing these two sets of 1985 demands, the feed component increases
16,000 tons, but the food and industrial component falls 213,000 tons.
Estimated total demand for fruit and vegetables increases 9.5 percent,
but in general the changes are small.

TABLE 8.22.   1985 and 2000 Production Projections for Oceania (1,000 metric tons)

Commodity	1985 Production	2000 Production
Cereals	18,740	21,262
Raw sugar	2,191	2,262
Root crops	638	484
Pulses	52	53
Fruit, veg.	3,267	3,857
Oil crops	66	90

TABLE 8.23. 1985 and 2000 Production-Demand Comparisons for Oceania, Assuming Low Population Growth Rates (1,000 metric tons)

1985 Comparisons, Assuming Constant Per Capita Income and **1985 Comparisons, Assuming Historical Income Growth Rates**

	Domestic Disappearance (Constant Per Capita Income)				Domestic Disappearance (Historical Income Growth Rates)			
Commodity	Food and Industrial	Feed	Total	Excess Demand	Food and Industrial	Feed	Total	Excess Demand
Cereals	2,816	2,310	5,126	−13,613	2,595	2,327	4,922	−13,817
Raw sugar	993	...	993	−1,198	999	...	999	−1,192
Root crops	1,037	...	1,037	399	1,037	...	1,037	399
Pulses	54	2	56	4	58	2	60	8
Fruit, veg.	2,828	...	2,828	−439	3,126	...	3,126	−140
Oil crops	74	28	102	36	77	28	105	39
Meat	2,279	...	2,279	...	2,288	...	2,288	...
Milk	8,505	2,809	11,314	...	8,633	2,827	11,460	...
Eggs	194	...	194	...	197	...	197	...

2000 Comparisons, Assuming Constant Per Capita Income and **2000 Comparisons, Assuming Historical Income Growth Rates**

	Domestic Disappearance (Constant Per Capita Income)				Domestic Disappearance (Historical Income Growth Rates)			
Commodity	Food and Industrial	Feed	Total	Excess Demand	Food and Industrial	Feed	Total	Excess Demand
Cereals	3,315	2,706	6,021	−15,240	2,938	2,737	5,675	−15,586
Raw sugar	1,170	...	1,170	−1,091	1,182	...	1,182	−1,079
Root crops	1,225	...	1,225	740	1,225	...	1,225	740
Pulses	63	2	65	13	72	2	74	21
Fruit, veg.	3,325	...	3,325	−531	3,929	...	3,929	72
Oil crops	86	33	119	29	91	33	124	34
Meat	2,686	...	2,686	...	2,704	...	2,704	...
Milk	10,049	3,358	13,407	...	10,267	3,394	13,661	...
Eggs	229	...	229	...	236	...	236	...

TABLE 8.24. 1985 and 2000 Production-Demand Comparisons for Oceania, Assuming Medium Population Growth Rates (1,000 metric tons)

| | 1985 Comparisons, Assuming Constant Per Capita Income | | | | 1985 Comparisons, Assuming Historical Income Growth Rates | | | |
| | Domestic Disappearance | | | | Domestic Disappearance | | | |
Commodity	Food and Industrial	Feed	Total	Excess Demand	Food and Industrial	Feed	Total	Excess Demand
Cereals	2,957	2,425	5,382	−13,358	2,744	2,441	5,185	−13,554
Raw sugar	1,043	...	1,043	−1,148	1,048	...	1,048	−1,142
Root crops	1,089	2	1,089	451	1,089	2	1,089	451
Pulses	56	...	58	7	60	...	62	11
Fruit, veg.	2,969	...	2,969	−297	3,251	...	3,251	−16
Oil crops	77	29	106	41	80	29	109	44
Meat	2,393	...	2,393	...	2,402	...	2,402	...
Milk	8,936	2,956	11,892	...	9,061	2,973	12,034	...
Eggs	204	...	204	...	207	...	207	...

| | 2000 Comparisons, Assuming Constant Per Capita Income | | | | 2000 Comparisons, Assuming Historical Income Growth Rates | | | |
| | Domestic Disappearance | | | | Domestic Disappearance | | | |
Commodity	Food and Industrial	Feed	Total	Excess Demand	Food and Industrial	Feed	Total	Excess Demand
Cereals	3,733	3,038	6,771	−14,490	3,350	3,069	6,419	−14,842
Raw sugar	1,318	...	1,318	−943	1,330	...	1,330	−932
Root crops	1,381	...	1,381	897	1,381	...	1,381	897
Pulses	71	3	74	21	80	3	83	29
Fruit, veg.	3,742	...	3,742	−114	4,324	...	4,324	467
Oil crops	97	37	134	44	102	37	139	49
Meat	3,026	...	3,026	...	3,044	...	3,044	...
Milk	11,339	3,814	15,153	...	11,562	3,849	15,411	...
Eggs	259	...	259	...	265	...	265	...

TABLE 8.25. 1985 and 2000 Production-Demand Comparisons for Oceania, Assuming High Population Growth Rates (1,000 metric tons)

1985 Comparisons, Assuming Constant Per Capita Income — Domestic Disappearance

Commodity	Food and Industrial	Feed	Total	Excess Demand
Cereals	3,083	2,528	5,611	−13,128
Raw sugar	1,087	...	1,087	−1,103
Root crops	1,136	2	1,136	498
Pulses	59	...	61	9
Fruit, veg.	3,096	...	3,096	−171
Oil crops	81	30	111	45
Meat	2,496	...	2,496	...
Milk	9,318	3,084	12,402	...
Eggs	212	...	212	...

1985 Comparisons, Assuming Historical Income Growth Rates — Domestic Disappearance

Commodity	Food and Industrial	Feed	Total	Excess Demand
Cereals	2,879	2,543	5,422	−13,318
Raw sugar	1,092	...	1,092	−1,098
Root crops	1,136	2	1,136	498
Pulses	63	...	65	13
Fruit, veg.	3,361	...	3,361	94
Oil crops	83	31	114	48
Meat	2,504	...	2,504	...
Milk	9,439	3,100	12,539	...
Eggs	215	...	215	...

2000 Comparisons, Assuming Constant Per Capita Income — Domestic Disappearance

Commodity	Food and Industrial	Feed	Total	Excess Demand
Cereals	4,003	3,254	7,257	−14,004
Raw sugar	1,414	...	1,414	−847
Root crops	1,482	3	1,482	998
Pulses	77	...	80	26
Fruit, veg.	4,012	...	4,012	155
Oil crops	104	39	143	53
Meat	3,245	...	3,245	...
Milk	12,167	4,102	16,269	...
Eggs	278	...	278	...

2000 Comparisons, Assuming Historical Income Growth Rates — Domestic Disappearance

Commodity	Food and Industrial	Feed	Total	Excess Demand
Cereals	3,621	3,284	6,906	−14,355
Raw sugar	1,425	...	1,425	−837
Root crops	1,482	3	1,482	998
Pulses	85	...	88	35
Fruit, veg.	4,575	...	4,575	718
Oil crops	109	40	149	59
Meat	3,263	...	3,263	...
Milk	12,392	4,136	16,528	...
Eggs	283	...	283	...

Referring to Tables 8.23 and 8.25, we see that variability between low and high variant population estimates creates 1985 demand estimates which are about 10 percent higher under the latter than under the former assumptions when per capita income is held constant.

The estimates for 2000 under benchmark assumptions show the projected surplus of cereals to continue expansion. The estimated surplus for 2000 shown in Table 8.24 is 14.5 million tons. Estimated cereals production is then 121 percent above the 1960 level while estimated demand grows 90 percent. A sugar surplus of 0.9 million tons is projected, together with a root crops deficit at the same level.

Variability in results under other population and income assumptions is not large, and in all cases, the large potential growth of cereals production relative to demand is the dominant characteristic of the projected comparisons.

To summarize, very large increases in potential production relative to demand are evident in all of the projections for Oceania. Population increases ranging from about 70 to 100 percent by 2000 are major determinants of projected demand, and the impact of income increases is severely damped because present per capita consumption is at near-saturation levels. A substantial expansion in possible crop area is projected, but increasing yields are also important in explaining the production trend.

CHAPTER 9

Production-Demand Comparisons
for Africa and Asia

FOOD DEMAND AND SUPPLY PROSPECTS among any set of countries as inclusive as the Africa and Asia grouping are bound to show great diversity. But aside from intercountry variability, there has been great diversity of opinion about the outlook for individual countries within this group. Much of the pessimistic writing of the early and mid-1960s having to do with stagnating economic conditions, uncontrolled population expansion, and imminent food crises was directed at the developing countries of Africa and Asia. Presently the green revolution captures the attention of many who write about food and agriculture in developing countries. Certain of these same Asian countries are now cited as leading examples where, in a short time, developments in improving agricultural productivity may be most "revolutionary," and food needs or even surpluses will be assured. Of course, neither the pessimism nor the optimism has been reserved exclusively for the developing countries of Africa and Asia. But, because of the large and rapidly growing population concentrations in some of these countries, their present low levels of food intake, and the intensity of current agricultural development efforts there, the discussion of food problems, their consequences, and their solutions has particular relevance in this part of the world.

In this chapter, as in earlier ones, we make no attempt to translate the possible effects of the green revolution into specific numerical estimates of future food production and excess demand. Indeed, we have substantial doubts about the possibility of making any kind of useful production predictions beyond, say, five years into the future, given the highly uncertain conditions in many of today's developing countries. Thus we continue with our attempt to assess the effects on food commodity balances of projected population and income together with a continuation of secular production trends as measured over the postwar years. From this we believe it will be possible to derive a fuller understanding of the nature, size, and time dimension of the problems to be overcome by the green revolution.

Before considering the results, we remind the reader of two points having to do with the way cropland estimates entered the projections. First, no attempt was made to estimate crop area and yield trends for most African countries. Most of the production estimates reported are the result of estimating and extending production trends directly.

It will also be recalled that in several cases it was found desirable to establish two alternative upper bounds on cropland expansion. The measurement concepts underlying these were discussed in Chapter 6. Production estimates and excess demand estimates were determined for both low and high limits on cropland expansion whenever the total of all crop area trends for a country reached the lower limit before 2000. This occurred in one or more countries in each of four Asian regions, and accordingly, two sets of production and excess demand estimates are given for these four. With these reminders, we begin by taking up the comparisons for North Africa.

COMPARISONS FOR NORTH AFRICA

Table 9.1 shows that in 1960 this region exported about 30 percent, or 3 million tons, of fruit and vegetables production as well as minor quantities of pulses and oil crops. Deficits were reported for other product classes, and the 2.4 million ton excess demand for cereals was most significant. Feed was a minor component of demand; only 6 percent of cereals demand was for this use.

Demand structure throughout nearly all of Africa reflects the low income and consumption status common for the area. All estimated income elasticities for the seven countries of North Africa are positive. Elasticity estimates for cereals, root crops, pulses, and fruit and vegetables are, respectively, 0.31, 0.24, 0.39, and 0.50. The estimate for milk is 0.75, and estimates for sugar, oil crops, and meat are in the range 0.84 to 0.87. For eggs demand, the income elasticity estimate is 1.16.

Crops for which production trends are estimated account for nearly

TABLE 9.1. 1960 Production-Demand Comparisons for North Africa (1,000 metric tons)

Commodity	Domestic Disappearance			Production	Excess Demand
	Food and Industrial	Feed	Total		
Cereals	16,235	1,040	17,275	14,869	2,406
Raw sugar	1,194	. . .	1,194	392	802
Root crops	1,165	. . .	1,165	1,089	76
Pulses	1,065	212	1,277	1,308	−31
Fruit, veg.	7,913	. . .	7,913	10,972	−3,059
Oil crops	1,250	12	1,262	1,396	−134
Meat	1,279	. . .	1,279	1,261	18
Milk	6,762	. . .	6,762	6,213	549
Eggs	208	. . .	208	205	3

TABLE 9.2. 1985 and 2000 Production Projections for North Africa (1,000 metric tons)

Commodity	1985 Production	2000 Production
Cereals	21,379	23,749
Raw sugar	1,063	1,350
Root crops	2,533	3,128
Pulses	1,857	2,105
Fruit, veg.	19,627	24,181
Oil crops	1,896	2,169

all production of cereals, root crops, and pulses, and coverage is least adequate for oil crops and fruit and vegetables.

The UAR is the only country for which area trends were estimated in all of Africa. Total area expansion for that country was bounded between 1970 and 1985 at a level 15.6 percent above the estimated 1960 level. Production projections given in Table 9.2 suggest that food output is by no means stagnant, but as we shall see, the gains are not adequate relative to projected demand.

Table 9.4 shows a distinct worsening of the production-demand comparisons in the intermediate projection years. By 1985 medium population growth alone causes the cereals demand projection to be about 50 percent above projected production, and a 12.1 million ton deficit results. Slightly increased production surpluses are projected for root crops and fruit and vegetables, but other product classes show worsening deficits.

Under the same population assumption, but assuming a continuing income trend, Table 9.4 shows that deficits are projected for all commodities in 1985. Relative to 1985 benchmark projections, demand for cereals increases 13.5 percent. The feed component of domestic demand increases 53.4 percent, but it is only a small part of the total. Demands for oil crops and meat rise 48.7 and 52.5 percent, respectively, compared to 1985 benchmark levels.

Assuming low versus high population, with constant per capita income, results in 1985 demand levels which lie 6.9 percent below and 2.5 percent above 1985 benchmark levels, as seen in Tables 9.3 and 9.5. However, even under low population assumptions, demand is distinctly higher relative to production than in 1960.

Qualitatively, all results for 2000 are very similar: demand outstrips production. Assuming low population growth and constant per capita income, Table 9.3 shows slightly larger surpluses for fruit and vegetables and root crops than in 1960, but an 18 million ton cereals deficit is projected. At the other extreme where we adopt a high population and increasing income assumption, Table 9.5 shows a projected 36.7 million ton cereals deficit, or 61 percent of total demand, and deficits occur for all other commodities. It will be recalled that the elasticity for cereals demand was among the lowest for the region, but the effect of

TABLE 9.3. 1985 and 2000 Production-Demand Comparisons for North Africa, Assuming Low Population Growth Rates (1,000 metric tons)

1985 Comparisons

| | 1985 Comparisons, Assuming Constant Per Capita Income | | | | 1985 Comparisons, Assuming Historical Income Growth Rates | | | |
| | Domestic Disappearance | | | | Domestic Disappearance | | | |
Commodity	Food and Industrial	Feed	Total	Excess Demand	Food and Industrial	Feed	Total	Excess Demand
Cereals	29,355	1,973	31,328	9,949	32,879	3,176	36,055	14,656
Raw sugar	2,263	...	2,263	1,200	3,232	...	3,232	2,169
Root crops	2,174	...	2,174	−358	2,433	...	2,433	−99
Pulses	1,823	399	2,222	364	2,228	719	2,947	1,090
Fruit, veg.	14,763	...	14,763	−4,863	19,615	...	19,615	−11
Oil crops	2,173	24	2,197	301	3,401	48	3,449	1,553
Meat	2,224	...	2,224	...	3,575	...	3,575	...
Milk	11,855	...	11,855	...	16,244	...	16,244	...
Eggs	374	...	374	...	720	...	720	...

2000 Comparisons

| | 2000 Comparisons, Assuming Constant Per Capita Income | | | | 2000 Comparisons, Assuming Historical Income Growth Rates | | | |
| | Domestic Disappearance | | | | Domestic Disappearance | | | |
Commodity	Food and Industrial	Feed	Total	Excess Demand	Food and Industrial	Feed	Total	Excess Demand
Cereals	39,108	2,648	41,756	18,006	45,147	5,785	50,932	27,183
Raw sugar	3,039	...	3,039	1,689	5,152	...	5,152	3,802
Root crops	2,900	...	2,900	−227	3,447	...	3,447	319
Pulses	2,414	532	2,946	841	3,221	1,407	4,628	2,523
Fruit, veg.	19,712	...	19,712	−4,468	30,367	...	30,367	6,186
Oil crops	2,832	32	2,864	695	5,997	98	6,095	3,926
Meat	2,950	...	2,950	...	6,704	...	6,704	...
Milk	15,678	...	15,678	...	24,774	...	24,774	...
Eggs	499	...	499	...	1,607	...	1,607	...

TABLE 9.4. 1985 and 2000 Production-Demand Comparisons for North Africa, Assuming Medium Population Growth Rates (1,000 metric tons)

1985 Comparisons, Assuming Constant Per Capita Income

Commodity	Domestic Disappearance			Excess Demand
	Food and Industrial	Feed	Total	
Cereals	31,383	2,116	33,499	12,120
Raw sugar	2,427	...	2,427	1,364
Root crops	2,331	...	2,331	−201
Pulses	1,940	428	2,368	511
Fruit, veg.	15,822	...	15,822	−3,804
Oil crops	2,323	26	2,349	453
Meat	2,371	...	2,371	...
Milk	12,650	...	12,650	...
Eggs	399	...	399	...

1985 Comparisons, Assuming Historical Income Growth Rates

Commodity	Domestic Disappearance			Excess Demand
	Food and Industrial	Feed	Total	
Cereals	34,777	3,246	38,023	16,645
Raw sugar	3,336	...	3,336	2,274
Root crops	2,577	...	2,577	44
Pulses	2,324	738	3,062	1,205
Fruit, veg.	20,483	...	20,483	856
Oil crops	3,444	49	3,493	1,596
Meat	3,617	...	3,617	...
Milk	16,798	...	16,798	...
Eggs	713	...	713	...

2000 Comparisons, Assuming Constant Per Capita Income

Commodity	Domestic Disappearance			Excess Demand
	Food and Industrial	Feed	Total	
Cereals	44,901	3,068	47,969	24,219
Raw sugar	3,521	...	3,521	2,171
Root crops	3,352	...	3,352	224
Pulses	2,741	616	3,357	1,251
Fruit, veg.	22,769	...	22,769	−1,411
Oil crops	3,240	37	3,277	1,108
Meat	3,361	...	3,361	...
Milk	17,903	...	17,903	...
Eggs	572	...	572	...

2000 Comparisons, Assuming Historical Income Growth Rates

Commodity	Domestic Disappearance			Excess Demand
	Food and Industrial	Feed	Total	
Cereals	50,951	6,069	57,020	33,271
Raw sugar	5,576	...	5,576	4,225
Root crops	3,890	...	3,890	762
Pulses	3,525	1,485	5,010	2,905
Fruit, veg.	33,480	...	33,480	9,299
Oil crops	6,127	104	6,231	4,062
Meat	6,839	...	6,839	...
Milk	26,778	...	26,778	...
Eggs	1,568	...	1,568	...

TABLE 9.5. 1985 and 2000 Production-Demand Comparisons for North Africa, Assuming High Population Growth Rates (1,000 metric tons)

1985 Comparisons

| | 1985 Comparisons, Assuming Constant Per Capita Income | | | | 1985 Comparisons, Assuming Historical Income Growth Rates | | | |
| | Domestic Disappearance | | | | Domestic Disappearance | | | |
Commodity	Food and Industrial	Feed	Total	Excess Demand	Food and Industrial	Feed	Total	Excess Demand
Cereals	32,187	2,159	34,346	12,967	35,494	3,266	38,760	17,381
Raw sugar	2,476	...	2,476	1,414	3,362	...	3,362	2,299
Root crops	2,378	...	2,378	−154	2,619	...	2,619	86
Pulses	2,005	436	2,441	584	2,374	743	3,117	1,260
Fruit, veg.	16,153	...	16,153	−3,473	20,741	...	20,741	1,114
Oil crops	2,372	26	2,398	502	3,456	49	3,505	1,609
Meat	2,444	...	2,444	...	3,641	...	3,641	...
Milk	13,003	...	13,003	...	17,031	...	17,031	...
Eggs	410	...	410	...	711	...	711	...

2000 Comparisons

| | 2000 Comparisons, Assuming Constant Per Capita Income | | | | 2000 Comparisons, Assuming Historical Income Growth Rates | | | |
| | Domestic Disappearance | | | | Domestic Disappearance | | | |
Commodity	Food and Industrial	Feed	Total	Excess Demand	Food and Industrial	Feed	Total	Excess Demand
Cereals	48,387	3,282	51,669	27,919	54,218	6,200	60,418	36,669
Raw sugar	3,768	...	3,768	2,418	5,761	...	5,761	4,411
Root crops	3,583	...	3,583	455	4,110	...	4,110	982
Pulses	2,988	657	3,645	1,540	3,730	1,521	5,251	3,145
Fruit, veg.	24,371	...	24,371	189	34,992	...	34,992	10,811
Oil crops	3,459	40	3,499	1,329	6,189	106	6,295	4,126
Meat	3,650	...	3,650	...	6,933	...	6,933	...
Milk	19,343	...	19,343	...	27,898	...	27,898	...
Eggs	619	...	619	...	1,551	...	1,551	...

TABLE 9.6. 1960 Production-Demand Comparisons for West Central Africa (1,000 metric tons)

Commodity	Domestic Disappearance			Production	Excess Demand
	Food and Industrial	Feed	Total		
Cereals	9,336	591	9,927	9,164	763
Raw sugar	370	. . .	370	137	233
Root crops	39,827	. . .	39,827	40,059	—232
Pulses	1,072	. . .	1,072	1,738	—666
Fruit, veg.	12,633	. . .	12,633	12,710	—77
Oil crops	1,157	. . .	1,157	3,455	—2,298
Meat	521	. . .	521	488	33
Milk	729	7	736	620	116
Eggs	68	. . .	68	68	. . .

these population and income assumptions is a 250 percent increase in cereals demand above the 1960 level. Cereals production increases only 60 percent in the 40-year projection period when present trends are extended. Larger production increases are estimated for other product classes, but demand estimates are higher also.

The summary of the projections for North Africa can be brief. Present rates of agricultural production growth are less than necessary to meet almost any foreseeable pattern of demand, and continuation of past production trends would result in steadily worsening production-demand comparisons throughout the period under study.

COMPARISONS FOR WEST CENTRAL AFRICA

Table 9.6 shows that West Central Africa produced substantial quantities of oil crops in excess of domestic demand in 1960. Root crops, pulses, and fruit and vegetables were also produced in surplus quantities, but domestic demand exceeded production of cereals and sugar by modest amounts. Feed demand was small—only 6 percent of total cereals demand. Cereals, root crops, and fruit and vegetables were the most important crops used domestically.

Income elasticity estimates for West Central Africa are similar to those for North Africa. Average root crops elasticity for the 11 countries is 0.19, and values in the range 0.40 to 0.56 are estimated for cereals, pulses, and fruit and vegetables. A higher value, 0.74, is estimated for oil crops, and the income elasticities estimated for sugar and the three livestock products are 1.21 to 1.47. All of the above values are simple averages of estimates for the 11 countries. However, Nigeria has over half the region's population, and elasticity estimates for livestock products demand in that country are significantly lower than the average.

Production trends have been estimated for crops accounting for nearly all production in each product class. Table 9.7 shows only mod-

TABLE 9.7. 1985 and 2000 Production Projections for West Central Africa (1,000 metric tons)

Commodity	1985 Production	2000 Production
Cereals	14,077	16,766
Raw sugar	239	288
Root crops	59,944	69,393
Pulses	3,440	4,284
Fruit, veg.	17,465	19,893
Oil crops	4,663	5,208

est production increases compared to population growth. Cereals, root crops, and fruit and vegetables increase only 55 to 80 percent between 1960 and 2000.

Projections to 1985 under benchmark assumptions are presented in Table 9.9. They show pulses and oil crops production increasing slightly faster than demand. Increasing deficits are projected for all other product classes. By 1985 a 5 million ton deficit is projected for cereals, a 6.4 million ton deficit for fruit and vegetables, and an 18.8 million ton deficit for root crops.

Projected income increases are very small for Nigeria and Congo (Kinshasa), the region's two largest countries, and consequently, the combination of high income and medium population produces little increase in demand above that reflected in the benchmark projections. Comparing the two sets of 1985 projections shown in Table 9.9, demands for cereals, root crops, and fruit and vegetables increase, respectively, 10.2, 2.7, and 10.6 percent.

Tables 9.8 and 9.10 show that low and high population assumptions, together with constant per capita income, result in 1985 demands 5.2 percent below and 6.3 percent above benchmark levels.

Production-demand comparisons for 2000 are similar to those for 1985. Table 9.9 shows that projections to 2000 under the benchmark variant produce a 12.9 million ton cereals deficit, a 55.2 million ton root crops deficit, and a 17.2 million ton deficit of fruit and vegetables. Small surpluses are projected for pulses and oil crops. Assuming high population and continued income growth, Table 9.10 shows that cereals, root crops, and fruit and vegetables production estimates are only about half the corresponding demand estimates, and respective deficits of 18.2, 74.0, and 24.3 million tons are projected.

The synopsis of the projections for West Central Africa is much the same as for North Africa. Demand outstrips production under all population and income assumptions. But there is another, more pessimistic conclusion to be drawn from these comparisons. While the contributions of population to demand growth are large (Nigeria's high variant projection for 2000 quadruples the 1960 population), per capita income increases are modest. If genuine economic development occurs

TABLE 9.8. 1985 and 2000 Production-Demand Comparisons for West Central Africa, Assuming Low Population Growth Rates (1,000 metric tons)

| | 1985 Comparisons, Assuming Constant Per Capita Income | | | | 1985 Comparisons, Assuming Historical Income Growth Rates | | | |
| | Domestic Disappearance | | | | Domestic Disappearance | | | |
Commodity	Food and Industrial	Feed	Total	Excess Demand	Food and Industrial	Feed	Total	Excess Demand
Cereals	16,931	1,206	18,137	4,060	19,055	1,335	20,390	6,313
Raw sugar	666	...	666	428	1,375	...	1,375	1,136
Root crops	74,650	...	74,650	14,706	77,261	...	77,261	17,317
Pulses	1,909	...	1,909	–1,530	2,115	...	2,115	–1,324
Fruit, veg.	22,583	...	22,583	5,119	25,509	...	25,509	8,045
Oil crops	2,111	...	2,111	–2,551	2,608	...	2,608	–2,054
Meat	965	...	965	...	1,431	...	1,431	...
Milk	1,313	9	1,322	...	2,017	20	2,037	...
Eggs	121	...	121	...	181	...	181	...

| | 2000 Comparisons, Assuming Constant Per Capita Income | | | | 2000 Comparisons, Assuming Historical Income Growth Rates | | | |
| | Domestic Disappearance | | | | Domestic Disappearance | | | |
Commodity	Food and Industrial	Feed	Total	Excess Demand	Food and Industrial	Feed	Total	Excess Demand
Cereals	24,867	1,850	26,717	9,950	28,217	2,163	30,380	13,613
Raw sugar	980	...	980	692	2,111	...	2,111	1,824
Root crops	111,278	...	111,278	41,885	115,813	...	115,813	46,420
Pulses	2,784	...	2,784	–1,500	3,100	...	3,100	–1,184
Fruit, veg.	33,038	...	33,038	13,145	38,016	...	38,016	18,122
Oil crops	3,106	...	3,106	–2,101	3,874	...	3,874	–1,333
Meat	1,431	...	1,431	...	2,165	...	2,165	...
Milk	1,918	11	1,929	...	3,041	27	3,068	...
Eggs	176	...	176	...	268	...	268	...

TABLE 9.9. 1985 and 2000 Production-Demand Comparisons for West Central Africa, Assuming Medium Population Growth Rates (1,000 metric tons)

1985 Comparisons, Assuming Constant Per Capita Income

Commodity	Food and Industrial	Feed	Total	Excess Demand
Cereals	17,820	1,268	19,088	5,011
Raw sugar	701	...	701	462
Root crops	78,739	...	78,739	18,795
Pulses	2,019	...	2,019	−1,420
Fruit, veg.	23,859	...	23,859	6,894
Oil crops	2,227	...	2,227	−2,435
Meat	1,017	...	1,017	...
Milk	1,385	10	1,395	...
Eggs	128	...	128	...

1985 Comparisons, Assuming Historical Income Growth Rates

Commodity	Food and Industrial	Feed	Total	Excess Demand
Cereals	19,705	1,343	21,048	6,971
Raw sugar	1,340	...	1,340	1,102
Root crops	80,887	...	80,887	20,943
Pulses	2,200	...	2,200	−1,239
Fruit, veg.	26,393	...	26,393	8,928
Oil crops	2,663	...	2,663	−1,999
Meat	1,421	...	1,421	...
Milk	2,006	20	2,026	...
Eggs	180	...	180	...

2000 Comparisons, Assuming Constant Per Capita Income

Commodity	Food and Industrial	Feed	Total	Excess Demand
Cereals	27,642	2,055	29,697	12,931
Raw sugar	1,087	...	1,087	799
Root crops	124,578	...	124,578	55,185
Pulses	3,138	...	3,138	−1,145
Fruit, veg.	37,140	...	37,140	17,246
Oil crops	3,477	...	3,477	−1,730
Meat	1,593	...	1,593	...
Milk	2,144	12	2,156	...
Eggs	196	...	196	...

2000 Comparisons, Assuming Historical Income Growth Rates

Commodity	Food and Industrial	Feed	Total	Excess Demand
Cereals	30,219	2,188	32,407	15,640
Raw sugar	2,007	...	2,007	1,719
Root crops	127,628	...	127,628	58,235
Pulses	3,370	...	3,370	−913
Fruit, veg.	40,857	...	40,857	20,963
Oil crops	4,049	...	4,049	−1,158
Meat	2,137	...	2,137	...
Milk	3,007	27	3,034	...
Eggs	265	...	265	...

TABLE 9.10. 1985 and 2000 Production-Demand Comparisons for West Central Africa, Assuming High Population Growth Rates (1,000 metric tons)

1985 Comparisons

Commodity	1985 Comparisons, Assuming Constant Per Capita Income — Domestic Disappearance				1985 Comparisons, Assuming Historical Income Growth Rates — Domestic Disappearance			
	Food and Industrial	Feed	Total	Excess Demand	Food and Industrial	Feed	Total	Excess Demand
Cereals	18,941	1,351	20,292	6,215	20,478	1,352	21,830	7,753
Raw sugar	745	...	745	506	1,302	...	1,302	1,063
Root crops	83,761	...	83,761	23,817	85,211	...	85,211	25,267
Pulses	2,144	...	2,144	−1,295	2,293	...	2,293	−1,146
Fruit, veg.	25,349	...	25,349	7,884	27,327	...	27,327	9,863
Oil crops	2,367	...	2,367	−2,295	2,727	...	2,727	−1,935
Meat	1,081	...	1,081	...	1,411	...	1,411	...
Milk	1,471	10	1,481	...	1,992	20	2,012	...
Eggs	136	...	136	...	179	...	179	...

2000 Comparisons

Commodity	2000 Comparisons, Assuming Constant Per Capita Income — Domestic Disappearance				2000 Comparisons, Assuming Historical Income Growth Rates — Domestic Disappearance			
	Food and Industrial	Feed	Total	Excess Demand	Food and Industrial	Feed	Total	Excess Demand
Cereals	31,642	2,367	34,009	17,243	32,758	2,222	34,980	18,214
Raw sugar	1,243	...	1,243	956	1,884	...	1,884	1,596
Root crops	143,328	...	143,328	73,935	143,422	...	143,422	74,030
Pulses	3,608	...	3,608	−675	3,704	...	3,704	−579
Fruit, veg.	42,681	...	42,681	22,788	44,190	...	44,190	24,296
Oil crops	3,993	...	3,993	−1,214	4,274	...	4,274	−933
Meat	1,827	...	1,827	...	2,108	...	2,108	...
Milk	2,457	13	2,470	...	2,964	28	2,993	...
Eggs	224	...	224	...	262	...	262	...

TABLE 9.11. 1960 Production-Demand Comparisons for East Africa (1,000 metric tons)

Commodity	Domestic Disappearance			Production	Excess Demand
	Food and Industrial	Feed	Total		
Cereals	7,138	275	7,413	7,394	19
Raw sugar	363	...	363	268	95
Root crops	4,441	45	4,486	4,538	—52
Pulses	647	...	647	690	—43
Fruit, veg.	5,439	...	5,439	5,390	49
Oil crops	286	61	347	420	—73
Meat	713	...	713	749	—36
Milk	1,449	133	1,582	1,388	194
Eggs	23	...	23	22	1

in this region, the income elasticities presently in evidence ensure that food demand will be much greater.

COMPARISONS FOR EAST AFRICA

Table 9.11 shows that East Africa was essentially self-sufficient in each of the six crop product classes in 1960. Minor excess demands were reported for cereals, fruit and vegetables, and sugar, and production exceeded demand by a small margin in each of the other product classes. Cereals, root crops, and fruit and vegetables were again the dominant commodities in both production and domestic disappearance, and the feed component of domestic demand was small.

The 1960 income elasticity of demand for root crops in East Africa is estimated to be 0.22, while for cereals, pulses, and fruit and vegetables the estimates range from 0.41 to 0.54. Demands for oil crops and the three classes of livestock products are projected, using base period elasticity estimates in the range 0.78 to 1.07. A value of 1.26 is estimated as the income elasticity for sugar.

Crops for which time trends are estimated account for high proportions of production in each product class except fruit and vegetables where the coverage is estimated to be about one-half.

With the exception of sugar, the production projections for East Africa presented in Table 9.12 suggest a low rate of increase much like that of West Central Africa.

TABLE 9.12. 1985 and 2000 Production Projections for East Africa (1,000 metric tons)

Commodity	1985 Production	2000 Production
Cereals	11,042	12,984
Raw sugar	1,221	1,734
Root crops	5,527	6,043
Pulses	891	965
Fruit, veg.	6,999	7,911
Oil crops	821	996

Under the benchmark assumptions of medium population and constant per capita income, 1985 demands exceed production for four of the six product classes as shown in Table 9.14. Cereals, root crops, and fruit and vegetables deficits of 2.3, 2.2, and 2.1 million tons, respectively, are projected. The cereals deficit is 17.4 percent of domestic demand; the deficit of root crops is 28.3 percent; and the fruit and vegetables deficit is 23.1 percent of demand.

The second set of 1985 projections given in Table 9.14 shows that the effect of combining high income and medium population assumptions is to increase 1985 demands for cereals, root crops, and fruit and vegetables 14.9, 10.6, and 43.5 percent, respectively, above benchmark levels. Sugar demand, with a 1960 income elasticity of 1.26, increases 113 percent. Deficits are projected for all six product classes in 1985. Each cereals, root crops, and fruit and vegetables demand exceeds production by between 3 and 6 million tons.

The range between 1985 demands under low and high population assumptions with constant per capita income is 11 percent.

Projections to 2000 under benchmark assumptions show even larger deficits for the three major product classes, together with small to moderate surpluses of sugar and oil crops. About 65 percent of projected 2000 demand for cereals, root crops, and fruit and vegetables can be provided by domestic production according to estimates presented in Table 9.14.

Again demand outpaces production under all assumptions as to future population and income growth. With low population and constant per capita income (Table 9.13), projected deficits for the three major crops range from 3.8 to 5.1 million tons, and moderate sugar and oil crops surpluses are projected. At the other extreme, assumptions of high income and population growth result in projected 2000 deficits for all commodities as shown in Table 9.15. Production projections are less than half of demand for cereals, root crops, and fruit and vegetables, and deficits of 13.4, 8.3, and 13.5 million tons, respectively, are estimated.

Clearly, the summary for East Africa must be similar to those for the preceding two African regions. Expansionary forces on the demand side far exceed those on the production side insofar as they are reflected in the projections.

COMPARISONS FOR THE REPUBLIC OF SOUTH AFRICA

Estimates presented in Table 9.16 show that in 1960 South Africa was a surplus producer in each crop commodity class except pulses, where a very small deficit was reported. Cereals and fruit and vegetables were the major commodities produced and consumed, and the feed component of cereals demand was about 30 percent of the total.

TABLE 9.13. 1985 and 2000 Production-Demand Comparisons for East Africa, Assuming Low Population Growth Rates (1,000 metric tons)

1985 Comparisons, Assuming Constant Per Capita Income

Commodity	Food and Industrial	Feed	Total	Excess Demand
	Domestic Disappearance			
Cereals	12,174	494	12,668	1,626
Raw sugar	617	..	617	−603
Root crops	7,239	72	7,311	1,783
Pulses	1,104	..	1,104	213
Fruit, veg.	8,649	..	8,649	1,650
Oil crops	486	121	607	−213
Meat	1,248	..	1,248	..
Milk	2,516	240	2,756	..
Eggs	42	..	42	..

1985 Comparisons, Assuming Historical Income Growth Rates

Commodity	Food and Industrial	Feed	Total	Excess Demand
	Domestic Disappearance			
Cereals	14,077	698	14,775	3,733
Raw sugar	1,411	..	1,411	190
Root crops	7,994	151	8,145	2,618
Pulses	1,443	..	1,443	552
Fruit, veg.	12,652	..	12,652	5,652
Oil crops	893	156	1,049	227
Meat	1,968	..	1,968	..
Milk	3,652	270	3,922	..
Eggs	71	..	71	..

2000 Comparisons, Assuming Constant Per Capita Income

Commodity	Food and Industrial	Feed	Total	Excess Demand
	Domestic Disappearance			
Cereals	17,332	722	18,054	5,070
Raw sugar	875	..	875	−859
Root crops	10,052	97	10,149	4,106
Pulses	1,569	..	1,569	604
Fruit, veg.	11,731	..	11,731	3,820
Oil crops	686	183	869	−126
Meat	1,803	..	1,803	..
Milk	3,615	352	3,967	..
Eggs	61	..	61	..

2000 Comparisons, Assuming Historical Income Growth Rates

Commodity	Food and Industrial	Feed	Total	Excess Demand
	Domestic Disappearance			
Cereals	20,616	1,255	21,871	8,886
Raw sugar	2,587	..	2,587	853
Root crops	11,303	315	11,618	5,575
Pulses	2,207	..	2,207	1,242
Fruit, veg.	18,409	..	18,409	10,498
Oil crops	1,413	272	1,685	690
Meat	3,405	..	3,405	..
Milk	6,473	416	6,889	..
Eggs	130	..	130	..

TABLE 9.14. 1985 and 2000 Production-Demand Comparisons for East Africa, Assuming Medium Population Growth Rates (1,000 metric tons)

1985 Comparisons

	Assuming Constant Per Capita Income — Domestic Disappearance				Assuming Historical Income Growth Rates — Domestic Disappearance			
Commodity	Food and Industrial	Feed	Total	Excess Demand	Food and Industrial	Feed	Total	Excess Demand
Cereals	12,854	523	13,377	2,335	14,673	705	15,378	4,335
Raw sugar	652	...	652	−568	1,392	...	1,392	172
Root crops	7,631	75	7,706	2,180	8,373	152	8,525	2,999
Pulses	1,166	...	1,166	275	1,488	...	1,488	597
Fruit, veg.	9,105	...	9,105	2,106	13,068	...	13,068	6,099
Oil crops	513	128	641	−179	894	157	1,051	230
Meat	1,319	...	1,319	...	1,974	...	1,974	...
Milk	2,658	254	2,912	...	3,706	274	3,980	...
Eggs	44	...	44	...	70	...	70	...

2000 Comparisons

	Assuming Constant Per Capita Income — Domestic Disappearance				Assuming Historical Income Growth Rates — Domestic Disappearance			
Commodity	Food and Industrial	Feed	Total	Excess Demand	Food and Industrial	Feed	Total	Excess Demand
Cereals	19,377	811	20,188	7,203	22,500	1,278	23,778	10,794
Raw sugar	977	...	977	−756	2,515	...	2,515	780
Root crops	11,178	108	11,286	5,242	12,420	320	12,740	6,697
Pulses	1,754	...	1,754	789	2,357	...	2,357	1,392
Fruit, veg.	12,979	...	12,979	5,068	19,662	...	19,662	11,751
Oil crops	766	207	973	−22	1,415	276	1,691	694
Meat	2,022	...	2,022	...	3,429	...	3,429	...
Milk	4,049	396	4,445	...	6,666	430	7,096	...
Eggs	69	...	69	...	128	...	128	...

TABLE 9.15. 1985 and 2000 Production-Demand Comparisons for East Africa, Assuming High Population Growth Rates (1,000 metric tons)

1985 Comparisons, Assuming Constant Per Capita Income

Commodity	Domestic Disappearance			Excess Demand
	Food and Industrial	Feed	Total	
Cereals	13,677	557	14,234	3,192
Raw sugar	693	...	693	−527
Root crops	8,108	80	8,188	2,661
Pulses	1,240	...	1,240	349
Fruit, veg.	9,661	...	9,661	2,662
Oil crops	545	137	682	−138
Meat	1,405	...	1,405	...
Milk	2,830	270	3,100	...
Eggs	47	...	47	...

1985 Comparisons, Assuming Historical Income Growth Rates

Commodity	Domestic Disappearance			Excess Demand
	Food and Industrial	Feed	Total	
Cereals	15,367	712	16,079	5,037
Raw sugar	1,371	...	1,371	150
Root crops	8,826	154	8,980	3,453
Pulses	1,541	...	1,541	650
Fruit, veg.	13,560	...	13,560	6,561
Oil crops	896	158	1,054	233
Meat	1,981	...	1,981	...
Milk	3,767	279	4,046	...
Eggs	70	...	70	...

2000 Comparisons, Assuming Constant Per Capita Income

Commodity	Domestic Disappearance			Excess Demand
	Food and Industrial	Feed	Total	
Cereals	22,472	947	23,419	10,435
Raw sugar	1,182	...	1,132	−602
Root crops	12,872	124	12,996	6,952
Pulses	2,033	...	2,033	1,068
Fruit, veg.	14,843	...	14,843	6,932
Oil crops	886	244	1,130	135
Meat	2,355	...	2,355	...
Milk	4,707	463	5,170	...
Eggs	80	...	80	...

2000 Comparisons, Assuming Historical Income Growth Rates

Commodity	Domestic Disappearance			Excess Demand
	Food and Industrial	Feed	Total	
Cereals	25,123	1,309	26,432	13,447
Raw sugar	2,423	...	2,423	689
Root crops	14,046	326	14,372	8,329
Pulses	2,566	...	2,566	1,601
Fruit, veg.	21,415	...	21,415	13,504
Oil crops	1,418	280	1,697	702
Meat	3,461	...	3,461	...
Milk	6,914	447	7,361	...
Eggs	126	...	126	...

TABLE 9.16. 1960 Production-Demand Comparisons for Republic of South Africa (1,000 metric tons)

Commodity	Domestic Disappearance			Production	Excess Demand
	Food and Industrial	Feed	Total		
Cereals	3,067	1,343	4,410	5,205	—795
Raw sugar	636	...	636	982	—346
Root crops	330	39	369	372	—3
Pulses	70	...	70	68	2
Fruit, veg.	2,194	...	2,194	2,715	—521
Oil crops	50	75	125	223	—98
Meat	687	...	687	696	—9
Milk	2,068	321	2,389	2,363	26
Eggs	53	...	53	61	—8

The 1960 estimated demand structure for this country differs somewhat from the rest of Africa. The estimated income elasticity of demand for cereals is —0.09, estimates for sugar and root crops are each 0.1, and the elasticity estimate for pulses is 0.2. Elasticity estimates for meat and fruit and vegetables are about 0.4, those for milk and eggs are each 0.5, and the elasticity estimate for oil crops, 0.8, is the highest recorded for the country. Compared to the rest of Africa, these elasticities reflect South Africa's higher average income level and dietary sufficiency.

Crops for which time trends are estimated account for an estimated 81 percent of fruit and vegetables production, and for nearly all production in every other product class. Table 9.17 shows that cereals production is projected to increase only about 50 percent above the 1960 level by 2000. However, much more rapid growth is projected for sugar and for fruit and vegetables.

Estimates presented in Table 9.19 show that increased surpluses are projected in 1985 for fruit and vegetables and sugar under benchmark assumptions. However, the balance for cereals changes from an 0.8 million ton surplus in 1960 to a 1.3 million ton deficit in 1985. Estimated 1985 cereals production is 40 percent greater than in 1960, while demand increases 95 percent.

The 1985 high variant income projection, when combined with medium population, results in a per capita income estimate 41 percent higher. Livestock products demand increases about 17 percent over

TABLE 9.17. 1985 and 2000 Production Projections for Republic of South Africa (1,000 metric tons)

Commodity	1985 Production	2000 Production
Cereals	7,280	7,801
Raw sugar	2,158	2,839
Root crops	613	726
Pulses	68	68
Fruit, veg.	5,195	6,660
Oil crops	232	232

1985 benchmark levels, fruit and vegetables demand increases about 14 percent, but cereals demand increases only 2.4 percent. Thus the over-all comparison between production and demand is little different from benchmark results.

Alternative population estimates are closely grouped, and there-fore, so are the associated demand estimates which appear in Tables 9.18 and 9.20. The 1985 low variant estimate is only 2.6 percent below the estimate for the medium assumption, and the high estimate is only 4.2 percent above.

By 2000, the projected cereals deficit grows to 5.1 million tons un-der benchmark assumptions as shown in Table 9.17. Production-demand comparisons for other commodity classes are not drastically different from 1960.

Under high income and population assumptions, population more than triples by 2000, and per capita income rises 53 percent relative to 1960. Estimates given in Table 9.20 show that sugar production is still 0.6 million tons greater than demand, but cereals demand exceeds pro-duction by 7.3 million tons and deficits are projected for the other four product classes.

Essentially the same result is obtained for 2000 when the high pop-ulation assumption is retained, but per capita income is assumed con-stant at the 1960 level. Even under assumptions of low population and constant per capita income, a 3.8 million ton deficit of cereals is pro-jected as shown in Table 9.18. However, a 1.2 million ton surplus of sugar is also estimated, along with a 0.9 million ton surplus of fruit and vegetables.

In summary, production increases projected for the Republic of South Africa are substantial, especially for sugar and fruit and vegetables. But production of corn, the major subsistence crop, increases only about 50 percent over the 40-year projection period. Population growth domi-nates all of these, however. Projected population increases between 1960 and 2000 range from 163 to 233 percent, and it is only under the most conservative population and income assumptions that estimated demand in 2000 is held to levels near projected production.

Before proceeding to consider the remainder of the regions, it should be observed that though it has not generally been possible to take account of the limitations on crop area in Africa, it seems likely that most results would have changed little if the study had included explicit area trends and cropland expansion constraints. It has been seen earlier that potential cropland estimates for tropical South America were far in excess of projected cropland use in 2000. Moreover, it will be recalled that trends in total crop area for countries in that region were rising rapidly. Population density per unit of area in Latin America is much lower than in most underdeveloped areas, but in Africa, it is lower still. Thus the similarity in soils between parts of Africa and South America, together with the low population densities in both areas, suggests that

TABLE 9.18. 1985 and 2000 Production-Demand Comparisons for the Republic of South Africa, Assuming Low Population Growth Rates (1,000 metric tons)

Commodity	1985 Comparisons, Assuming Constant Per Capita Income — Domestic Disappearance				1985 Comparisons, Assuming Historical Income Growth Rates — Domestic Disappearance				2000 Comparisons, Assuming Constant Per Capita Income — Domestic Disappearance				2000 Comparisons, Assuming Historical Income Growth Rates — Domestic Disappearance			
	Food and Industrial	Feed	Total	Excess Demand	Food and Industrial	Feed	Total	Excess Demand	Food and Industrial	Feed	Total	Excess Demand	Food and Industrial	Feed	Total	Excess Demand
Cereals	5,836	2,557	8,393	1,113	5,640	2,963	8,603	1,323	8,053	3,528	11,581	3,780	7,576	4,481	12,057	4,255
Raw sugar	1,210	...	1,210	−947	1,248	...	1,248	−908	1,670	...	1,670	−1,168	1,753	...	1,753	−1,085
Root crops	627	74	701	87	650	85	735	123	865	102	967	241	922	129	1,051	326
Pulses	134	...	134	66	144	...	144	76	185	...	185	117	209	...	209	141
Fruit, veg.	4,174	...	4,174	−1,020	4,795	...	4,795	−399	5,761	...	5,761	−898	7,279	...	7,279	619
Oil crops	95	142	237	5	123	164	287	55	131	196	327	95	199	248	447	216
Meat	1,307	...	1,307	...	1,484	...	1,484	...	1,803	...	1,803	...	2,198	...	2,198	...
Milk	3,935	612	4,547	...	4,666	709	5,375	...	5,430	844	6,274	...	7,219	1,072	8,291	...
Eggs	100	...	100	...	119	...	119	...	138	...	138	...	184	...	184	...

TABLE 9.19. 1985 and 2000 Production-Demand Comparisons for the Republic of South Africa, Assuming Medium Population Growth Rates (1,000 metric tons)

1985 Comparisons, Assuming Constant Per Capita Income

Commodity	Domestic Disappearance			Excess Demand
	Food and Industrial	Feed	Total	
Cereals	5,992	2,625	8,617	1,336
Raw sugar	1,243	...	1,243	−914
Root crops	644	76	720	106
Pulses	137	...	137	69
Fruit, veg.	4,286	...	4,286	−908
Oil crops	97	146	243	11
Meat	1,342	...	1,342	...
Milk	4,040	628	4,668	...
Eggs	103	...	103	...

1985 Comparisons, Assuming Historical Income Growth Rates

Commodity	Domestic Disappearance			Excess Demand
	Food and Industrial	Feed	Total	
Cereals	5,805	3,014	8,819	1,539
Raw sugar	1,279	...	1,279	−877
Root crops	666	87	753	140
Pulses	147	...	147	79
Fruit, veg.	4,878	...	4,878	−316
Oil crops	124	167	291	59
Meat	1,512	...	1,512	...
Milk	4,738	721	5,459	...
Eggs	121	...	121	...

2000 Comparisons, Assuming Constant Per Capita Income

Commodity	Domestic Disappearance			Excess Demand
	Food and Industrial	Feed	Total	
Cereals	8,981	3,935	12,916	5,115
Raw sugar	1,863	...	1,863	−975
Root crops	965	113	1,078	353
Pulses	206	...	206	138
Fruit, veg.	6,425	...	6,425	−234
Oil crops	146	218	364	132
Meat	2,011	...	2,011	...
Milk	6,056	941	6,997	...
Eggs	154	...	154	...

2000 Comparisons, Assuming Historical Income Growth Rates

Commodity	Domestic Disappearance			Excess Demand
	Food and Industrial	Feed	Total	
Cereals	8,537	4,835	13,372	5,571
Raw sugar	1,943	...	1,943	−895
Root crops	1,018	139	1,157	432
Pulses	228	...	228	160
Fruit, veg.	7,837	...	7,837	1,178
Oil crops	210	268	478	246
Meat	2,392	...	2,392	...
Milk	7,721	1,157	8,878	...
Eggs	196	...	196	...

TABLE 9.20. 1985 and 2000 Production-Demand Comparisons for the Republic of South Africa, Assuming High Population Growth Rates (1,000 metric tons)

	1985 Comparisons, Assuming Constant Per Capita Income — Domestic Disappearance				1985 Comparisons, Assuming Historical Income Growth Rates — Domestic Disappearance			
Commodity	Food and Industrial	Feed	Total	Excess Demand	Food and Industrial	Feed	Total	Excess Demand
Cereals	6,244	2,735	8,979	1,699	6,073	3,095	9,168	1,887
Raw sugar	1,295	...	1,295	−862	1,329	...	1,329	−828
Root crops	671	79	750	136	691	89	780	167
Pulses	143	...	143	75	152	...	152	84
Fruit, veg.	4,466	...	4,466	−728	5,010	...	5,010	−184
Oil crops	101	152	253	21	126	172	298	66
Meat	1,398	...	1,398	...	1,557	...	1,557	...
Milk	4,210	654	4,864	...	4,851	740	5,591	...
Eggs	107	...	107	...	123	...	123	...

	2000 Comparisons, Assuming Constant Per Capita Income — Domestic Disappearance				2000 Comparisons, Assuming Historical Income Growth Rates — Domestic Disappearance			
Commodity	Food and Industrial	Feed	Total	Excess Demand	Food and Industrial	Feed	Total	Excess Demand
Cereals	10,193	4,465	14,658	6,857	9,804	5,267	15,071	7,270
Raw sugar	2,114	...	2,114	−724	2,188	...	2,188	−650
Root crops	1,095	129	1,224	498	1,142	152	1,294	568
Pulses	234	...	234	166	253	...	253	185
Fruit, veg.	7,291	...	7,291	631	8,526	...	8,526	1,866
Oil crops	165	248	413	181	221	292	513	282
Meat	2,283	...	2,283	...	2,629	...	2,629	...
Milk	6,873	1,068	7,941	...	8,327	1,260	9,587	...
Eggs	175	...	175	...	212	...	212	...

in most African countries ample cropland is available to support the production projections which have been estimated. If there are exceptions to this judgment, they would be the countries of North Africa where large expanses of desert terrain are dominant.

COMPARISONS FOR WEST ASIA

This region is made quite heterogeneous by inclusion of Cyprus and Israel, two relatively wealthy countries, with six others which are definitely underdeveloped. Although Cyprus and Israel are very small relative to the regional total (together, they had 4 percent of the region's 1960 population), their demand structure is quite unlike the other six countries' demand structures, and it will be discussed separately. Average 1960 income elasticities of demand for cereals in Cyprus and Israel are estimated to be —0.23, and zero income elasticities are estimated for root crops and pulses. Estimated elasticities for the other six commodities lie in the range 0.25 to 0.42.

Elasticity estimates for cereals and root crops in the remaining six countries are about 0.25; pulses and fruit and vegetables elasticities are estimated to be about 0.45; the average estimate for oil crops is 0.57; and sugar's elasticity is estimated to be 0.67. The three livestock product classes have elasticity estimates in the range 0.83 to 1.10.

The 1960 production-demand comparisons for West Asia are presented in Table 9.21. These estimates as well as later projections are strongly influenced by circumstances in Turkey and Iran which together account for 73 percent of the total population of the region. Cereals and fruit and vegetables are seen to be the most important crops in both production and domestic disappearance, and the feed component of cereals demand amounts to 6.6 million tons, or 31 percent. Cereals production is 88 percent of demand, and a deficit of 2.5 million tons is estimated. Sugar production amounts to only slightly more than half of demand, but the commodity class is not a major one for the region. Other surpluses and deficits are small relative to production.

TABLE 9.21. 1960 Production-Demand Comparisons for West Asia (1,000 metric tons)

Commodity	Domestic Disappearance			Production	Excess Demand
	Food and Industrial	Feed	Total		
Cereals	14,798	6,601	21,399	18,879	2,250
Raw sugar	1,432	...	1,432	738	694
Root crops	1,707	36	1,743	1,762	—19
Pulses	700	8	708	766	—58
Fruit, veg.	14,195	805	15,000	16,124	—1,124
Oil crops	852	200	1,052	893	159
Meat	1,061	...	1,061	1,048	13
Milk	6,699	391	7,090	6,881	209
Eggs	169	...	169	189	—20

Crops for which time trends have been estimated in this study account for only about 40 percent of fruit and vegetables production, but coverage in all other product classes is nearly complete.

Production estimates and production-demand comparisons were made for both low and high variant constraints on cropland expansion. For four of the eight countries, two constraints were specified and land expansion was bounded under the low variant constraint. Area expansion was bounded for Iran between 1970 and 1985 under the low variant constraint and between 1985 and 2000 under the high variant constraint. Area expansion in Israel was unbounded through 2000 under the high variant constraint; but under the low variant constraint, further expansion in total area was limited between 1970 and 1985. Jordan's total crop area trend was also unbounded for the high variant constraint; but for the low variant constraint, the boundary was reached before 1970. Finally, Syria's total area expansion was limited before 1970 when a low variant constraint was specified, and between 1970 and 1985 when a high variant bound was specified. Only single variants were specified for Cyprus and Turkey, and total area expansion for both was limited between 1970 and 1985. Two variants were specified for both Iraq and Lebanon, but in both cases, area was unbounded through 2000 under the low variant constraint.

For the region as a whole, total area trends projected to 2000 under the low variant constraints yield estimates of cropland use which are 16.6 percent above the estimated cropland total for 1960; but when high variant constraints are specified, total projected crop area for 2000 is 27.3 percent above the 1960 level. However, some countries exhibit much more expansion than indicated above. Crop area in Turkey accounts for over half the regional total in 1960, and the single constraint on area expansion estimated for that country is less than 1 percent above the estimated area in use in 1960.

The production projections for West Asia presented in Table 9.22 show only limited cereals increases. Low restrictions on cropland expansion result in a 40 percent increase by 2000. Even the high variant

TABLE 9.22.   1985 and 2000 Production Projections for West Asia (1,000 metric tons)

Commodity	1985 Production[a]		2000 Production[a]	
	(L)	(H)	(L)	(H)
Cereals	25,402	26,384	26,682	29,031
Raw sugar	1,700	1,733	2,166	2,306
Root crops	3,716	3,740	4,571	4,650
Pulses	911	952	948	1,025
Fruit, veg.	30,407	30,563	38,294	38,704
Oil crops	1,357	1,360	1,601	1,606

[a] Projections designated (L) are made under a low variant upper bound on cropland expansion. Those designated (H) are made under a high variant bound on cropland expansion.

restraints yield only a 54 percent increase. Output of sugar, root crops, and fruit and vegetables expands more rapidly, but the limited expansion of cereals production will be a major factor conditioning the resulting production-demand comparisons.

Projections for the region as a whole are dominated by production and demand estimates for Turkey and Iran, while results for Israel and Cyprus are quite different from others in the region. Projections for these two will not be singled out for detailed examination, but as a qualifying statement to the discussion which follows, it should be recognized that projected production-demand comparisons for Israel and Cyprus indicate that production will rise about as rapidly as demand regardless of the assumptions adopted. Increasing surpluses of fruit and vegetables production are projected under all assumptions.

Table 9.24 shows that the most significant feature of projections through the intermediate years under benchmark assumptions is an increasing deficit of cereals. Estimated 1985 cereals production is only 3.9 percent greater under the high land constraint assumption than under the low assumption. For the low constraint assumption, estimated 1985 cereals production equals only 60.9 percent of demand, and a deficit of 16.3 million tons is estimated. On the whole, estimated 1985 excess demands in other product classes do not differ greatly from 1960 levels.

Estimated 1985 cereals demand under assumptions of high income and medium population, also shown in Table 9.24, is 18.6 percent greater than under the benchmark variant, and for a low constraint on cropland expansions, a 24.0 million ton deficit is projected. Livestock product demands under these assumptions are about 50 percent greater than under benchmark assumptions in 1985, and the estimated increase in feed demand is about 40 percent. Demands for sugar, fruit and vegetables, and oil crops all rise substantially, and deficits of 2.2, 5.7, and 1.2 million tons, respectively, are projected.

Table 9.23 shows 1985 demands which are 4.8 percent lower under low population and constant per capita income assumptions than under benchmark assumptions. All demands increase 3.1 percent above benchmark estimates when high population is assumed (Table 9.25).

Projections for West Asia indicate serious deficits in the cereals category by 2000 under all population, income, and land constraint assumptions, but the variability under different assumptions is still large. Under benchmark population and income assumptions, the high land constraint assumption results in 8.8 percent greater cereals production than is estimated for a low constraint, but neither production projection reaches 50 percent of estimated 2000 demand. Estimated cereals deficits shown in Table 9.24 are 32.0 million tons with a low constraint, and 29.6 million tons with a high constraint. Deficits are projected for all other product classes under both land expansion constraints, but they are less striking.

TABLE 9.23. 1985 and 2000 Production-Demand Comparisons for West Asia, Assuming Low Population Growth Rates (1,000 metric tons)

1985 Comparisons

| | Assuming Constant Per Capita Income | | | | | Assuming Historical Income Growth Rates | | | | |
| | Domestic Disappearance | | | Excess Demand | | Domestic Disappearance | | | Excess Demand | |
Commodity	Food and Industrial	Feed	Total	Low Land Bound	High Land Bound	Food and Industrial	Feed	Total	Low Land Bound	High Land Bound
Cereals	27,490	12,178	39,668	14,266	13,284	30,399	17,386	47,785	22,383	21,400
Raw sugar	2,640	...	2,640	940	907	3,805	...	3,805	2,105	2,072
Root crops	3,159	68	3,227	−488	−513	3,445	95	3,540	−175	−200
Pulses	1,315	17	1,332	421	380	1,607	25	1,632	722	681
Fruit, veg.	26,280	1,509	27,789	−2,617	−2,773	33,038	2,117	35,155	4,748	4,592
Oil crops	1,616	329	1,945	588	585	2,056	420	2,476	1,118	1,116
Meat	1,955	...	1,955	...	...	3,127	...	3,127	...	...
Milk	12,405	726	13,131	...	...	17,852	1,006	18,858	...	...
Eggs	297	...	297	...	...	509	...	509	...	...

2000 Comparisons

| | Assuming Constant Per Capita Income | | | | | Assuming Historical Income Growth Rates | | | | |
| | Domestic Disappearance | | | Excess Demand | | Domestic Disappearance | | | Excess Demand | |
Commodity	Food and Industrial	Feed	Total	Low Land Bound	High Land Bound	Food and Industrial	Feed	Total	Low Land Bound	High Land Bound
Cereals	36,793	16,234	53,027	26,346	23,996	41,813	25,333	67,146	40,465	38,115
Raw sugar	3,525	...	3,525	1,360	1,219	5,778	...	5,778	3,612	3,472
Root crops	4,214	90	4,304	−265	−344	4,696	130	4,826	256	177
Pulses	1,774	24	1,798	850	773	2,357	37	2,394	1,446	1,369
Fruit, veg.	35,069	2,019	37,088	−1,205	−1,614	46,737	2,928	49,665	11,371	10,962
Oil crops	2,194	415	2,609	1,008	1,003	2,942	563	3,505	1,904	1,899
Meat	2,608	...	2,608	...	...	4,953	...	4,953	...	...
Milk	16,590	967	17,557	...	...	27,601	1,379	28,980	...	...
Eggs	387	...	387	...	...	935	...	935	...	...

TABLE 9.24. 1985 and 2000 Production-Demand Comparisons for West Asia, Assuming Medium Population Growth Rates (1,000 metric tons)

1985 Comparisons, Assuming Constant Per Capita Income

Commodity	Domestic Disappearance			Excess Demand	
	Food and Industrial	Feed	Total	Low Land Bound	High Land Bound
Cereals	28,910	12,779	41,689	16,287	15,305
Raw sugar	2,778	...	2,778	1,078	1,045
Root crops	3,313	71	3,384	-331	-356
Pulses	1,382	18	1,400	489	448
Fruit, veg.	27,593	1,582	29,175	-1,232	-1,388
Oil crops	1,695	344	2,039	681	679
Meat	2,057		2,057	...	...
Milk	13,054	761	13,815	...	...
Eggs	312	...	312	...	...

1985 Comparisons, Assuming Historical Income Growth Rates

Commodity	Domestic Disappearance			Excess Demand	
	Food and Industrial	Feed	Total	Low Land Bound	High Land Bound
Cereals	31,772	17,674	49,446	24,045	23,063
Raw sugar	3,892	...	3,892	2,192	2,159
Root crops	3,586	96	3,682	-33	-57
Pulses	1,662	25	1,687	776	735
Fruit, veg.	34,090	2,144	36,234	5,827	5,671
Oil crops	2,102	432	2,534	1,176	1,174
Meat	3,166		3,166	...	...
Milk	18,281	1,020	19,301	...	...
Eggs	510	...	510	...	...

2000 Comparisons, Assuming Constant Per Capita Income

Commodity	Domestic Disappearance			Excess Demand	
	Food and Industrial	Feed	Total	Low Land Bound	High Land Bound
Cereals	40,716	17,963	58,679	31,997	29,647
Raw sugar	3,898	...	3,898	1,733	1,592
Root crops	4,665	100	4,765	194	115
Pulses	1,967	27	1,994	1,046	969
Fruit, veg.	38,803	2,237	41,040	2,745	2,336
Oil crops	2,440	453	2,893	1,292	1,287
Meat	2,884		2,884	...	...
Milk	18,351	1,071	19,422	...	...
Eggs	426	...	426	...	...

2000 Comparisons, Assuming Historical Income Growth Rates

Commodity	Domestic Disappearance			Excess Demand	
	Food and Industrial	Feed	Total	Low Land Bound	High Land Bound
Cereals	45,772	26,255	72,027	45,345	42,996
Raw sugar	6,057	...	6,057	3,891	3,751
Root crops	5,120	134	5,254	683	604
Pulses	2,529	38	2,567	1,619	1,542
Fruit, veg.	49,960	3,010	52,970	14,676	14,266
Oil crops	3,091	595	3,686	2,085	2,080
Meat	5,080		5,080	...	...
Milk	29,012	1,419	30,431	...	...
Eggs	933	...	933	...	...

TABLE 9.25. 1985 and 2000 Production-Demand Comparisons for West Asia, Assuming High Population Growth Rates (1,000 metric tons)

1985 Comparisons, Assuming Constant Per Capita Income

Commodity	Domestic Disappearance			Excess Demand	
	Food and Industrial	Feed	Total	Low Land Bound	High Land Bound
Cereals	29,831	18,142	42,973	17,571	16,589
Raw sugar	2,870	...	2,870	1,170	1,137
Root crops	3,403	73	3,476	−239	−264
Pulses	1,424	18	1,442	−532	491
Fruit, veg.	28,403	1,623	30,026	−380	−536
Oil crops	1,741	352	2,093	735	732
Meat	2,124	...	2,124	...	...
Milk	13,485	780	14,265	...	...
Eggs	321	...	321	...	...

1985 Comparisons, Assuming Historical Income Growth Rates

Commodity	Domestic Disappearance			Excess Demand	
	Food and Industrial	Feed	Total	Low Land Bound	High Land Bound
Cereals	32,658	17,845	50,503	25,101	24,119
Raw sugar	3,949	...	3,949	2,249	2,216
Root crops	3,669	96	3,765	49	24
Pulses	1,695	25	1,720	810	769
Fruit, veg.	34,720	2,158	36,878	6,471	6,315
Oil crops	2,128	438	2,566	1,209	1,206
Meat	3,193	...	3,193	...	...
Milk	18,561	1,027	19,588	...	...
Eggs	509	...	509	...	...

2000 Comparisons, Assuming Constant Per Capita Income

Commodity	Domestic Disappearance			Excess Demand	
	Food and Industrial	Feed	Total	Low Land Bound	High Land Bound
Cereals	44,465	19,557	64,022	37,340	34,990
Raw sugar	4,259	...	4,259	2,093	1,952
Root crops	5,075	109	5,184	613	534
Pulses	2,150	29	2,179	1,232	1,155
Fruit, veg.	42,279	2,433	44,712	6,418	6,009
Oil crops	2,664	486	3,150	1,549	1,545
Meat	3,149	...	3,149	...	...
Milk	20,053	1,163	21,216	...	...
Eggs	462	...	462	...	...

2000 Comparisons, Assuming Historical Income Growth Rates

Commodity	Domestic Disappearance			Excess Demand	
	Food and Industrial	Feed	Total	Low Land Bound	High Land Bound
Cereals	49,502	27,044	76,546	49,864	47,515
Raw sugar	6,321	...	6,321	4,156	4,015
Root crops	5,497	136	5,633	1,063	983
Pulses	2,686	39	2,725	1,777	1,700
Fruit, veg.	52,784	3,074	55,858	17,564	17,155
Oil crops	3,220	620	3,840	2,240	2,235
Meat	5,206	...	5,206	...	...
Milk	30,300	1,451	31,751	...	...
Eggs	931	...	931	...	...

Estimated cereals deficits in 2000 range from 24.0 million tons when low population, constant per capita income, and high variant land constraints are assumed (Table 9.23) to 49.9 million tons when high population, high income, and low variant land constraints are assumed (Table 9.25). Under these later assumptions, estimated 2000 cereals demand is 3.57 times 1960 demand, but 2000 production is only 1.41 times the 1960 level. Small surpluses are projected for fruit and vegetables and root crops under assumptions of less vigorous population and income trends, but under most assumptions deficits are projected for 2000 in every product class.

In summary, projected production-demand comparisons for West Asia are perhaps the most alarming of any appearing in this study. Estimated population growth rates are very high, and projected demand is quite responsive to increasing per capita income. Recent production trends have been influenced substantially by expansion of crop area, and yield increases have been less important. Estimates of available crop area suggest that future increases in cropland will be limited. Thus something approximating a green revolution in crop yields will be necessary if large future deficits are to be avoided, even if future demand grows at a relatively conservative pace.

COMPARISONS FOR INDIA

Table 9.26 shows that in 1960 India was nearly self-sufficient in every product class. The feed component of demand was very small except for oil crops where a 2.3 million ton feed demand was estimated. Cereals are the most important crops produced and consumed.

The 1960 structure of demand in India reflects the low average income and consumption levels found in that country. The lowest estimated income elasticity, 0.2, is associated with root crops demand. Cereals, pulses, and meat demands are estimated to have elasticities of 0.43, 0.50, and 0.54, respectively. The elasticity estimate for fruit and

TABLE 9.26. 1960 Production-Demand Comparisons for India (1,000 metric tons)

Commodity	Domestic Disappearance			Production	Excess Demand
	Food and Industrial	Feed	Total		
Cereals	83,130	454	83,584	82,604	980
Raw sugar	8,441	409	8,850	9,359	−509
Root crops	8,399	...	8,399	8,399	...
Pulses	11,997	480	12,477	12,474	3
Fruit, veg.	23,619	...	23,619	23,619	...
Oil crops	1,914	2,344	4,258	4,818	−560
Meat	600	...	600	600	...
Milk	21,233	...	21,233	20,942	291
Eggs	126	...	126	125	1

vegetables is 0.78; for sugar, it is 0.90; and for oil crops, it is 1.0. Estimates are highest for milk and eggs, 1.6 and 1.5, respectively.

Crops for which time trends are estimated account for nearly all production of cereals and oil crops, but for sugar, root crops, and pulses, the estimated coverage is only about two-thirds. The coverage estimate for fruit and vegetables is very small, about 10 percent.

Crop areas are projected subject to low and high variant constraints on total cropland expansion. Under the low variant constraint, total area expands 21.5 percent above the 1960 level before becoming bounded between 1985 and 2000. Total crop area remains unbounded through 2000 under the high variant constraint, and projected area in 2000 is 27.2 percent greater than the 1960 value. Table 9.27 shows that projected production of cereals, sugar, and oil crops is more than twice their 1960 levels by 2000, even under the low constraint on cropland expansion. The difference between the production projections under low and high cropland restrictions amounts to only about 5 percent. Production of other commodities shows less growth; and fruit and vegetables production remains constant at the 1960 level.

Table 9.29 shows that under assumptions of constant per capita income and medium population growth, projected production generally keeps pace with demand rather well. Cereals demand exceeds production by only 1.6 million tons in 1985. A 15.6 million ton deficit is projected for fruit and vegetables, along with a 6.6 million ton annual deficit of pulses. However, sugar and oil crops surpluses are projected for 1985, and these are well above corresponding surplus levels estimated for 1960.

With medium population and continued income growth, the resulting per capita income estimate is only 28 percent above the level reflected in the 1985 benchmark projections. In the resulting projections, also shown in Table 9.29, cereals demand increases 9.5 percent, pulses demand increases 13.2 percent, and demand for vegetables and fruit increases 19.0 percent. Demands for meat, milk, and eggs increase 14.1

TABLE 9.27. 1985 and 2000 Production Projections for India (1,000 metric tons)

Commodity	1985 Production	2000 Production[a]	
		(L)	(H)
Cereals	137,045	167,141	175,641
Raw sugar	18,685	23,888	24,840
Root crops	11,969	13,570	14,111
Pulses	14,119	14,528	15,106
Fruit, veg.	23,619	23,619	23,619
Oil crops	8,377	9,805	10,223

[a] Projections designated (L) are made under a low variant upper bound on cropland expansion. Those designated (H) are made under a high variant bound on cropland expansion.

percent, 38.9 percent, and 44.0 percent, respectively, and corresponding increases are indicated for feed demand. However, feed demand is such a small portion of total demand for crop products that the relative impact on total demand is not great. The projections indicate total 1985 deficits of 14.7 and 23.0 million tons of cereals and fruit and vegetables, respectively.

Alternative population estimates for 1985 differ by about 11 percent, and corresponding variations are estimated for demand as shown in Tables 9.28 and 9.30.

Next, consider the benchmark projections for 2000 shown in Table 9.29. With medium population, constant per capita income, and a low variant land expansion constraint, deficits of 8.2, 11.6, and 25.9 million tons of cereals, pulses, and fruit and vegetables, respectively, are projected. When these population and income assumptions are retained, but a high variant land constraint is assumed, a 0.3 million ton production surplus of cereals is estimated. The cereals production estimate for the high land constraint assumption represents a 5.1 percent increase over the estimate associated with low land constraints.

Projections under assumptions of high population and income are given in Table 9.30. For these assumptions, 2000 population is 126 percent above the level for 1960, but per capita income is only 46 percent greater. However, assumptions of low land expansion potentials then result in an estimated 48.6 million ton cereals deficit, or 22.5 percent of estimated demand. Deficits are estimated for all other product classes, including 45.9 million tons for fruit and vegetables and 19.6 million tons for pulses.

Assuming a high land expansion constraint, the cereals deficit is reduced by 8.5 million tons, but it and most others are still large.

Table 9.28 gives projections for assumptions of low population and continued income growth. The cereals deficit is reduced substantially relative to results with high population and income, though the projected deficit of fruit and vegetables remains large. With a low variant land restriction, the estimated cereals deficit is 24.3 million tons, while if the high land restriction is adopted, it is reduced to 15.9 million tons.

India's low and high variant population estimates for 2000 differ by 143 million, and, of course, the potential difference in food demand is immense.

In summary, projected production-demand comparisons for India are not as unfavorable as those for some other underdeveloped areas when measured by relative growth in production and demand. Indeed, under assumptions of low population and constant per capita income, a substantial surplus of cereals is estimated, and other deficits are not extremely large. However, the very large population of this country and its potential for further growth under less conservative population assumptions combine to effect some projected deficits which are very large in absolute value. This picture must be expanded further to recognize

TABLE 9.28. 1985 and 2000 Production-Demand Comparisons for India, Assuming Low Population Growth Rates (1,000 metric tons)

1985 Comparisons, Assuming Constant Per Capita Income

Commodity	Domestic Disappearance			Excess Demand
	Food and Industrial	Feed	Total	
Cereals	130,415	713	131,128	−5,917
Raw sugar	13,243	642	13,885	−4,799
Root crops	13,176	...	13,176	1,208
Pulses	18,821	753	19,574	5,454
Fruit, veg.	37,054	...	37,054	13,435
Oil crops	3,002	3,677	6,679	−1,697
Meat	941	...	941	...
Milk	33,311	...	33,311	...
Eggs	198	...	198	...

1985 Comparisons, Assuming Historical Income Growth Rates

Commodity	Domestic Disappearance			Excess Demand
	Food and Industrial	Feed	Total	
Cereals	144,935	1,042	145,977	8,933
Raw sugar	16,814	939	17,753	−930
Root crops	13,966	...	13,966	1,997
Pulses	21,640	1,101	22,741	8,622
Fruit, veg.	45,715	...	45,715	22,096
Oil crops	4,051	5,378	9,429	1,052
Meat	1,107	...	1,107	...
Milk	49,282	...	49,282	...
Eggs	310	...	310	...

2000 Comparisons, Assuming Constant Per Capita Income

Commodity	Domestic Disappearance			Excess Demand	
	Food and Industrial	Feed	Total	Low Land Bound	High Land Bound
Cereals	161,072	880	161,952	−5,187	−13,687
Raw sugar	16,356	793	17,149	−6,738	−7,690
Root crops	16,274	...	16,274	2,704	2,163
Pulses	23,245	930	24,175	9,647	9,068
Fruit, veg.	45,764	...	45,764	22,145	22,145
Oil crops	3,708	4,541	8,249	−1,555	−1,973
Meat	1,163	...	1,163	...	...
Milk	41,141	...	41,141	...	...
Eggs	244	...	244	...	...

2000 Comparisons, Assuming Historical Income Growth Rates

Commodity	Domestic Disappearance			Excess Demand	
	Food and Industrial	Feed	Total	Low Land Bound	High Land Bound
Cereals	189,912	1,615	191,527	24,386	15,887
Raw sugar	24,283	1,455	25,738	1,850	898
Root crops	18,027	...	18,027	4,457	3,916
Pulses	29,504	1,705	31,209	16,681	16,103
Fruit, veg.	64,988	...	64,988	41,369	41,369
Oil crops	6,353	8,331	14,684	4,879	4,461
Meat	1,555	...	1,555	...	...
Milk	76,591	...	76,591	...	...
Eggs	548	...	548	...	...

TABLE 9.29. 1985 and 2000 Production-Demand Comparisons for India, Assuming Medium Population Growth Rates (1,000 metric tons)

	1985 Comparisons, Assuming Constant Per Capita Income				1985 Comparisons, Assuming Historical Income Growth Rates			
	Domestic Disappearance				Domestic Disappearance			
Commodity	Food and Industrial	Feed	Total	Excess Demand	Food and Industrial	Feed	Total	Excess Demand
Cereals	137,916	754	138,670	1,625	150,744	1,037	151,781	14,736
Raw sugar	14,004	679	14,683	−4,000	17,076	935	18,011	−673
Root crops	13,934	...	13,934	1,965	14,614	...	14,614	2,645
Pulses	19,903	796	20,699	6,580	22,329	1,095	23,424	9,305
Fruit, veg.	39,185	...	39,185	15,566	46,635	...	46,635	23,016
Oil crops	3,175	3,888	7,063	−1,313	4,051	5,350	9,401	1,025
Meat	995	...	995	...	1,135	...	1,135	...
Milk	35,227	...	35,227	...	48,965	...	48,965	...
Eggs	209	...	209	...	301	...	301	...

	2000 Comparisons, Assuming Constant Per Capita Income					2000 Comparisons, Assuming Historical Income Growth Rates				
	Domestic Disappearance			Excess Demand		Domestic Disappearance			Excess Demand	
Commodity	Food and Industrial	Feed	Total	Low Land Bound	High Land Bound	Food and Industrial	Feed	Total	Low Land Bound	High Land Bound
Cereals	174,423	953	175,376	8,235	−263	202,025	1,630	203,655	36,515	28,015
Raw sugar	17,711	859	18,570	−5,317	−6,269	25,026	1,469	26,495	2,607	1,655
Root crops	17,623	...	17,623	4,053	3,512	19,240	...	19,240	5,670	5,129
Pulses	25,171	1,007	26,178	11,650	11,072	30,947	1,721	32,668	18,141	17,562
Fruit, veg.	49,558	...	49,558	25,989	25,939	67,297	...	67,297	43,678	43,678
Oil crops	4,015	4,918	8,933	−872	−1,289	6,353	8,410	14,763	4,958	4,540
Meat	1,259	...	1,259	...	...	1,613	...	1,613	...	...
Milk	44,551	...	44,551	...	...	77,263	...	77,263	...	...
Eggs	264	...	264	...	...	526	...	526	...	...

TABLE 9.30. 1985 and 2000 Production-Demand Comparisons for India, Assuming High Population Growth Rates (1,000 metric tons)

1985 Comparisons, Assuming Constant Per Capita Income

| Commodity | Domestic Disappearance | | | Excess Demand |
	Food and Industrial	Feed	Total	
Cereals	144,663	791	145,454	8,409
Raw sugar	14,689	712	15,401	−3,282
Root crops	14,616	...	14,616	2,647
Pulses	20,877	835	21,712	7,592
Fruit, veg.	41,102	...	41,102	17,483
Oil crops	3,330	4,079	7,409	−967
Meat	1,044	...	1,044	...
Milk	36,950	...	36,950	...
Eggs	219	...	219	...

1985 Comparisons, Assuming Historical Income Growth Rates

| Commodity | Domestic Disappearance | | | Excess Demand |
	Food and Industrial	Feed	Total	
Cereals	155,734	1,029	156,763	19,718
Raw sugar	17,280	928	18,208	−476
Root crops	15,189	...	15,189	3,220
Pulses	22,922	1,087	24,009	9,890
Fruit, veg.	47,385	...	47,385	23,766
Oil crops	4,051	5,311	9,362	985
Meat	1,161	...	1,161	...
Milk	48,537	...	48,537	...
Eggs	294	...	294	...

2000 Comparisons, Assuming Constant Per Capita Income

| Commodity | Domestic Disappearance | | | Excess Demand | |
	Food and Industrial	Feed	Total	Low Land Bound	High Land Bound
Cereals	188,446	1,030	189,476	22,335	13,835
Raw sugar	19,135	928	20,063	−3,824	−4,776
Root crops	19,040	...	19,040	5,469	4,929
Pulses	27,195	1,087	28,282	13,755	13,176
Fruit, veg.	53,542	...	53,542	29,923	29,923
Oil crops	4,338	5,313	9,651	−154	−571
Meat	1,360	...	1,360	...	...
Milk	48,133	...	48,133	...	...
Eggs	286	...	286	...	...

2000 Comparisons, Assuming Historical Income Growth Rates

| Commodity | Domestic Disappearance | | | Excess Demand | |
	Food and Industrial	Feed	Total	Low Land Bound	High Land Bound
Cereals	214,150	1,637	215,777	48,647	40,147
Raw sugar	25,707	1,475	27,182	3,294	2,342
Root crops	20,493	...	20,493	6,922	6,382
Pulses	32,384	1,729	34,113	19,585	19,006
Fruit, veg.	69,478	...	69,478	45,859	45,859
Oil crops	6,353	8,447	14,800	4,994	4,577
Meat	1,671	...	1,671	...	...
Milk	77,519	...	77,519	...	...
Eggs	506	...	506	...	...

189

TABLE 9.31. 1960 Production-Demand Comparisons for Other South Asia (1,000 metric tons)

Commodity	Domestic Disappearance			Production	Excess Demand
	Food and Industrial	Feed	Total		
Cereals	24,117	...	24,117	21,543	2,574
Raw sugar	1,634	...	1,634	1,393	241
Root crops	1,050	...	1,050	987	63
Pulses	1,236	...	1,236	1,179	57
Fruit, veg.	7,787	...	7,787	7,714	73
Oil crops	224	304	528	594	—66
Meat	607	...	607	607	...
Milk	6,359	...	6,359	6,221	138
Eggs	45	...	45	44	1

that present food consumption and income levels are very low, and that per capita income increases embodied in the demand estimates are very meager. Any real economic development would increase the projected deficits for 2000 and earlier years by a wide margin if production expanded only at historical rates.

COMPARISONS FOR OTHER SOUTH ASIA

Table 9.31 shows that 1960 cereals production was about 89 percent of demand, and that a 2.6 million ton deficit was estimated. Except for an estimated 65,000 ton surplus of oil crops, small deficits were estimated for all other product classes. The only feed demand estimated for the region was 0.3 million tons of oil crops. As is true of India, cereals are by far the dominant commodity class.

Root crops and pulses demands in this region have estimated income elasticities of about 0.2, and estimated elasticities for cereals, fruit and vegetables, sugar, and oil crops are 0.44, 0.68, 0.91, and 1.0, respectively. Demand elasticities for both meat and milk are estimated to be 1.2, and the estimate for eggs is 1.5.

Estimated crop coverage is essentially the same as for India. High proportions of cereals and oil crops production are accounted for by crops for which time trends have been estimated, but the estimated coverage of sugar, root crops, and pulses is only about half, and for fruit and vegetables it is less than 5 percent.

Both low and high variant constraints on cropland expansion were estimated for the two countries of this region, but Ceylon's area expansion was unbounded under the low variant constraint. Area expansion in Pakistan was bounded between 1985 and 2000 under the low constraint, but the bound was not reached by 2000 when the high constraint was specified. For the region, estimated total crop area in 2000 is 20.2 percent greater than 1960 when low constraints are specified, and 29.3 percent greater than 1960 when high constraints are introduced.

TABLE 9.32.   1985 and 2000 Production Projections for Other South Asia (1,000 metric tons)

Commodity	1985 Production	2000 Production[a]	
		(L)	(H)
Cereals	33,957	39,311	42,201
Raw sugar	2,864	3,378	3,643
Root crops	1,100	1,063	1,116
Pulses	1,337	1,319	1,421
Fruit, veg.	25,546	31,448	33,866
Oil crops	594	569	594

[a] Projections designated (L) are made under a low variant upper bound on cropland expansion. Those designated (H) are made under a high variant bound on cropland expansion.

Resulting production projections are given in Table 9.32. Continuation of past trends would about double cereals production by 2000, but the two sets of production projections for 2000 differ by about 7 percent. Sugar production increases at a substantially greater rate, but little growth is projected for root crops, pulses, and oil crops. The seemingly large increase in fruit and vegetables production must be interpreted with caution because, as noted above, it is based on very incomplete data.

Table 9.34 shows that a substantial cereals deficit is projected for 1985 when medium population growth and constant per capita incomes are assumed. The deficit, 10.6 million tons, is 23.7 percent of estimated demand. A surplus of 11.2 million tons of fruit and vegetables is also projected but this estimate must be qualified since, as specified earlier, it is based on time trend representing only a very small part of total production.

Under assumptions of continued income growth and medium population, average per capita incomes for the region increase about 70 percent above 1985 benchmark levels. Estimated 1985 demand increases above benchmark levels are 17.0 percent for cereals and 36.9 percent for vegetables and fruit. Total 1985 demand for cereals is then estimated to be 53 percent greater than production, and an 18.1 million ton deficit results.

Population growth is rapid under all projections. In 1985 demand is 6 percent lower with low population and constant income than under benchmark assumptions, while demand under high population assumptions is 5.2 percent greater than under the benchmark variant (Tables 9.33 and 9.35). However, a 7.9 million ton cereals deficit is projected for 1985 even under assumptions of low population growth and constant per capita income.

Table 9.34 shows that benchmark demand assumptions result in relatively large deficits for 2000 in all product classes except fruit and vegetables regardless of which bound is specified for cropland expansion.

TABLE 9.33. 1985 and 2000 Production-Demand Comparisons for Other South Asia, Assuming Low Population Growth Rates (1,000 metric tons)

1985 Comparisons

Commodity	1985 Comparisons, Assuming Constant Per Capita Income — Domestic Disappearance			Excess Demand	1985 Comparisons, Assuming Historical Income Growth Rates — Domestic Disappearance			Excess Demand
	Food and Industrial	Feed	Total		Food and Industrial	Feed	Total	
Cereals	41,894	...	41,894	7,937	49,662	...	49,662	15,705
Raw sugar	2,850	...	2,850	−13	4,286	...	4,286	1,422
Root crops	1,855	...	1,855	755	2,030	...	2,030	930
Pulses	2,142	...	2,142	805	2,403	...	2,403	1,066
Fruit, veg,	13,519	...	13,519	−12,026	19,078	...	19,078	−6,466
Oil crops	394	541	935	341	667	813	1,480	886
Meat	1,051	...	1,051	...	2,278	...	2,278	...
Milk	11,007	...	11,007	...	18,173	...	18,173	...
Eggs	79	...	79	...	178	...	178	...

2000 Comparisons

Commodity	2000 Comparisons, Assuming Constant Per Capita Income — Domestic Disappearance			Excess Demand		2000 Comparisons, Assuming Historical Income Growth Rates — Domestic Disappearance			Excess Demand	
	Food and Industrial	Feed	Total	Low Land Bound	High Land Bound	Food and Industrial	Feed	Total	Low Land Bound	High Land Bound
Cereals	54,491	...	54,491	15,180	12,290	69,020	...	69,020	29,710	26,820
Raw sugar	3,702	...	3,702	325	59	6,941	...	6,941	3,563	3,298
Root crops	2,399	...	2,399	1,337	1,284	2,805	...	2,805	1,742	1,689
Pulses	2,788	...	2,788	1,469	1,367	3,373	...	3,373	2,054	1,952
Fruit, veg,	17,587	...	17,587	−13,860	−16,278	29,985	...	29,985	−1,462	−3,880
Oil crops	511	698	1,209	640	615	1,298	1,379	2,677	2,109	2,083
Meat	1,369	...	1,369	...	...	5,144	...	5,144	...	...
Milk	14,333	...	14,333	...	...	29,154	...	29,154	...	...
Eggs	102	...	102	...	...	432	...	432	...	...

TABLE 9.34. 1985 and 2000 Production-Demand Comparisons for Other South Asia, Assuming Medium Population Growth Rates (1,000 metric tons)

1985 Comparisons, Assuming Constant Per Capita Income

Commodity	Domestic Disappearance			Excess Demand
	Food and Industrial	Feed	Total	
Cereals	44,532	...	44,532	10,576
Raw sugar	3,031	...	3,031	167
Root crops	1,975	...	1,975	876
Pulses	2,276	...	2,276	939
Fruit, veg.	14,369	...	14,369	−11,176
Oil crops	419	577	996	402
Meat	1,117	...	1,117	...
Milk	11,696	...	11,696	...
Eggs	84	...	84	...

1985 Comparisons, Assuming Historical Income Growth Rates

Commodity	Domestic Disappearance			Excess Demand
	Food and Industrial	Feed	Total	
Cereals	52,083	...	52,083	18,126
Raw sugar	4,386	...	4,386	1,522
Root crops	2,137	...	2,137	1,037
Pulses	2,524	...	2,524	1,187
Fruit, veg.	19,673	...	19,673	−5,871
Oil crops	667	821	1,488	894
Meat	2,244	...	2,244	...
Milk	18,598	...	18,598	...
Eggs	173	...	173	...

2000 Comparisons, Assuming Constant Per Capita Income

Commodity	Domestic Disappearance			Excess Demand	
	Food and Industrial	Feed	Total	Low Land Bound	High Land Bound
Cereals	59,270	...	59,270	19,960	17,070
Raw sugar	4,028	...	4,028	650	385
Root crops	2,612	...	2,612	1,549	1,496
Pulses	3,032	...	3,032	1,713	1,611
Fruit, veg.	19,129	...	19,129	−12,318	−14,736
Oil crops	556	760	1,316	747	722
Meat	1,489	...	1,489	...	...
Milk	15,588	...	15,588	...	...
Eggs	111	...	111	...	...

2000 Comparisons, Assuming Historical Income Growth Rates

Commodity	Domestic Disappearance			Excess Demand	
	Food and Industrial	Feed	Total	Low Land Bound	High Land Bound
Cereals	74,207	...	74,207	34,896	32,006
Raw sugar	7,240	...	7,240	3,862	3,596
Root crops	3,009	...	3,009	1,946	1,893
Pulses	3,614	...	3,614	2,295	2,193
Fruit, veg.	31,506	...	31,506	57	−2,359
Oil crops	1,298	1,411	2,709	2,141	2,116
Meat	5,034	...	5,034	...	...
Milk	30,641	...	30,641	...	...
Eggs	414	...	414	...	...

TABLE 9.35. 1985 and 2000 Production-Demand Comparisons for Other South Asia, Assuming High Population Growth Rates (1,000 metric tons)

1985 Comparisons, Assuming Constant Per Capita Income

Commodity	Domestic Disappearance			Excess Demand
	Food and Industrial	Feed	Total	
Cereals	46,854	...	46,854	12,897
Raw sugar	3,189	...	3,189	325
Root crops	2,077	...	2,077	977
Pulses	2,395	...	2,395	1,058
Fruit, veg.	15,118	...	15,118	−10,426
Oil crops	441	606	1,047	454
Meat	1,176	...	1,176	...
Milk	12,307	...	12,307	...
Eggs	88	...	88	...

1985 Comparisons, Assuming Historical Income Growth Rates

Commodity	Domestic Disappearance			Excess Demand
	Food and Industrial	Feed	Total	
Cereals	54,161	...	54,161	20,204
Raw sugar	4,466	...	4,466	1,602
Root crops	2,226	...	2,226	1,126
Pulses	2,630	...	2,630	1,293
Fruit, veg.	20,171	...	20,171	−5,373
Oil crops	667	827	1,494	900
Meat	2,215	...	2,215	...
Milk	18,932	...	18,932	...
Eggs	169	...	169	...

2000 Comparisons, Assuming Constant Per Capita Income

Commodity	Domestic Disappearance			Excess Demand	
	Food and Industrial	Feed	Total	Low Land Bound	High Land Bound
Cereals	64,356	...	64,356	25,045	22,155
Raw sugar	4,375	...	4,375	997	732
Root crops	2,840	...	2,840	1,778	1,725
Pulses	3,292	...	3,292	1,973	1,870
Fruit, veg.	20,770	...	20,770	−10,678	−13,095
Oil crops	604	827	1,431	863	837
Meat	1,616	...	1,616	...	...
Milk	16,920	...	16,920	...	...
Eggs	121	...	121	...	...

2000 Comparisons, Assuming Historical Income Growth Rates

Commodity	Domestic Disappearance			Excess Demand	
	Food and Industrial	Feed	Total	Low Land Bound	High Land Bound
Cereals	79,564	...	79,564	40,253	37,364
Raw sugar	7,530	...	7,530	4,153	3,887
Root crops	3,223	...	3,223	2,161	2,108
Pulses	3,866	...	3,866	2,547	2,445
Fruit, veg.	33,028	...	33,028	1,579	−837
Oil crops	1,298	1,441	2,739	2,171	2,145
Meat	4,929	...	4,929	...	...
Milk	32,076	...	32,076	...	...
Eggs	397	...	397	...	...

Cereals production is 7.4 percent greater when high variant land restrictions are specified than under low variant assumptions, but the lowest projected deficit is still 17.1 million tons, or 29 percent of projected demand.

Table 9.35 presents production-demand comparisons for 2000 under assumptions of high population and income growth. Population for the region is thus presumed to be 2.67 times the 1960 level, and average per capita income increases 2.3 times. The projected deficit of cereals is 40.3 million tons under low land restrictions and 37.4 million tons under high land restrictions. Except for fruit and vegetables, estimated production in each other commodity class is less than one-half of demand.

Even under assumptions of low population growth and constant per capita income (Table 9.33), cereals deficits of 15.1 and 12.3 million tons are projected for 2000 under low and high variant land expansion constraints, respectively.

To summarize, future production-demand comparisons for Other South Asia are dominated by expansionary forces on the demand side. Of course, this is not to say that postwar trends in production have been absent. Continued trends for cereals production will nearly double output by 2000. Both crop areas and yields increase, but the combined effect is inadequate to keep pace with demand. Here again we have a situation where failure of the green revolution to significantly improve most yield trends would result in a large gap between future production and demand.

COMPARISONS FOR JAPAN

Table 9.36 shows that 1960 demand exceeded production in all product classes, and that the largest deficits were reported for cereals, sugar, and oil crops. Cereals, root crops, and fruit and vegetables are the major components in Japanese production and demand, and cereals and root crops are also the most important commodities fed to livestock.

TABLE 9.36. 1960 Production-Demand Comparisons for Japan (1,000 metric tons)

Commodity	Domestic Disappearance			Production	Excess Demand
	Food and Industrial	Feed	Total		
Cereals	21,166	3,054	24,220	19,919	4,301
Raw sugar	1,477	...	1,477	195	1,282
Root crops	7,864	2,294	10,158	10,094	64
Pulses	401	59	460	370	90
Fruit, veg.	12,959	...	12,959	12,943	16
Oil crops	1,483	708	2,191	765	1,426
Meat	600	...	600	575	25
Milk	2,004	172	2,176	1,905	271
Eggs	552	...	552	560	−8

96

2. PRODUCTION AND DEMAND PROSPECTS

TABLE 9.37. 1985 and 2000 Production Projections for Japan (1,000 metric tons)

Commodity	1985 Production	2000 Production
Cereals	26,934	32,293
Raw sugar	526	705
Root crops	12,896	13,733
Pulses	476	538
Fruit, veg.	34,024	47,847
Oil crops	674	900

However, the feed component in cereals demand is only 12.6 percent of the total, and root crops fed to livestock are 22.6 percent.

The estimated structure of demand for Japan in 1960 is quite different from its neighbors. Two negative income elasticities are estimated: −0.1 for root crops and −0.06 for cereals. Pulses and fruit and vegetables have estimated elasticities of 0.1 and 0.4, respectively. Estimates for sugar, milk, and eggs lie in the range 0.7 to 0.8, and values of 1.0 and 1.12 are estimated for oil crops and meat.

Crops for which trends have been estimated account for nearly all cereals, root crops, and oil crops production, but coverage for sugar and pulses is only about 80 percent, and for fruit and vegetables, about 30 percent.

The trend in total crop area for Japan is distinctly negative, and by 2000, estimated total area in crops is only 80 percent of the estimated level in 1960. With the exception of fruit and vegetables, the production increases for Japan shown in Table 9.37 are not large. Much of the increase shown is due to further increases in crop yields which were already relatively high in 1960.

Estimates in Table 9.39 indicate that under benchmark assumptions, Japan's agricultural production rises at about the same rate as demand. The cereals deficit declines slightly through 1985, and an 18.1 million ton surplus of fruit and vegetables is estimated.

With medium population and continuing income trends, 1985 per capita income rises to 3.86 times the level under the benchmark assumptions, but only moderate increases in crop commodity demands are projected. Relative to 1985 demand under benchmark assumptions, demands for meat, milk, and eggs increase 151, 109, and 81 percent, respectively, and demand for feed increases 119 percent. However, total cereals demand increases only 11 percent, and total demand for root crops increases 21 percent. Projected 1985 deficits of 6.2 million tons of cereals, 3.0 million tons of sugar, and 5.5 million tons of oil crops are estimated.

Low and high variant population estimates for 1985 differ by 11 percent, and corresponding differences between demand estimates are observed in Tables 9.38 and 9.40 in the projections for constant per capita income.

TABLE 9.38. 1985 and 2000 Production-Demand Comparisons for Japan, Assuming Low Population Growth Rates (1,000 metric tons)

	1985 Comparisons, Assuming Constant Per Capita Income — Domestic Disappearance				1985 Comparisons, Assuming Historical Income Growth Rates — Domestic Disappearance			
Commodity	Food and Industrial	Feed	Total	Excess Demand	Food and Industrial	Feed	Total	Excess Demand
Cereals	25,053	3,615	28,668	1,733	23,949	8,032	31,981	5,047
Raw sugar	1,748	...	1,748	1,222	3,449	...	3,449	2,923
Root crops	9,307	2,716	12,023	-871	8,634	6,035	14,669	1,773
Pulses	475	70	545	69	541	155	696	220
Fruit, veg.	15,338	...	15,338	-18,685	20,711	...	20,711	-18,312
Oil crops	1,755	838	2,593	1,919	4,194	1,861	6,055	5,381
Meat	710	...	710	...	1,814	...	1,814	...
Milk	2,372	204	2,576	...	5,009	453	5,462	...
Eggs	653	...	653	...	1,191	...	1,191	...

	2000 Comparisons, Assuming Constant Per Capita Income — Domestic Disappearance				2000 Comparisons, Assuming Historical Income Growth Rates — Domestic Disappearance			
Commodity	Food and Industrial	Feed	Total	Excess Demand	Food and Industrial	Feed	Total	Excess Demand
Cereals	26,189	3,779	29,968	-2,323	24,927	9,454	34,381	2,088
Raw sugar	1,828	...	1,828	1,123	4,044	...	4,044	3,340
Root crops	9,730	2,839	12,569	-1,163	8,961	7,103	16,064	2,331
Pulses	496	73	569	32	582	183	765	227
Fruit, veg.	16,034	...	16,034	-31,812	22,287	...	22,287	-25,558
Oil crops	1,835	876	2,711	1,810	5,014	2,191	7,205	6,305
Meat	742	...	742	...	2,182	...	2,182	...
Milk	2,480	213	2,693	...	5,917	533	6,450	...
Eggs	683	...	683	...	1,320	...	1,320	...

TABLE 9.39. 1985 and 2000 Production-Demand Comparisons for Japan, Assuming Medium Population Growth Rates (1,000 metric tons)

	1985 Comparisons, Assuming Constant Per Capita Income — Domestic Disappearance				1985 Comparisons, Assuming Historical Income Growth Rates — Domestic Disappearance			
Commodity	Food and Industrial	Feed	Total	Excess Demand	Food and Industrial	Feed	Total	Excess Demand
Cereals	26,027	3,756	29,783	2,848	24,895	8,226	33,121	6,187
Raw sugar	1,816	...	1,816	1,290	3,534	...	3,534	3,009
Root crops	9,669	2,822	12,491	−403	8,979	6,180	15,159	2,264
Pulses	493	73	566	90	560	159	719	243
Fruit, veg.	15,935	...	15,935	−18,089	21,433	...	21,433	−12,590
Oil crops	1,823	870	2,693	2,020	4,287	1,906	6,193	5,520
Meat	737	...	737	...	1,853	...	1,853	...
Milk	2,464	212	2,676	...	5,128	464	5,592	...
Eggs	679	...	679	...	1,228	...	1,228	...

	2000 Comparisons, Assuming Constant Per Capita Income — Domestic Disappearance				2000 Comparisons, Assuming Historical Income Growth Rates — Domestic Disappearance			
Commodity	Food and Industrial	Feed	Total	Excess Demand	Food and Industrial	Feed	Total	Excess Demand
Cereals	27,795	4,011	31,806	−486	26,472	9,841	36,313	4,021
Raw sugar	1,940	...	1,940	1,235	4,212	...	4,212	3,507
Root crops	10,326	3,014	13,340	−392	9,520	7,394	16,914	3,181
Pulses	527	78	605	67	615	190	805	268
Fruit, veg.	17,017	...	17,017	−30,829	23,551	...	23,551	−24,294
Oil crops	1,947	929	2,876	1,976	5,206	2,280	7,486	6,586
Meat	787	...	787	...	2,263	...	2,263	...
Milk	2,632	226	2,858	...	6,155	555	6,710	...
Eggs	725	...	725	...	1,388	...	1,388	...

TABLE 9.40. 1985 and 2000 Production-Demand Comparisons for Japan, Assuming High Population Growth Rates (1,000 metric tons)

1985 Comparisons, Assuming Constant Per Capita Income / **1985 Comparisons, Assuming Historical Income Growth Rates**

Commodity	Food and Industrial	Feed	Total	Excess Demand	Food and Industrial	Feed	Total	Excess Demand
Cereals	27,754	4,005	31,759	4,824	26,574	8,558	35,132	8,198
Raw sugar	1,937	...	1,937	1,411	3,682	...	3,682	3,156
Root crops	10,311	3,009	13,320	424	9,591	6,430	16,021	3,125
Pulses	526	77	603	127	594	165	759	283
Fruit, veg.	16,992	...	16,992	−17,031	22,699	...	22,699	−11,324
Oil crops	1,944	928	2,872	2,199	4,447	1,983	6,430	5,756
Meat	786	...	786	...	1,920	...	1,920	...
Milk	2,628	226	2,854	...	5,334	483	5,817	...
Eggs	724	...	724	...	1,292	...	1,292	...

(Domestic Disappearance: Food and Industrial, Feed, Total)

2000 Comparisons, Assuming Constant Per Capita Income / **2000 Comparisons, Assuming Historical Income Growth Rates**

Commodity	Food and Industrial	Feed	Total	Excess Demand	Food and Industrial	Feed	Total	Excess Demand
Cereals	31,503	4,546	36,049	3,756	30,049	10,692	40,741	8,449
Raw sugar	2,198	...	2,198	1,494	4,581	...	4,581	3,876
Root crops	11,704	3,416	15,120	1,386	10,818	8,033	18,851	5,118
Pulses	597	88	685	147	690	207	897	359
Fruit, veg.	19,287	...	19,287	−28,558	26,427	...	26,427	−21,418
Oil crops	2,207	1,053	3,260	2,360	5,624	2,477	8,101	7,201
Meat	892	...	892	...	2,440	...	2,440	...
Milk	2,983	256	3,239	...	6,677	603	7,280	...
Eggs	822	...	822	...	1,542	...	1,542	...

(Domestic Disappearance: Food and Industrial, Feed, Total)

Medium population and constant per capita income assumptions result in a small surplus of cereals for Japan by 2000 as shown in Table 9.39. Again, a large surplus of fruit and vegetables is projected.

Under the most pressing demand assumptions, high population, and increasing income, deficits are projected for all commodity classes except fruit and vegetables. Cereals and oil crops deficits of 8.4 and 7.2 million tons, respectively, are the most notable results reported in Table 9.40 for 2000.

Table 9.38 shows that the low population growth rate assumption, together with an assumption of increasing per capita income, results in a projected cereals deficit in 2000 which is less than was realized in 1960. The projected deficit for oil crops remains large, and deficits for sugar, root crops, and pulses are also above 1960 levels. However, the 25.5 million ton estimated fruit and vegetables surplus is an offsetting factor.

To summarize, the projections indicate that Japan's agricultural production may keep pace with demand reasonably well, despite a rather striking 20 percent projected decline in crop area by 2000. Of course, projected yields rise; but most important, projected population growth is very modest. Large increases in projected income cause moderate increases in demand for sugar and fruit and vegetables, but its positive effect on feed demand for cereals and root crops is partially offset by the negative effect on demand for direct consumption. Japan seems likely to continue in its present role as an importer of cereals and oil crops, though this tendency may be less pronounced in the case of cereals than in oil crops.

COMPARISONS FOR OTHER EAST ASIA

Table 9.41 shows that production exceeded demand in each of the six crop product classes in 1960. The largest surplus recorded was 3.1 million tons of cereals, but this was only 6 percent of production. Cereals were the most important product class in both production and consump-

TABLE 9.41.   1960 Production-Demand Comparisons for Other East Asia (1,000 metric tons)

Commodity	Domestic Disappearance			Production	Excess Demand
	Food and Industrial	Feed	Total		
Cereals	47,499	1,081	48,580	51,683	—3,103
Raw sugar	2,368	. . .	2,368	3,982	—1,614
Root crops	20,625	1,745	22,370	24,310	—1,940
Pulses	901	. . .	901	1,044	—143
Fruit, veg.	26,452	79	26,532	26,758	—226
Oil crops	2,212	821	3,033	4,133	—1,100
Meat	1,333	. . .	1,333	1,314	19
Milk	1,117	. . .	1,117	440	677
Eggs	550	. . .	550	557	—7

tion, but large quantities of root crops and fruit and vegetables were also reported. Cereals and root crops were the most important commodities fed to livestock, but the feed component was small relative to total demand in each case.

The estimated average 1960 income elasticity of demand for root crops is 0.09. Estimates for cereals, pulses, and fruit and vegetables are 0.29, 0.33, and 0.52, respectively. All other estimates are greater than 1.0. Those for sugar, oil crops, meat, and eggs lie in the range 1.02 to 1.16; and milk has the highest estimated elasticity, 1.44.

Crops for which time trends are estimated account for nearly all production of cereals, sugar, root crops, and oil crops, but only about 25 percent of pulses production is covered, and estimated coverage for fruit and vegetables is less than 10 percent.

Both low and high variant cropland expansion constraints are estimated for all countries of the region except South Korea. Total cropland expansion in the estimates for South Korea is bounded between 1985 and 2000. Estimated total crop areas for Burma, Cambodia, China (Taiwan), Malaya, and Indonesia are all unbounded through 2000 under their low variant constraints. Land expansion for both the Philippines and South Vietnam is bounded between 1970 and 1985 when low variant constraints are specified, but expansion is unbounded under high variant land restrictions. Thailand's land expansion is bounded by 1985 under both high and low cropland constraints. For the entire region, projections of total crop area in 2000 are 51.9 percent greater than 1960 under low variant constraints, and 62.2 percent higher than 1960 when high variant constraints are assumed. Overall, the production projections given in Table 9.42 suggest that Other East Asia's food production would about double by 2000 under the high cropland constraint. However, the differential effects of the two upper bounds on land result in about 6 percent more cereals production in 2000 under the high land constraint than under the low restraint.

Production and demand trends advance at nearly equal rates

TABLE 9.42. 1985 and 2000 Production Projections for Other East Asia (1,000 metric tons)

Commodity	1985 Production[a]		2000 Production[a]	
	(L)	(H)	(L)	(H)
Cereals	90,346	93,779	111,215	118,279
Raw sugar	6,111	6,236	6,989	7,141
Root crops	46,345	48,097	56,231	58,317
Pulses	1,447	1,502	1,607	1,702
Fruit, veg.	52,670	53,650	69,076	70,241
Oil crops	6,665	6,675	8,005	8,014

[a] Projections designated (L) are made under a low variant upper bound on cropland expansion. Those designated (H) are made under a high variant bound on cropland expansion.

through 1985 under benchmark assumptions as shown in Table 9.44. By 1985 the 3.1 million ton cereals surplus estimated for 1960 falls to a 2.5 million ton deficit when a low constraint on cropland expansion is assumed. If the high constraint is adopted, a 0.9 million ton surplus is projected. Estimated cereals production in 1985 is higher by 3.4 million tons, or 3.8 percent, under high rather than low land constraints. Excess demands for other commodities are near their 1960 levels.

The effect of assuming continuing income trends rather than constant per capita incomes is to increase the projected 1985 cereals demand by 11.8 percent. The three livestock product demands increase between 52 and 77 percent, and feed demand rises about 50 percent. When a low land constraint is assumed along with the above population and income assumptions, a 13.5 million ton cereals deficit is estimated. Deficits then are estimated for all other product classes except root crops, and the surplus estimated for this class is only 0.3 million tons. Demand estimates for 1985 under assumptions of low population and constant per capita income are 4.1 percent below benchmark levels (Table 9.43) while high population assumptions result in demand increases of 3.4 percent (Table 9.45).

By 2000, cereals production under low and high land constraints is projected at 111.2 and 118.3 million tons, respectively. In both cases, the estimates are more than double the 51.7 million tons production of 1960. However, benchmark cereals demand is estimated to be 134 million tons, and therefore, deficits of 22.9 or 15.8 million tons are estimated, depending upon the land constraint assumption. A slight surplus is projected for sugar, but small to moderate deficits are estimated for each other product class.

The only assumptions for population, income, and land constraints which result in estimated excess demands for 2000 near 1960 levels are low population, constant per capita income, and high land expansion constraints (Table 9.43). At the other extreme, high population and high income assumptions result in estimated cereals deficits of 56.3 or 49.2 million tons, depending upon assumptions about land expansion constraints (Table 9.45). These population and income assumptions result in projected deficits for all other product classes as well.

In summarizing the projections for Other East Asia, it should be pointed out that considerable intercountry variability exists. Projected excess demands for the Philippines are very high, while Burma and Thailand exhibit growing surpluses under most variants. For the region as a whole, even the large crop area increases projected, together with moderate yield increases, do not advance production levels as rapidly as demand under most population and income assumptions. However, high variant estimates of cropland constraints based on the special soils study indicate that almost twice the crop area projected for 2000 may be developed if necessary.

TABLE 9.43. 1985 and 2000 Production-Demand Comparisons for Other East Asia, Assuming Low Population Growth Rates (1,000 metric tons)

1985 Comparisons

| | 1985 Comparisons, Assuming Constant Per Capita Income | | | | | 1985 Comparisons, Assuming Historical Income Growth Rates | | | | |
| | Domestic Disappearance | | | Excess Demand | | Domestic Disappearance | | | Excess Demand | |
Commodity	Food and Industrial	Feed	Total	Low Land Bound	High Land Bound	Food and Industrial	Feed	Total	Low Land Bound	High Land Bound
Cereals	87,089	2,169	89,258	−1,088	−4,520	97,322	3,343	100,665	10,318	6,886
Raw sugar	4,419	...	4,419	−1,691	−1,816	6,930	...	6,930	819	694
Root crops	37,160	2,958	40,118	−6,226	−7,979	38,663	5,825	44,488	−1,856	−3,609
Pulses	1,678	...	1,678	231	176	1,884	...	1,884	437	382
Fruit, veg.	49,182	132	49,314	−3,355	−4,335	60,093	266	60,359	7,690	6,710
Oil crops	4,048	1,565	5,613	−1,051	−1,061	7,011	2,390	9,401	2,736	2,726
Meat	2,499	...	2,499	...	...	4,098	...	4,098	...	...
Milk	2,151	...	2,151	...	...	3,971	...	3,971	...	...
Eggs	1,028	...	1,028	...	...	1,666	...	1,666	...	...

2000 Comparisons

| | 2000 Comparisons, Assuming Constant Per Capita Income | | | | | 2000 Comparisons, Assuming Historical Income Growth Rates | | | | |
| | Domestic Disappearance | | | Excess Demand | | Domestic Disappearance | | | Excess Demand | |
Commodity	Food and Industrial	Feed	Total	Low Land Bound	High Land Bound	Food and Industrial	Feed	Total	Low Land Bound	High Land Bound
Cereals	118,724	3,137	121,861	10,646	3,582	139,315	6,248	145,563	34,349	27,284
Raw sugar	6,119	...	6,119	−869	−1,021	13,242	...	13,242	6,253	6,101
Root crops	50,682	3,651	54,333	−1,897	−3,983	54,031	9,324	63,355	7,124	5,039
Pulses	2,331	...	2,331	724	629	2,816	...	2,816	1,209	1,114
Fruit, veg.	68,129	159	68,288	−787	−1,952	94,023	421	94,444	25,367	24,202
Oil crops	5,556	2,190	7,746	−258	−267	15,951	4,321	20,272	12,267	12,258
Meat	3,449	...	3,449	...	...	7,721	...	7,721	...	...
Milk	3,034	...	3,034	...	...	8,179	...	8,179	...	...
Eggs	1,428	...	1,428	...	...	3,357	...	3,357	...	...

TABLE 9.44. 1985 and 2000 Production-Demand Comparisons for Other East Asia, Assuming Medium Population Growth Rates (1,000 metric tons)

1985 Comparisons

| | 1985 Comparisons, Assuming Constant Per Capita Income | | | | | 1985 Comparisons, Assuming Historical Income Growth Rates | | | | |
| | Domestic Disappearance | | | Excess Demand | | Domestic Disappearance | | | Excess Demand | |
Commodity	Food and Industrial	Feed	Total	Low Land Bound	High Land Bound	Food and Industrial	Feed	Total	Low Land Bound	High Land Bound
Cereals	90,622	2,258	92,880	2,532	−897	100,472	3,352	103,824	13,478	10,046
Raw sugar	4,591	...	4,591	−1,519	−1,644	6,947	...	6,947	836	711
Root crops	38,624	3,169	41,793	−4,551	−6,304	40,008	6,055	46,063	−281	−2,034
Pulses	1,743	...	1,743	296	241	1,938	...	1,938	491	436
Fruit, veg.	51,075	142	51,217	−1,452	−2,432	61,335	278	61,613	8,943	7,963
Oil crops	4,213	1,630	5,843	−822	−832	6,973	2,394	9,367	2,702	2,692
Meat	2,603	...	2,603	...	...	4,104	...	4,104	...	...
Milk	2,230	...	2,230	...	...	3,964	...	3,964	...	...
Eggs	1,067	...	1,067	...	...	1,661	...	1,661	...	...

2000 Comparisons

| | 2000 Comparisons, Assuming Constant Per Capita Income | | | | | 2000 Comparisons, Assuming Historical Income Growth Rates | | | | |
| | Domestic Disappearance | | | Excess Demand | | Domestic Disappearance | | | Excess Demand | |
Commodity	Food and Industrial	Feed	Total	Low Land Bound	High Land Bound	Food and Industrial	Feed	Total	Low Land Bound	High Land Bound
Cereals	130,580	3,505	134,085	22,871	15,806	150,746	6,340	157,086	45,872	38,807
Raw sugar	6,757	...	6,757	−230	−382	13,374	...	13,374	6,385	6,233
Root crops	55,860	4,237	60,097	3,866	1,780	58,952	10,278	69,230	12,999	10,913
Pulses	2,575	...	2,575	967	873	3,034	...	3,034	1,427	1,333
Fruit, veg.	75,087	186	75,273	6,197	5,032	99,422	467	99,889	30,813	29,648
Oil crops	6,128	2,435	8,563	559	549	15,796	4,353	20,149	12,144	12,135
Meat	3,832	...	3,832	...	...	7,758	...	7,758	...	...
Milk	3,354	...	3,854	...	...	8,203	...	8,203	...	...
Eggs	1,575	...	1,575	...	...	3,330	...	3,330	...	...

TABLE 9.45. 1985 and 2000 Production-Demand Comparisons for Other East Asia, Assuming High Population Growth Rates (1,000 metric tons)

1985 Comparisons, Assuming Constant Per Capita Income

Commodity	Domestic Disappearance			Excess Demand	
	Food and Industrial	Feed	Total	Low Land Bound	High Land Bound
Cereals	93,702	2,324	96,026	5,680	2,247
Raw sugar	4,726	...	4,726	−1,385	−1,510
Root crops	39,773	3,294	43,067	−3,277	−5,030
Pulses	1,793	...	1,793	346	291
Fruit, veg.	52,611	147	52,758	88	−891
Oil crops	4,350	1,677	6,027	−637	−647
Meat	2,678	...	2,678	...	...
Milk	2,291	...	2,291	...	...
Eggs	1,095	...	1,095	...	...

1985 Comparisons, Assuming Historical Income Growth Rates

Commodity	Domestic Disappearance			Excess Demand	
	Food and Industrial	Feed	Total	Low Land Bound	High Land Bound
Cereals	103,242	3,355	106,597	16,251	12,818
Raw sugar	6,954	...	6,954	843	718
Root crops	41,058	6,182	47,240	896	−856
Pulses	1,979	...	1,979	532	477
Fruit, veg.	62,339	284	62,623	9,953	8,973
Oil crops	6,918	2,394	9,312	2,646	2,636
Meat	4,102	...	4,102	...	...
Milk	3,958	...	3,958	...	...
Eggs	1,658	...	1,658	...	...

2000 Comparisons, Assuming Constant Per Capita Income

Commodity	Domestic Disappearance			Excess Demand	
	Food and Industrial	Feed	Total	Low Land Bound	High Land Bound
Cereals	141,278	3,790	145,068	33,853	26,789
Raw sugar	7,272	...	7,272	283	131
Root crops	60,060	4,601	64,661	8,430	6,344
Pulses	2,769	...	2,769	1,162	1,067
Fruit, veg.	80,881	202	81,083	12,007	10,842
Oil crops	6,621	2,622	9,243	1,239	1,229
Meat	4,121	...	4,121	...	...
Milk	3,606	...	3,606	...	...
Eggs	1,689	...	1,689	...	...

2000 Comparisons, Assuming Historical Income Growth Rates

Commodity	Domestic Disappearance			Excess Demand	
	Food and Industrial	Feed	Total	Low Land Bound	High Land Bound
Cereals	161,092	6,386	167,478	56,263	49,199
Raw sugar	13,434	...	13,434	6,446	6,294
Root crops	62,897	10,831	73,728	17,497	15,411
Pulses	3,205	...	3,205	1,597	1,503
Fruit, veg.	103,811	493	104,304	35,228	34,063
Oil crops	15,425	4,365	19,790	11,785	11,776
Meat	7,764	...	7,764	...	...
Milk	8,207	...	8,207	...	...
Eggs	3,313	...	3,313	...	...

205

CHAPTER 10

A Synthesis of
Regional Production-Demand Comparisons

WE NOW PRESENT a brief summary of the projections discussed in Chapters 7 through 9. Production-demand comparisons are considered for aggregates of low, medium, and high income countries, and for the 96-country total. These serve as a useful vehicle to integrate data and analyses presented previously and to emphasize some significant contrasts among parts of the world with differing levels of economic development.

Production is projected by the methods explained in earlier chapters. Food demand is based on the two major variables, population and income (as reflected through income elasticities of demand). Hence, the demand functions are not fully specified in the sense of simultaneously determined prices and quantities; nor are supply and demand relationships specified and estimated within this framework. Production and demand, in the sense used here, are estimated entirely separately. We could discuss our demand projections as *food requirements,* but this terminology would be even less appropriate than the term *food demand.* This is true because projections reflect income variables as well as population growth. The major difficulty is our use of the terms *surplus* and *deficits* of production over demand. If we had completely specified demand relations, price flexibilities would allow the greater supplies to be absorbed through market reaction. However, our methods do not allow these equilibrating processes (since they are rather impossible, given the nature of world and country data available). Our terms *surplus* and *deficit* refer only to the extent that production, projected as it is, exceeds demand projected on the basis of population and income variables. We hope the reader will keep these technical considerations in mind and bear with us as we use the terms *demand, surplus,* and *deficit* as we do for purposes of simplicity.

A similar explanation or qualification applies to the relative *surpluses* or *deficits* projected for different crops. Since projections are not within a system of endogenously determined quantities and prices, the analysis does not automatically generate a shift of land and other resources from a crop for which a surplus is projected to one for which a

206

deficit is projected. In other words, projections of output and balances for the individual crop groups are mechanically and economically separate. Under typical responses of farmers throughout the world and under the equilibrating forces of markets (or even through the lagged response of planners) we would expect a large deficit for a particular crop to be overcome through shift of resources from crops of large outputs relative to demand (just as we would expect a large supply of a particular crop to be absorbed through price reactions which eliminate physical surpluses in markets). It is not, however, our purpose to estimate equilibrating conditions of various countries or groups (and generally we believe it to be impossible with the data available). Our purpose is simply to project independently levels of commodity supply and demand under the given specifications. Even without showing details depicting market clearing and equilibrium attainment, these projections provide broad indicators of where relative problems of production may occur in the future. They do not define equilibrium conditions of supply and demand for a single country or a group of countries.

BASE PERIOD COMPARISONS

We begin by considering the base period production-demand comparisons and the production projections. The 1960 food balance for the low income countries shown in Table 10.1 indicates that production exceeded food requirements for all product classes except cereals and milk. The cereals deficit, 7.8 million tons, was only about 3 percent of demand. Oil crops production exceeded demand by about one-third, but other excess supplies were small fractions of production.

Projected aggregate production for the low income countries in 2000 is about double the 1960 level for all commodity groups except pulses (Table 10.2). The difference associated with alternative upper bounds on cropland expansion amounts to about 5 percent for cereals in 2000, but lesser differences are reported for other crops.

TABLE 10.1. 1960 Production-Demand Comparisons for Low Income Countries (1,000 metric tons)

| | Domestic Disappearance | | | | |
Commodity	Food and Industrial	Feed	Total	Production	Excess Demand
Cereals	217,161	15,570	232,731	224,920	7,811
Raw sugar	20,426	409	20,835	23,179	—2,344
Root crops	98,127	6,787	104,914	107,343	—2,429
Pulses	19,907	700	20,607	21,756	—1,149
Fruit, veg.	115,136	1,425	116,561	122,905	—6,344
Oil crops	8,396	3,897	12,293	16,600	—4,307
Meat	9,204	...	9,204	9,263	—59
Milk	52,422	1,048	53,470	51,148	2,322
Eggs	1,610	...	1,610	1,610	...

TABLE 10.2. 1985 and 2000 Production Projections for Low Income Countries (1,000 metric tons)

Commodity	1985 Production[a]		2000 Production[a]	
	(L)	(H)	(L)	(H)
Cereals	366,700	371,104	440,478	461,147
Raw sugar	44,573	44,727	56,187	57,644
Root crops	179,739	181,513	217,710	220,442
Pulses	27,704	27,801	29,998	30,851
Fruit, veg.	205,390	206,518	249,025	252,938
Oil crops	27,331	27,343	32,827	33,281

[a] Projections designated (L) are made under a low variant upper bound on cropland expansion. Those designated (H) are made under a high variant bound on cropland expansion.

The medium income countries also had a 1960 deficit of 8.2 million tons, or about 6 percent of requirements, in the important cereals group and a 2.3 million ton deficit in oil crops (Table 10.3). Sugar, root crops, and fruit and vegetables were the most important commodities in surplus. The sugar surplus was about 40 percent of production.

Overall, the projected production increases for the medium income countries suggest a somewhat less rapid growth of food supplies than for the low income countries. The projection in Table 10.4 for sugar, pulses, and fruit and vegetables shows production at least doubling by the year 2000. However, projected cereals production for 2000 is only 64 percent greater than in 1960. An 82 percent increase is projected for oil crops, and root crops production increases 35 percent.

The pooled 1960 food balance for the high income countries is shown in Table 10.5. Surpluses are shown for milk, root crops, eggs, and most notably, for cereals, where a net 25 million ton excess supply is estimated. Deficits were estimated for the other five commodity groups. The high income countries were most dependent on the medium and low income countries for sugar supplies, but significant quantities of fruit and vegetables and oil crops were also required.

TABLE 10.3. 1960 Production-Demand Comparisons for Medium Income Countries (1,000 metric tons)

Commodity	Domestic Disappearance			Production	Excess Demand
	Food and Industrial	Feed	Total		
Cereals	85,126	51,713	136,839	128,625	8,214
Raw sugar	9,944	...	9,944	16,430	—6,486
Root crops	43,358	33,657	77,015	81,750	—4,735
Pulses	3,263	960	4,223	4,047	176
Fruit, veg.	80,216	2,315	82,531	92,206	—9,495
Oil crops	8,629	3,167	11,796	9,507	2,289
Meat	11,471	...	11,471	12,173	—702
Milk	41,154	17,006	58,160	57,329	831
Eggs	2,506	...	2,506	2,536	—30

TABLE 10.4. 1985 and 2000 Production Projections for Medium Income Countries
(1,000 metric tons)

Commodity	1985 Production	2000 Production
Cereals	180,889	210,621
Raw sugar	28,059	34,964
Root crops	99,662	110,087
Pulses	6,480	8,346
Fruit, veg.	175,487	224,989
Oil crops	14,384	17,300

Production projections for the high income countries are given in Table 10.6. A decline is projected for root crops production, and very large proportionate increases are projected for pulses and oil crops. However, projected cereals production for 2000 is 2.3 times 1960 production, and the cereals tonnage projected for 2000 is over 4 times greater than any other commodity group.

The base period food balances and production projections for the 96-country total (Tables 10.7 and 10.8) are perhaps of less interest than the others. The excess food "requirements" or demand for 1960 give a crude measure of the extent of the interface between the countries under study and those which have been excluded. Generally, the excess demands and excess supplies are on the order of 1 percent of production or less. The root crops surplus and the oil crops surplus are the only exceptions. The production projections shown in Table 10.8 take on more meaning in the context of production-demand comparisons, and we turn now to a consideration of these.

Table 10.19 shows that for the 96-nation total, large and growing surpluses are estimated through 1985 when medium population and constant per capita income are assumed. A 145 million ton surplus of cereals is projected under the high land restriction assumption, or about 12 percent of production. A deficit is estimated only in the case of root crops. However, an examination of the 1985 benchmark projections for

TABLE 10.5. 1960 Production-Demand Comparisons for High Income Countries
(1,000 metric tons)

Commodity	Domestic Disappearance Food and Industrial	Feed	Total	Production	Excess Demand
Cereals	128,463	201,901	330,364	355,355	−24,991
Raw sugar	26,965	61	27,026	18,059	8,967
Root crops	105,136	66,967	172,103	173,829	−1,726
Pulses	3,924	2,148	6,072	4,909	1,163
Fruit, veg.	114,764	70	114,834	97,800	17,034
Oil crops	6,676	17,937	24,613	20,723	3,890
Meat	42,585	...	42,585	42,369	216
Milk	157,656	55,743	213,399	222,407	−9,008
Eggs	8,628	...	8,628	8,675	−47

low, medium, and high income aggregates shows that the surplus is concentrated in the high income nations. Growing cereals deficits are projected for the low and medium income countries. The estimated deficit for the low income countries is 60.0 or 55.6 million tons, depending on whether low or high bounds are assumed for cropland expansion (Table 10.10).

TABLE 10.6. 1985 and 2000 Production Projections for High Income Countries (1,000 metric tons)

Commodity	1985 Production	2000 Production
Cereals	639,716	828,425
Raw sugar	34,569	41,151
Root crops	152,632	130,526
Pulses	28,583	40,130
Fruit, veg.	154,336	179,024
Oil crops	62,940	93,391

TABLE 10.7. 1960 Production-Demand Comparisons for 96-Country Total (1,000 metric tons)

	Domestic Disappearance				
Commodity	Food and Industrial	Feed	Total	Production	Excess Demand
Cereals	430,748	269,182	699,930	708,900	—8,970
Raw sugar	57,334	470	57,804	57,668	136
Root crops	246,620	107,410	354,030	362,922	—8,892
Pulses	27,095	3,808	30,903	30,712	191
Fruit, veg.	310,114	3,810	313,924	312,731	1,193
Oil crops	23,701	25,001	48,702	46,830	1,872
Meat	63,260	. . .	63,260	63,805	—545
Milk	251,232	73,797	325,029	330,884	—5,855
Eggs	12,744	. . .	12,744	12,821	—77

TABLE 10.8. 1985 and 2000 Production Projections for 96-Country Total (1,000 metric tons)

Commodity	1985 Production[a]		2000 Production[a]	
	(L)	(H)	(L)	(H)
Cereals	1,187,286	1,191,701	1,479,368	1,500,175
Raw sugar	107,196	107,354	132,248	133,759
Root crops	432,029	433,807	458,295	461,053
Pulses	62,767	62,863	78,474	79,327
Fruit, veg.	535,205	536,341	652,958	656,950
Oil crops	104,652	104,665	143,513	143,970

[a] Projections designated (L) are made under a low variant upper bound on cropland expansion. Those designated (H) are made under a high variant bound on cropland expansion.

TABLE 10.9. 1985 and 2000 Production-Demand Comparisons for Low Income Countries, Assuming Low Population Growth Rates (1,000 metric tons)

1985 Comparisons, Assuming Constant Per Capita Income

Commodity	Domestic Disappearance			Excess Demand	
	Food and Industrial	Feed	Total	Low Land Bound	High Land Bound
Cereals	373,291	29,193	402,484	35,784	31,380
Raw sugar	35,360	642	36,002	-8,569	-8,724
Root crops	177,962	12,224	190,186	10,447	8,672
Pulses	33,049	1,169	34,218	6,513	6,417
Fruit, veg.	204,419	2,688	207,107	1,717	589
Oil crops	14,784	6,593	21,377	-5,952	-5,964
Meat	16,649	...	16,649	...	...
Milk	89,851	1,975	91,826	...	...
Eggs	2,942	...	2,942	...	...

1985 Comparisons, Assuming Historical Income Growth Rates

Commodity	Domestic Disappearance			Excess Demand	
	Food and Industrial	Feed	Total	Low Land Bound	High Land Bound
Cereals	420,426	41,574	462,000	95,300	90,896
Raw sugar	46,943	939	47,882	3,310	3,156
Root crops	188,041	18,757	206,798	27,060	25,285
Pulses	37,925	1,845	39,770	12,066	11,969
Fruit, veg.	255,352	3,837	259,189	53,799	52,671
Oil crops	22,036	9,717	31,753	4,422	4,410
Meat	25,694	...	25,694	...	...
Milk	132,737	2,667	135,404	...	...
Eggs	4,810	...	4,810	...	...

2000 Comparisons, Assuming Constant Per Capita Income

Commodity	Domestic Disappearance			Excess Demand	
	Food and Industrial	Feed	Total	Low Land Bound	High Land Bound
Cereals	489,132	39,144	528,276	87,798	67,129
Raw sugar	45,970	793	46,763	-9,423	-10,880
Root crops	247,830	15,652	263,482	45,772	43,040
Pulses	42,456	1,486	43,942	13,943	13,091
Fruit, veg.	273,860	3,591	277,451	28,425	24,512
Oil crops	19,841	8,518	28,359	-4,467	-4,921
Meat	22,324	...	22,324	...	...
Milk	116,522	2,676	119,198	...	...
Eggs	3,961	...	3,961	...	...

2000 Comparisons, Assuming Historical Income Growth Rates

Commodity	Domestic Disappearance			Excess Demand	
	Food and Industrial	Feed	Total	Low Land Bound	High Land Bound
Cereals	578,051	62,612	640,663	200,185	179,516
Raw sugar	72,442	1,455	73,897	17,710	16,253
Root crops	265,919	27,453	293,372	75,662	72,930
Pulses	52,791	3,149	55,940	25,942	25,098
Fruit, veg.	378,806	5,737	384,543	135,518	131,606
Oil crops	39,816	15,745	55,561	22,734	22,279
Meat	43,637	...	43,637	...	...
Milk	207,795	3,897	211,692	...	...
Eggs	9,194	...	9,194	...	...

TABLE 10.10. 1985 and 2000 Production-Demand Comparisons for Low Income Countries, Assuming Medium Population Growth Rates (1,000 metric tons)

1985 Comparisons, Assuming Constant Per Capita Income

Commodity	Domestic Disappearance			Excess Demand	
	Food and Industrial	Feed	Total	Low Land Bound	High Land Bound
Cereals	395,206	31,450	426,656	59,956	55,552
Raw sugar	37,862	679	38,541	−6,031	−6,185
Root crops	189,629	13,482	203,111	23,373	21,598
Pulses	35,183	1,242	36,425	8,720	8,624
Fruit, veg.	217,229	2,886	220,115	14,725	13,597
Oil crops	15,637	6,985	22,622	−4,708	−4,720
Meat	17,916	...	17,916	...	...
Milk	95,979	2,136	98,115	...	...
Eggs	3,143	...	3,143	...	...

1985 Comparisons, Assuming Historical Income Growth Rates

Commodity	Domestic Disappearance			Excess Demand	
	Food and Industrial	Feed	Total	Low Land Bound	High Land Bound
Cereals	439,299	42,891	482,190	115,490	111,086
Raw sugar	48,431	935	49,366	4,793	4,639
Root crops	198,779	19,764	218,543	38,804	37,030
Pulses	39,536	1,858	41,394	13,690	13,594
Fruit, veg.	264,482	3,934	268,416	63,026	61,898
Oil crops	22,218	9,753	31,971	4,640	4,628
Meat	26,215	...	26,215	...	...
Milk	135,088	2,763	137,851	...	...
Eggs	4,838	...	4,838	...	...

2000 Comparisons, Assuming Constant Per Capita Income

Commodity	Domestic Disappearance			Excess Demand	
	Food and Industrial	Feed	Total	Low Land Bound	High Land Bound
Cereals	542,962	45,706	588,668	148,189	127,521
Raw sugar	42,295	859	53,154	−3,032	−4,489
Root crops	282,855	19,196	302,051	84,340	81,609
Pulses	47,355	1,649	49,004	19,006	18,154
Fruit, veg.	309,649	4,186	313,835	64,810	60,897
Oil crops	22,212	9,437	31,649	−1,177	−1,632
Meat	25,928	...	25,928	...	...
Milk	131,360	3,156	134,516	...	...
Eggs	4,561	...	4,561	...	...

2000 Comparisons, Assuming Historical Income Growth Rates

Commodity	Domestic Disappearance			Excess Demand	
	Food and Industrial	Feed	Total	Low Land Bound	High Land Bound
Cereals	629,461	66,777	696,238	255,761	235,092
Raw sugar	77,061	1,469	78,530	22,343	20,886
Root crops	298,819	30,737	329,556	111,846	109,115
Pulses	56,976	3,244	60,220	30,222	29,369
Fruit, veg.	408,701	6,095	414,796	165,771	161,858
Oil crops	40,347	16,053	56,400	23,573	23,119
Meat	45,316	...	45,316	...	...
Milk	217,784	4,226	222,010	...	...
Eggs	9,250	...	9,250	...	...

TABLE 10.11. 1985 and 2000 Production-Demand Comparisons for Low Income Countries, Assuming High Population Growth Rates (1,000 metric tons)

1985 Comparisons, Assuming Constant Per Capita Income

Commodity	Domestic Disappearance			Excess Demand	
	Food and Industrial	Feed	Total	Low Land Bound	High Land Bound
Cereals	412,484	32,659	445,143	78,443	74,039
Raw sugar	39,524	712	40,236	−4,335	−4,490
Root crops	199,268	14,103	213,371	33,632	31,858
Pulses	36,859	1,289	38,148	10,444	10,347
Fruit, veg.	226,362	2,986	229,348	23,958	22,830
Oil crops	16,272	7,294	23,566	−3,764	−3,776
Meat	18,696	...	18,696	...	...
Milk	100,226	2,223	102,449	...	...
Eggs	3,262	...	3,262	...	...

1985 Comparisons, Assuming Historical Income Growth Rates

Commodity	Domestic Disappearance			Excess Demand	
	Food and Industrial	Feed	Total	Low Land Bound	High Land Bound
Cereals	453,516	43,490	497,006	130,306	125,902
Raw sugar	49,181	928	50,109	5,536	5,381
Root crops	207,341	20,237	227,578	47,839	46,065
Pulses	40,715	1,855	42,570	14,867	14,770
Fruit, veg.	270,427	3,980	274,407	69,017	67,889
Oil crops	22,300	9,741	32,041	4,711	4,699
Meat	26,449	...	26,449	...	...
Milk	136,063	2,807	138,870	...	...
Eggs	4,843	...	4,843	...	...

2000 Comparisons, Assuming Constant Per Capita Income

Commodity	Domestic Disappearance			Excess Demand	
	Food and Industrial	Feed	Total	Low Land Bound	High Land Bound
Cereals	590,384	49,528	639,912	199,434	178,765
Raw sugar	56,509	928	57,437	1,250	−206
Root crops	314,219	20,680	334,899	117,189	114,457
Pulses	51,500	1,774	53,274	23,276	22,423
Fruit, veg.	337,512	4,532	342,044	93,019	89,106
Oil crops	24,270	10,205	34,475	1,647	1,193
Meat	28,250	...	28,250	...	...
Milk	142,476	3,440	145,916	...	...
Eggs	4,927	...	4,927	...	...

2000 Comparisons, Assuming Historical Income Growth Rates

Commodity	Domestic Disappearance			Excess Demand	
	Food and Industrial	Feed	Total	Low Land Bound	High Land Bound
Cereals	672,488	68,611	741,099	300,621	279,952
Raw sugar	79,311	1,475	80,786	24,600	23,142
Root crops	326,477	32,118	358,595	140,885	138,154
Pulses	60,200	3,288	63,488	33,490	32,637
Fruit, veg.	429,571	6,259	435,830	186,806	182,893
Oil crops	40,462	16,190	56,652	23,826	23,371
Meat	45,955	...	45,955	...	...
Milk	223,400	4,367	227,767	...	...
Eggs	9,219	...	9,219	...	...

213

TABLE 10.12. 1985 and 2000 Production-Demand Comparisons for Medium Income Countries, Assuming Low Population Growth Rates (1,000 metric tons)

1985 Comparisons

Commodity	1985 Comparisons, Assuming Constant Per Capita Income — Domestic Disappearance				1985 Comparisons, Assuming Historical Income Growth Rates — Domestic Disappearance			
	Food and Industrial	Feed	Total	Excess Demand	Food and Industrial	Feed	Total	Excess Demand
Cereals	110,481	62,929	173,410	−7,478	102,364	94,583	196,947	16,058
Raw sugar	13,919	...	13,919	−14,139	19,208	...	19,208	−8,850
Root crops	53,585	42,218	95,803	−3,858	46,161	57,451	103,611	3,949
Pulses	4,779	1,070	5,849	−630	5,206	1,837	7,043	564
Fruit, veg.	100,021	2,482	102,503	−72,983	131,009	3,625	134,634	−40,852
Oil crops	10,108	4,106	14,214	−169	15,143	6,560	21,703	7,319
Meat	15,320	...	15,320	...	21,996	...	21,996	...
Milk	55,025	20,019	75,044	...	70,131	30,114	100,245	...
Eggs	3,181	...	3,181	...	4,932	...	4,932	...

2000 Comparisons

Commodity	2000 Comparisons, Assuming Constant Per Capita Income — Domestic Disappearance				2000 Comparisons, Assuming Historical Income Growth Rates — Domestic Disappearance			
	Food and Industrial	Feed	Total	Excess Demand	Food and Industrial	Feed	Total	Excess Demand
Cereals	125,329	68,270	193,599	−17,021	113,726	114,656	228,382	17,762
Raw sugar	16,589	...	16,589	−18,374	24,715	...	24,715	−10,248
Root crops	58,098	45,360	103,458	−6,628	45,714	67,459	113,173	3,085
Pulses	5,896	1,104	7,000	−1,345	6,667	2,328	8,995	648
Fruit, veg.	111,177	2,572	113,749	−111,238	158,489	4,250	162,739	−62,248
Oil crops	10,812	4,650	15,462	−1,837	18,172	8,213	26,385	9,085
Meat	17,700	...	17,700	...	28,446	...	28,446	...
Milk	63,557	21,280	84,837	...	86,967	37,076	124,043	...
Eggs	3,573	...	3,573	...	6,363	...	6,363	...

TABLE 10.13. 1985 and 2000 Production-Demand Comparisons for Medium Income Countries, Assuming Medium Population Growth Rates (1,000 metric tons)

1985 Comparisons, Assuming Constant Per Capita Income

Commodity	Domestic Disappearance			Excess Demand
	Food and Industrial	Feed	Total	
Cereals	115,530	65,479	181,009	120
Raw sugar	14,647	…	14,647	−13,411
Root crops	55,966	43,912	99,878	216
Pulses	5,047	1,110	6,157	−322
Fruit, veg.	104,582	2,567	107,149	−68,337
Oil crops	10,537	4,304	14,841	457
Meat	16,075	…	16,075	…
Milk	57,715	20,777	78,492	…
Eggs	3,327	…	3,327	…

1985 Comparisons, Assuming Historical Income Growth Rates

Commodity	Domestic Disappearance			Excess Demand
	Food and Industrial	Feed	Total	
Cereals	107,375	97,275	204,650	23,762
Raw sugar	19,951	…	19,951	−8,108
Root crops	48,557	59,244	107,801	8,140
Pulses	5,469	1,877	7,346	867
Fruit, veg.	135,539	3,715	139,254	−36,233
Oil crops	15,581	6,767	22,348	7,964
Meat	22,723	…	22,723	…
Milk	72,777	30,889	103,666	…
Eggs	5,074	…	5,074	…

2000 Comparisons, Assuming Constant Per Capita Income

Commodity	Domestic Disappearance			Excess Demand
	Food and Industrial	Feed	Total	
Cereals	137,471	74,063	211,534	913
Raw sugar	18,515	…	18,515	−16,448
Root crops	63,480	49,364	112,844	2,757
Pulses	6,628	1,187	7,815	−531
Fruit, veg.	121,862	2,735	124,597	−100,391
Oil crops	11,737	5,119	16,856	−444
Meat	19,576	…	19,576	…
Milk	70,415	22,983	93,398	…
Eggs	3,915	…	3,915	…

2000 Comparisons, Assuming Historical Income Growth Rates

Commodity	Domestic Disappearance			Excess Demand
	Food and Industrial	Feed	Total	
Cereals	125,766	121,830	247,596	36,976
Raw sugar	26,893	…	26,893	−8,070
Root crops	50,768	72,415	123,178	13,091
Pulses	7,423	2,439	9,862	1,516
Fruit, veg.	170,190	4,463	174,653	−50,335
Oil crops	19,249	8,764	28,013	10,712
Meat	30,533	…	30,533	…
Milk	94,315	39,250	133,565	…
Eggs	6,755	…	6,755	…

TABLE 10.14. 1985 and 2000 Production-Demand Comparisons for Medium Income Countries, Assuming High Population Growth Rates (1,000 metric tons)

Commodity	1985 Comparisons, Assuming Constant Per Capita Income				1985 Comparisons, Assuming Historical Income Growth Rates			
	Domestic Disappearance				Domestic Disappearance			
	Food and Industrial	Feed	Total	Excess Demand	Food and Industrial	Feed	Total	Excess Demand
Cereals	120,505	67,976	188,481	7,592	112,300	99,907	212,207	31,318
Raw sugar	15,236	...	15,236	−12,822	20,555	...	20,555	−7,503
Root crops	58,356	45,667	104,023	4,631	50,964	61,099	112,063	12,402
Pulses	5,238	1,153	6,391	−89	5,656	1,918	7,574	1,094
Fruit, veg.	108,895	2,649	111,544	−63,942	139,938	3,802	143,740	−31,747
Oil crops	10,974	4,486	15,460	1,077	16,045	6,966	23,011	8,627
Meat	16,692	...	16,692	...	23,327	...	23,327	...
Milk	59,925	21,540	81,465	...	74,944	31,655	106,599	...
Eggs	3,467	...	3,467	...	5,222	...	5,222	...

Commodity	2000 Comparisons, Assuming Constant Per Capita Income				2000 Comparisons, Assuming Historical Income Growth Rates			
	Domestic Disappearance				Domestic Disappearance			
	Food and Industrial	Feed	Total	Excess Demand	Food and Industrial	Feed	Total	Excess Demand
Cereals	150,399	80,073	230,472	19,851	138,443	129,171	267,614	56,993
Raw sugar	20,206	...	20,206	−14,758	28,820	...	28,820	−6,143
Root crops	69,073	53,606	122,679	12,592	56,017	77,656	133,673	23,586
Pulses	7,213	1,275	8,488	141	8,023	2,555	10,578	2,231
Fruit, veg.	132,511	2,887	135,398	−89,590	181,901	4,660	186,561	−38,427
Oil crops	12,711	5,579	18,290	990	20,449	9,334	29,783	12,482
Meat	21,289	...	21,289	...	32,437	...	32,437	...
Milk	76,527	24,724	101,251	...	100,802	41,405	142,207	...
Eggs	4,266	...	4,266	...	7,181	...	7,181	...

TABLE 10.15. 1985 and 2000 Production-Demand Comparisons for High Income Countries, Assuming Low Population Growth Rates (1,000 metric tons)

	1985 Comparisons, Assuming Constant Per Capita Income — Domestic Disappearance				1985 Comparisons, Assuming Historical Income Growth Rates — Domestic Disappearance			
Commodity	Food and Industrial	Feed	Total	Excess Demand	Food and Industrial	Feed	Total	Excess Demand
Cereals	161,319	252,225	413,544	−226,171	134,383	299,691	434,074	−205,642
Raw sugar	33,388	73	33,461	−1,106	39,255	88	39,343	4,774
Root crops	130,258	78,707	208,965	56,333	97,585	108,755	206,340	53,707
Pulses	4,850	2,747	7,597	−20,985	5,024	3,867	8,891	−19,691
Fruit, veg.	140,684	70	140,704	−13,632	167,545	91	167,636	13,300
Oil crops	8,199	22,177	30,376	−32,563	9,733	26,564	36,297	−26,642
Meat	52,930	...	52,930	...	66,343	...	66,343	...
Milk	195,067	66,615	261,682	...	210,128	89,264	299,392	...
Eggs	10,698	...	10,698	...	12,044	...	12,044	...

	2000 Comparisons, Assuming Constant Per Capita Income — Domestic Disappearance				2000 Comparisons, Assuming Historical Income Growth Rates — Domestic Disappearance			
Commodity	Food and Industrial	Feed	Total	Excess Demand	Food and Industrial	Feed	Total	Excess Demand
Cereals	177,616	277,691	455,307	−373,117	140,088	353,174	493,262	−335,162
Raw sugar	36,562	79	36,641	−4,509	44,683	103	44,786	3,635
Root crops	142,494	84,227	226,721	96,155	97,966	128,229	226,195	95,670
Pulses	5,304	3,039	8,343	−31,786	5,532	4,615	10,147	−29,983
Fruit, veg.	153,264	70	153,334	−25,689	194,455	96	194,551	15,526
Oil crops	8,950	24,296	33,246	−60,143	11,123	31,252	42,375	−51,014
Meat	58,128	...	58,128	...	79,577	...	79,577	...
Milk	213,485	71,582	285,067	...	231,535	105,475	337,010	...
Eggs	11,740	...	11,740	...	13,562	...	13,562	...

TABLE 10.16. 1985 and 2000 Production-Demand Comparisons for High Income Countries, Assuming Medium Population Growth Rates (1,000 metric tons)

	1985 Comparisons, Assuming Constant Per Capita Income — Domestic Disappearance				1985 Comparisons, Assuming Historical Income Growth Rates — Domestic Disappearance			
Commodity	Food and Industrial	Feed	Total	Excess Demand	Food and Industrial	Feed	Total	Excess Demand
Cereals	169,969	268,798	438,767	−200,948	143,063	315,565	458,628	−181,087
Raw sugar	35,329	77	35,406	838	41,202	92	41,294	6,725
Root crops	136,678	81,956	218,634	66,001	103,931	112,049	215,980	63,347
Pulses	5,122	2,877	7,999	−20,583	5,297	3,997	9,294	−19,288
Fruit, veg.	148,529	72	148,601	−5,735	175,139	93	175,232	20,895
Oil crops	8,682	23,589	32,271	−30,667	10,210	27,919	38,129	−24,810
Meat	56,146	...	56,146	...	69,371	...	69,371	...
Milk	206,372	69,500	275,872	...	221,813	92,150	313,963	...
Eggs	11,364	...	11,364	...	12,720	...	12,720	...

	2000 Comparisons, Assuming Constant Per Capita Income — Domestic Disappearance				2000 Comparisons, Assuming Historical Income Growth Rates — Domestic Disappearance			
Commodity	Food and Industrial	Feed	Total	Excess Demand	Food and Industrial	Feed	Total	Excess Demand
Cereals	199,857	321,068	520,925	−307,498	161,238	397,293	558,531	−269,892
Raw sugar	41,570	90	41,660	509	49,872	115	49,987	8,835
Root crops	158,687	91,742	250,429	119,903	112,961	136,763	249,724	119,199
Pulses	5,998	3,375	9,373	−30,756	6,226	4,981	11,207	−28,922
Fruit, veg.	178,345	73	173,418	−5,606	215,124	99	215,223	36,199
Oil crops	10,188	27,947	38,135	−55,255	12,414	34,979	47,393	−45,996
Meat	66,464	...	66,464	...	88,195	...	88,195	...
Milk	242,725	78,580	321,305	...	261,406	113,285	374,691	...
Eggs	13,465	...	13,465	...	15,323	...	15,323	...

TABLE 10.17. 1985 and 2000 Production-Demand Comparisons for High Income Countries, Assuming High Population Growth Rates (1,000 metric tons)

1985 Comparisons

	1985 Comparisons, Assuming Constant Per Capita Income — Domestic Disappearance				1985 Comparisons, Assuming Historical Income Growth Rates — Domestic Disappearance			
Commodity	Food and Industrial	Feed	Total	Excess Demand	Food and Industrial	Feed	Total	Excess Demand
Cereals	180,460	282,746	463,206	−176,509	153,734	328,855	482,589	−157,127
Raw sugar	37,225	81	37,306	2,738	43,072	95	43,167	8,598
Root crops	145,130	86,286	231,416	78,784	112,482	116,320	228,802	76,169
Pulses	5,405	3,087	8,492	−20,090	5,581	4,198	9,779	−18,802
Fruit, veg.	155,967	74	156,041	1,704	182,254	95	182,349	28,013
Oil crops	9,123	24,795	33,918	−29,021	10,636	29,068	39,704	−23,235
Meat	59,055	...	59,055	...	72,090	...	72,090	...
Milk	217,061	73,047	290,108	...	232,938	95,623	328,561	...
Eggs	11,951	...	11,951	...	13,310	...	13,310	...

2000 Comparisons

	2000 Comparisons, Assuming Constant Per Capita Income — Domestic Disappearance				2000 Comparisons, Assuming Historical Income Growth Rates — Domestic Disappearance			
Commodity	Food and Industrial	Feed	Total	Excess Demand	Food and Industrial	Feed	Total	Excess Demand
Cereals	220,602	345,196	565,798	−262,625	181,432	422,037	603,469	−224,955
Raw sugar	45,123	96	45,219	4,067	53,538	121	53,659	12,508
Root crops	175,819	100,521	276,340	145,814	129,335	146,318	275,653	145,127
Pulses	6,539	3,815	10,354	−29,775	6,772	5,439	12,211	−27,919
Fruit, veg.	187,221	75	187,296	8,271	229,366	102	229,468	50,443
Oil crops	10,994	30,058	41,052	−52,338	13,239	37,158	50,397	−42,993
Meat	71,718	...	71,718	...	93,625	...	93,625	...
Milk	262,734	85,993	348,727	...	282,631	121,311	403,942	...
Eggs	14,508	...	14,508	...	16,409	...	16,409	...

219

TABLE 10.18. 1985 and 2000 Production-Demand Comparisons for 96-Country Total, Assuming Low Population Growth Rates (1,000 metric tons)

1985 Comparisons

| | 1985 Comparisons, Assuming Constant Per Capita Income | | | | | 1985 Comparisons, Assuming Historical Income Growth Rates | | | | |
| | Domestic Disappearance | | | Excess Demand | | Domestic Disappearance | | | Excess Demand | |
Commodity	Food and Industrial	Feed	Total	Low Land Bound	High Land Bound	Food and Industrial	Feed	Total	Low Land Bound	High Land Bound
Cereals	645,090	344,346	989,436	-197,847	-202,262	657,173	435,846	1,093,019	-94,269	-98,684
Raw sugar	82,667	715	83,382	-23,813	-23,971	105,405	1,027	106,432	-762	-920
Root crops	361,805	133,149	494,954	62,924	61,147	331,785	184,962	516,747	84,719	82,941
Pulses	42,679	4,985	47,664	-15,103	-15,199	48,155	7,550	55,705	-7,061	-7,158
Fruit, veg.	445,072	5,240	450,312	-84,892	-86,027	553,906	7,552	561,458	26,253	25,117
Oil crops	33,091	32,876	65,967	-38,684	-38,696	46,912	42,841	89,753	-14,899	-14,911
Meat	84,899	...	84,898	...	...	114,031		114,031	...	...
Milk	339,941	88,608	428,549	...	...	412,994	122,045	535,039	...	...
Eggs	16,822	...	16,822	...	...	21,786	...	21,786	...	...

2000 Comparisons

| | 2000 Comparisons, Assuming Constant Per Capita Income | | | | | 2000 Comparisons, Assuming Historical Income Growth Rates | | | | |
| | Domestic Disappearance | | | Excess Demand | | Domestic Disappearance | | | Excess Demand | |
Commodity	Food and Industrial	Feed	Total	Low Land Bound	High Land Bound	Food and Industrial	Feed	Total	Low Land Bound	High Land Bound
Cereals	792,076	385,103	1,177,179	-302,191	-322,998	831,864	530,441	1,362,305	-177,069	-137,876
Raw sugar	99,120	872	99,992	-32,255	-33,765	141,838	1,559	143,397	11,149	9,638
Root crops	448,421	145,239	593,660	135,365	132,607	409,598	223,141	632,739	174,444	171,685
Pulses	53,656	5,629	59,285	-19,188	-20,041	64,989	10,092	75,081	-3,393	-4,245
Fruit, veg.	538,300	6,233	544,533	-108,424	-112,416	731,749	10,083	741,832	88,874	84,882
Oil crops	39,603	37,465	77,068	-66,445	-66,902	69,112	55,209	124,321	-19,192	-19,650
Meat	98,151		98,151	...	...	151,658		151,658	...	...
Milk	393,562	95,537	489,099	...	...	526,295	146,447	672,742	...	...
Eggs	19,273	...	19,273	...	...	29,119	...	29,119	...	...

220

TABLE 10.19. 1985 and 2000 Production-Demand Comparisons for 96-Country Total, Assuming Medium Population Growth Rates (1,000 metric tons)

1985 Comparisons

| | 1985 Comparisons, Assuming Constant Per Capita Income | | | | | 1985 Comparisons, Assuming Historical Income Growth Rates | | | | |
| | Domestic Disappearance | | | Excess Demand | | Domestic Disappearance | | | Excess Demand | |
Commodity	Food and Industrial	Feed	Total	Low Land Bound	High Land Bound	Food and Industrial	Feed	Total	Low Land Bound	High Land Bound
Cereals	680,704	365,725	1,046,429	−140,854	−145,269	689,736	455,730	1,145,466	−41,821	−46,236
Raw sugar	87,837	757	88,594	−18,601	−18,759	109,582	1,027	110,609	3,413	3,255
Root crops	382,272	139,350	521,622	89,592	87,815	351,267	191,056	542,323	110,294	108,517
Pulses	45,352	5,229	50,581	−12,185	−12,282	50,302	7,732	58,034	−4,732	−4,828
Fruit, veg.	470,339	5,524	475,863	−59,341	−60,476	575,158	7,741	582,899	47,695	46,559
Oil crops	34,856	34,878	69,734	−34,917	−34,929	48,009	44,438	92,447	−12,204	−12,216
Meat	90,135	...	90,135	...	...	118,308	...	118,308	...	...
Milk	360,065	92,412	452,477	...	...	429,676	125,801	555,477	...	...
Eggs	17,835	...	17,835	...	...	22,631	...	22,631	...	...

2000 Comparisons

| | 2000 Comparisons, Assuming Constant Per Capita Income | | | | | 2000 Comparisons, Assuming Historical Income Growth Rates | | | | |
| | Domestic Disappearance | | | Excess Demand | | Domestic Disappearance | | | Excess Demand | |
Commodity	Food and Industrial	Feed	Total	Low Land Bound	High Land Bound	Food and Industrial	Feed	Total	Low Land Bound	High Land Bound
Cereals	880,288	440,835	1,321,123	−158,248	−179,055	916,464	585,899	1,502,363	22,989	2,182
Raw sugar	112,378	949	113,327	−18,920	−20,430	153,824	1,584	155,408	23,159	21,649
Root crops	505,020	160,302	665,322	207,028	204,269	462,542	239,915	702,457	244,163	241,405
Pulses	59,981	6,210	66,191	−12,282	−13,135	70,625	10,664	81,289	2,815	1,963
Fruit, veg.	604,855	6,993	611,848	−41,109	−45,102	794,014	10,656	804,670	151,713	147,721
Oil crops	44,136	42,502	86,638	−56,874	−57,332	72,009	59,797	131,806	−11,708	−12,165
Meat	111,967	...	111,967	...	...	164,042	...	164,042	...	...
Milk	444,498	104,719	549,217	...	...	573,503	156,760	730,263	...	...
Eggs	21,940	...	21,940	...	...	31,328	...	31,328	...	...

221

TABLE 10.20. 1985 and 2000 Production-Demand Comparisons for 96-Country Total, Assuming High Population Growth Rates (1,000 metric tons)

1985 Comparisons

| | 1985 Comparisons, Assuming Constant Per Capita Income | | | | | 1985 Comparisons, Assuming Historical Income Growth Rates | | | | |
| | Domestic Disappearance | | | Excess Demand | | Domestic Disappearance | | | Excess Demand | |
Commodity	Food and Industrial	Feed	Total	Low Land Bound	High Land Bound	Food and Industrial	Feed	Total	Low Land Bound	High Land Bound
Cereals	713,448	383,378	1,096,826	−90,460	−94,875	719,549	472,250	1,191,799	4,507	92
Raw sugar	91,985	793	92,778	−14,416	−14,575	112,806	1,023	113,829	6,633	6,475
Root crops	402,753	146,056	548,809	116,780	115,002	370,786	197,655	568,441	136,412	134,635
Pulses	47,501	5,528	53,029	−9,736	−9,833	51,953	7,972	59,925	−2,841	−2,938
Fruit, veg.	491,222	5,709	496,931	−38,273	−39,409	592,618	7,876	600,494	65,289	64,153
Oil crops	36,370	36,575	72,945	−31,707	−31,720	48,980	45,775	94,755	−9,896	−9,909
Meat	94,441	...	94,441	...	...	121,865	...	121,865	...	...
Milk	377,210	96,809	474,019	...	...	443,944	130,085	574,029	...	...
Eggs	18,680	...	18,680	...	...	23,374	...	23,374	...	...

2000 Comparisons

| | 2000 Comparisons, Assuming Constant Per Capita Income | | | | | 2000 Comparisons, Assuming Historical Income Growth Rates | | | | |
| | Domestic Disappearance | | | Excess Demand | | Domestic Disappearance | | | Excess Demand | |
Commodity	Food and Industrial	Feed	Total	Low Land Bound	High Land Bound	Food and Industrial	Feed	Total	Low Land Bound	High Land Bound
Cereals	961,385	474,794	1,436,179	−43,193	−64,000	992,361	619,817	1,612,178	132,801	111,914
Raw sugar	121,836	1,024	122,860	−9,388	−10,898	161,668	1,596	163,264	31,015	29,505
Root crops	559,109	174,807	733,916	275,622	272,863	511,828	256,091	767,919	309,625	306,866
Pulses	65,253	6,864	72,117	−6,358	−7,210	74,994	11,282	86,276	7,802	6,949
Fruit, veg.	657,243	7,494	664,737	11,779	7,787	840,837	11,020	851,857	198,900	194,908
Oil crops	47,975	45,842	93,817	−49,696	−50,153	74,149	62,682	136,831	−6,682	−7,139
Meat	121,255	...	121,255	...	...	172,015	...	172,015	...	...
Milk	481,784	114,157	595,891	...	...	606,832	167,082	773,914	...	...
Eggs	23,700	...	23,700	...	...	32,808	...	32,808	...	...

Surpluses are projected only for sugar and oil crops. Table 10.13 shows a 0.1 million ton cereals deficit for the medium income countries under benchmark assumptions. However, large surpluses are projected for sugar and fruit and vegetables.

Table 10.16 shows that massive surpluses of cereals are projected for the high income nations in 1985 under benchmark assumptions. Production rises rapidly, while only modest demand increases are projected. The resulting surplus is about 201 million tons.

Referring again to Table 10.19 we see that by assuming high income rather than constant per capita income, estimated total cereals demand in 1985 for the 96 nations is increased 9.5 percent, or 99 million tons, above levels in the benchmark projections. For analogous comparisons, cereals demand in the low income nations increases 13.0 percent or 55.5 million tons (Table 10.10); estimated cereals demand in the medium income nations increases 13.1 percent or 23.6 million tons (Table 10.13); and cereals demand in the high income nations is estimated to increase 4.5 percent or 19.9 million tons (Table 10.16). Thus the 96-nation comparison is affected primarily by increased deficits in the low and medium income nations. The level of the 1985 cereals surplus for the 96 nations under these assumptions is 41.8 or 46.2 million tons, depending upon assumptions of cropland constraint.

Under medium population and continued income trends, positive excess demands exist in 1985 for all other product classes in the low income countries. Medium income countries show moderate deficits for root crops, pulses, and oil crops, but also a moderate sugar surplus and a relatively large surplus of fruit and vegetables. Relatively large deficits of root crops and surpluses of pulses and oil crops are projected (although in a market sense, we would expect forces oriented toward equilibrium to obscure these in market-directed shifts of resources).

For the 96-nation total, 1985 demands are 5.4 percent less under low population and constant per capita income than under benchmark conditions. Under high population, demand increases by 4.8 percent. However, the spread between low and high population estimates varies among the low, medium, and high income aggregates. For the low income countries the 1985 low variant population projections are 6 percent below the benchmark population estimates, while the high population estimates are 4.3 percent higher. The range for the medium income countries is from 4.4 percent below to 4.1 percent above the benchmark, and corresponding spreads for the high income countries are —5.7 percent and 5.6 percent.

None of the demand assumptions produces major, overall deficits for the 96-country total in 1985. With high population and continued income trends (Table 10.20) there is a relatively large deficit of root crops, but the excess demand for cereals is only a small fraction of total domestic disappearance.

Table 10.19 shows that, by 2000, conditions of medium population growth and constant per capita income result in an estimated cereals surplus for the 96-nation total of 158 or 179 million tons, depending on whether cropland expansion is based on the low or high variant constraints. The estimated surplus for the high income nations is about 307 million tons (Table 10.16) and demand in the medium income nations is about equal to production (Table 10.13). Table 10.10 shows deficits of either 148 or 128 million tons, depending upon which land constraint is used.

Estimated cereals production in 2000 is 2.1 times 1960 levels for the 96-nation aggregate, 2.3 times 1960 production for the high income nations, 1.6 times 1960 production for the medium income nations, and 2.0 times 1960 production for the low income nations. Cereals demand increases by the year 2000 under medium population and constant per capita income are greater, relative to 1960 levels, by 1.89 times for the 96-nation total, 1.58 times for the high income nations, 1.55 times for the medium income nations, and 2.53 times for the low income nations.

Since constant per capita income is used in the above demand comparisons, the ratios generated, for years 2000 and 1960, are the same as the ratios of medium variant population projections. Tables 10.10, 10.13, 10.16, and 10.19 also illustrate the potential effect of continued income trends rather than constant per capita income. Projected cereals demand in 2000 is then 2.15 times greater than 1960 demand for the 96-nation total. Corresponding ratios for the high, medium, and low income nations are 1.69, 1.81, and 2.99.

Finally, projections to 2000 under the highest and lowest demand assumptions will be considered. Table 10.18 shows that for low population growth and constant per capita income, cereals demand for the 96-nation total is only 78 percent of the production estimated under high variant land constraints. A 323 million ton surplus results. Root crops production lies below demand by about 134 million tons (with the qualification of equilibrating forces cited earlier), but surpluses are projected for all other commodities.

Excesses are projected for all commodities except root crops in the high income nations' production-demand comparisons, and the 373 million ton estimated surplus of cereals shown in Table 10.15 dominates the array of results. Every commodity is seen to be in surplus in the production-demand comparisons for the medium income nations (Table 10.12) and the 111 million ton estimated surplus of fruit and vegetables is most striking. In projections for the low income nations (Table 10.9), estimated production exceeds demand only for sugar and oil crops. Deficits are estimated for the other four product classes; and for cereals, the level is 67 or 88 million tons, depending on low or high cropland constraints. Thus projected production in the low income nations in 2000 is inadequate even to meet demands which embody no per capita income growth and only low variant population projections.

Table 10.20 shows that for the 96-nation total projected cereals production is only about 92 percent of demand in 2000 when high population and continued income trends are assumed. Under low variant cropland constraints, a 133 million ton cereals deficit is estimated. A small surplus of oil crops is estimated, but deficits occur in all other product classes. In addition to cereals shortages, large deficits are projected for root crops (310 million tons) and fruit and vegetables (199 million tons).

Deficits are estimated for three product classes, and surpluses are estimated for the other three in the corresponding projections for the high income nations (Table 10.17). The 225 million ton estimated surplus of cereals and the 145 million ton deficit of root crops stand out most notably. Production-demand comparisons for the medium income nations show a 57 million ton deficit of cereals (Table 10.14). Three deficits and two surpluses are found among the estimates for the other five product classes. Finally, projections for the low income nations under the highest demand assumptions (Table 10.11) follow a pattern which, by now, is familiar. Deficits are estimated for every product class, and in each case they are very large. The estimated cereals deficit is 280 or 301 million tons, depending upon land constraint assumptions. Production estimates range from 47 to 70 percent of the corresponding demands in the other five product classes.

Summary Comments

The production-demand comparisons are highly diverse in several ways. Estimated production trends for most high income countries are found to rise more rapidly than demand under plausible income, population, and land constraint assumptions. However, most low income countries' production trends rise less rapidly than demand. Projections to 2000 for the 96-nation total indicate both massive surpluses and massive deficits, depending upon assumptions; and corresponding wide ranges are found among the several production and demand variants for individual regions. Projected population growth is rampant in some regions and nearly stagnant in others. Still, some important conclusions do stand out in the midst of this diversity and are considered in succeeding paragraphs.

The analysis of production has shown few instances of stagnant trends. Projected yields for many developed countries, particularly the United States, are very high. Present yields are lower, and estimated yield trends are less vigorous in most underdeveloped countries, but even in these, substantial progress in improving yields over the postwar years has been apparent even before advent of the green revolution. In addition, cropland expansion has been important in explaining production trends in many developing countries.

Population growth accounts for most of the projected increase in demand in both low and high income countries. The impact of income increases on demand is not to be neglected, but over the 40-year projection period used, population effects are substantially greater than income effects.

Average per capita food intake in the high income countries is at near-saturation levels, and income elasticities are generally small. Increases in income result mainly in increased consumption of livestock products, but frequently are accompanied by offsetting declines in direct food consumption of crops. Thus, though projected population growth is not unduly high among high income countries, it is still more important as a determinant of future food demand than is income.

This same conclusion applies to the developing countries, but for different reasons. Consider the following comparisons. A 50 percent increase in food intake among the most poorly fed populations of the world could result in per capita nutrient consumption levels which compare reasonably well with consumption in the wealthy nations where food intake is at near-saturation levels. Thus, for a country where income and food consumption are low, an upper bound on potential food demand increases resulting from higher incomes might amount to (a) 50 percent of present demand, plus (b) an upward adjustment to allow for an increase in demand of primary products per unit of nutrients consumed, as livestock products are substituted for other foods in the diet. The amount of increase in incomes needed before such dietary substitutions have a large impact on demand seems to be much greater than is evident in the incomes projected as part of this study. But a doubling of projected population between 1960 and 2000 is a common occurrence under certain estimates for the underdeveloped countries, and in some cases, projected population for 2000 is over four times 1960 levels. Thus the greater importance of population growth as a demand determinant is apparent, even in the developing countries.

Because of the importance of population growth as a determinant of results reported, additional comments about alternative population variants are useful. As suggested in Chapter 1, falling mortality rates provide the major demographic factor affecting trends in population among the underdeveloped nations in recent years. However, future human fertility rates will undoubtedly be important factors affecting long-term balances of food production and demand in these nations. Demand projections presented for the three population alternatives illustrate the variability in food demand expected under different fertility trends. Actually, none of the population projections embodies assumptions of continued recent trends in mortality and fertility; in general, projections for the underdeveloped countries are based on assumptions that fertility will fall more rapidly than in the past.

Some experts believe that even these estimates are too high. Bogue argues that the political leadership in many underdeveloped countries has recently come to view family planning as a moral and sensible solution to their food and population problems.[1] He further argues that low-cost birth control technology is presently at hand, and that persons living in impoverished, overpopulated nations are receptive to family planning programs. Based on these postulates, he proposed that the world population growth rate may be reduced to zero by the year 2000. He constructs a set of population projections for the world by assuming that growth rates began falling in 1965, and that they fall linearly to zero by 2000. The resulting estimate of population for the world in 2000 is about 4.5 billion. He terms this approach "crude and subjective," and to allow for some possible contingencies, he submits for consideration an adjusted estimate of about 5 billion.

Estimates for world population in 2000 under the low, medium, and high variant projections used in this study are 5.3, 6.0, and 6.8 billion. Some lowering of fertility seems nearly certain as the number of surviving children, and hence, average family size, increases above prior levels. Thus population projections embodying assumptions of constant fertility and declining mortality have been excluded. This tendency toward smaller families will be augmented as birth control assistance becomes more widespread and parents gain access to knowledge and means to limit family size to desired levels. However, Bogue's projections reflect an assumption that this process will occur very quickly, and even with this assumption, his "adjusted" estimates for 2000 are only 6 percent lower than the low variant estimates used in this study. Thus projections such as Bogue's also have been excluded. Nevertheless, it is apparent that had they been included, the results would have been essentially the same. Continued income growth, together with these more modest population assumptions, would have yielded growing food surpluses in the developed countries and growing food deficits in the developing countries.

DIFFERENT OUTLOOK

The production-demand comparisons in this and preceding chapters were constructed to determine in some detail the implications for the future of available population projections along with continuation of postwar trends in income and food production. But we have noted before that the results are not to be regarded as the often dismal end product of an abstract process which will inexorably run its course through time. The outlook can be altered, but major policy decisions and commitments of resources will be required to bring about different results. Fortunately, there is good evidence that this process has begun,

---

1. Donald J. Bogue. "The Prospects for World Population Control." In *Alternatives for Balancing World Food Production and Needs.* Ames: Iowa State University Press, 1967, pp. 72–85.

and in the next chapter we consider the most topical aspect of these efforts: the green revolution. We conclude this chapter with a few general observations about the policy implications of the projections. We distinguish between the implications of the 1985 and 2000 projections, for in some respects they appear to be quite different.

Projections through 1985 suggest that some high income countries (most notably the United States) may well be pursuing policies designed to limit production for most of this period regardless of the course of future demand around the world. Most 1985 projections for the 96-nation total exhibit surpluses, and only the highest demand estimates produce overall deficits. Under high population and income assumptions and low constraints on land expansion, estimated cereals production for the 96-nation total falls short of demand by 4.5 million tons or 0.4 percent. If high land constraints are assumed, production almost exactly equals demand. However, the combined 1985 cereals deficit for the low and medium income countries is 161.6 or 157.2 million tons, depending upon land constraint assumptions.

The above projections are based on assumptions of income growth at past rates, and if they are realized, some additional imports into the low income countries can probably be financed commercially. However, deficits are projected for every other product class. If these were to materialize, it seems likely that a major share of the overall shortage would have to be made up by noncommercial shipments. While no precise evaluation has been made, it is clear that shipments this large would be difficult to sustain. Certainly the cost of financing such shipments would be immense, and in any case, such a policy would not lead to a viable and permanent solution.

But other alternatives exist for balancing food demand and supply in the low income countries. In many areas, particularly Latin America, Africa, and Other East Asia, available evidence indicates that cropland expansion will continue to be feasible for some time into the future. Estimates of agricultural land resources in Latin America are extremely large relative to any foreseeable requirements. To be sure, the natural productivity of many soils in this area is not high, and the economic and institutional problems involved in expanding the agricultural area could not be subjected to close study in this investigation. However, the potential for cultivating large additional areas of no lower productivity than many areas now in use is clearly present. In Asia, the potentials appear to be more limited; but still, they seem to be far from exhausted. The countries of Southeast Asia appear to have the greatest potential for adding new land, but additional multiple cropping and irrigation from available water sources seem possible throughout Asia. Again, these results do not imply that an all-out effort toward cropland expansion is the surest or most effective means of solving the food problems of the underdeveloped countries.

Currently, the most promising alternatives lie in policies designed to stimulate the rate of increase in crop yields in the developing countries. This, of course, is the central theme of the green revolution, but we shall defer discussion of this until Chapter 11.

The long-term policy implications appear somewhat different. Possibilities of meeting the projected food deficits of the poorer nations through increased trade seem remote. Of course an array of policies designed to increase food production will be required, and high level food shipments from the developed to the developing countries may still be necessary. However, it seems certain that population control, or more properly, fertility control, must play an important part in any long-term solution. A perspective on the possible gains from fertility control can be gotten from the following comparisons. For all low and medium income countries, high variant population projections in 2000 exceed the low variant projections by about 20 percent. Under high population and income assumptions, cereals deficits in the year 2000 for the low and medium income countries are about 350 million tons, while under low population and high income assumptions, the estimated deficit is about 210 million tons. Both figures are large, and even the lower suggests the need for greater efforts toward improving productivity.

However, it seems clear that population control must be given high priority in any realistic long-run solution. Continued rapid population growth seems certain to thwart any long-term efforts to balance food production and needs, much less to raise general levels of human well-being. With appropriate population control, the results of this study indicate that aggregate food problems of the world could disappear over time without a powerful and erupting green revolution. The developing countries have the power in their hands to steer world food balances toward a realistic convergence between supply and demand. Currently, they have a disproportionate underinvestment in efforts toward population control or birth reduction relative to increased agricultural output. In the year 2000 the world population can either praise or condemn the leaders of developing countries for the emphasis they did or did not place on birthrate reductions in the decades of the 1970s and 1980s. The solution for 2000 rests in their hands and it depends more on their efforts for population control and the "troops" they do not produce, than on the "troops" they assign to the green revolution.

CHAPTER 11

Outlook under the Green Revolution and Government Policy

PROJECTIONS OF PRODUCTION POTENTIAL and demand for the year 2000 suggest that with maintenance of a modest momentum in improvement of agriculture and the lower level of population growth the world food situation will not deteriorate over a 30-year period. Of course, the pressure of demand on food supplies also will depend on the rate at which per capita incomes increase over the next three decades. If they increase to the higher level used in previous chapters, food may have a higher real price than otherwise, but the cause will more nearly be the happiness of mankind than if the pressure of demand on supply is reflected mainly through population increase and moderate increases in income. In the one case food takes on a relative scarcity because man has enough income to buy more of it and drive its real price up; in the other case, the relative scarcity grows out of restrained supplies which hold populations in misery near a subsistence level.

The most recent change on the supply side of food has been the worldwide introduction and adoption of a set of techniques and auxiliary inputs that has facilitated the production of short-stemmed or Mexican wheat varieties. India would have suffered acute hunger and starvation of many million persons in the 1966 and 1967 droughts had not large imports of cereals been available through food aid programs such as that of the United States. But in a remarkable recovery, record crops of over 100 million tons of cereals were realized by 1970 and 1971. While weather was favorable, an important increment of this greater production was represented by the greater yields of the new wheat varieties. Similarly, a rapid upturn in yields and wheat production took place in Pakistan over a short period of time. By 1970 Indian officials were expressing hopes for self-sufficiency in food production by 1976 and Pakistan had already begun some exports.

Somewhat parallel developments have taken place in rice varieties and yields. The first major new variety, IR-8, was similar to the Mexican wheats with its shorter and stiffer stems. It also could absorb heavier doses of fertilizer without stem-breakage under a greater load of grain.

As for wheat, more recent modifications in varieties have added further to yield potentials and realization for rice. New varieties have made it possible for the Philippines to approach self-sufficiency in rice production. In general, however, adoption rates and yield increases from the new rice varieties have not been as spectacular as for wheat. The technology of paddy production is more complex than that of wheat. Also, the early rice varieties were not in conformity with preferences of consumers in countries such as India. Consequently, the market price and profitability of rice were depressed relative to native, local varieties. More recent emphasis in plant breeding has been to combine the yield potentials of the new varieties and the characteristics which appeal to consumers.

Of course, yield takeoff from new varieties and associated fertilization, pesticide, and tillage practices has been experienced in developed countries previously and broadly over more of the less developed world in recent years. The dates, if any can be discretely identified, at which yield takeoff occurred for selected crops and countries are sometimes placed at these approximations:

Wheat—Mexico, 1953; Pakistan and India, 1967; Afghanistan and Turkey, 1969; Iran and Morocco, 1970.

Rice—Taiwan, 1915 (under the influence of Japanese technology where the takeoff is dated at the turn of the century); the Philippines, 1965; Ceylon, 1967; India, 1969; Indonesia, 1970; and South Vietnam, 1966.

As mentioned earlier, progress has been much less for some crops and countries than for others. Progress for rice in southern India is markedly less than for wheat in northern India. Progress for India generally has been much less than for the Philippines in rice, and Thailand continues even to have falling yields for this crop. However, upward momentum for wheat is beginning to develop over much of the Middle East and in Asian countries. Progress has been nil for crops grown in drier regions of these same countries where water is not available and governments have placed no emphasis on new technologies.

The progress in adoption and successful cultivation of the new wheat varieties in northern India, Pakistan, and elsewhere over Asia has caused some experts on world food problems to nearly reverse their projections. For example, Brown's earlier forecast, along with many others of the late 1960s, presented an extremely gloomy outlook on the ability of the world aggregatively to meet the future food needs of growing populations in less developed countries.[1] However, in a recent book his outlook is entirely optimistic and he paints a rosy picture, based

1. Lester R. Brown. *Increasing World Food Output.* U.S. Department of Agriculture, Government Printing Office, Washington, D.C., 1965; Lester R. Brown. *Man, Land and Food.* U.S. Department of Agriculture, Government Printing Office, Washington, D.C., 1963.

on progress of the green revolution thus far, of the ability of developing countries and the world at large to increase food supplies by relevant magnitudes in future decades:

> For the first time in history, it is realistic to consider the eradication of hunger for the overwhelming majority of mankind. Breakthrough in cereal production and the new food technologies make that a realistic, attainable objective, particularly if the United States were to take the lead in a new worldwide effort. We would be showing the world that we had not lost confidence in ourselves or in those abroad who so desperately need help. . . . At present, the average diets in poor countries, which contain more than half of the world's people, are deficient in calories and protein. But enlightened agricultural policies, new food and agricultural technologies, and employment creating programs should make it possible to bring the diet of virtually every country above the nutritional minimum by 1980. This would eliminate most hunger and malnutrition, leaving only fringe groups to be reached by special efforts.[2]

Cochrane similarly is optimistic but in a more guarded manner and emphasizes the revolution in expectations more than the revolution in food production:

> The gap between expectations and fulfillment is narrower for food than for total level of living. And the total time span required to eradicate food and nutrition deficiencies in most developing countries is probably much shorter than for the total level of living. But the food problem is more acute and urgent of solution than the level of living problem. Hunger and malnutrition cannot wait; their crippling effects are immediate and direct. . . . Some gains in per capita availability of food supplies have been made in the 1950's and 1960's and more gains will be made in the 1970's and 1980's. But food supplies will not reach the level of wants and expectations in many, if not most, developed countries in this or the next decade.[3]

SOURCE

Future production possibilities posed from the initial successes of the green revolution will stem from additional research and improved communication of knowledge to farmers. Cultivators in less developed countries have shown remarkable ability to innovate and apply new technologies under favorable conditions of knowledge, prices, tenure, and capital. As we point out later, the secrets of progress and development of agriculture seem to be few, if government officials and administrators are bold and imaginative enough to provide an appropriate economic environment. The green revolution, as recent progress in high-

2. Lester R. Brown. *Seeds of Change.* New York: Praeger, 1970, p. 183.
3. Willard W. Cochrane. *World Food Problem, A Guardedly Optimistic View.* New York: Thomas Y. Crowell, 1969, p. 12.

yielding varieties and associated technologies has come to be known, emphasizes the need for development of new practices and inputs and appropriate combinations of these so that their interactions can be expressed in greater output. The high return on agricultural research for these purposes has been rather well quantified.[4] While this payoff is evident in the rate at which farmers have adopted hybrid corn and other new technologies in the United States, parallel progress was made as a result of early research and technical knowledge in Japan and other countries.[5] While the high level of technology within the setting of their domestic factor markets and relative prices has been held up as an example for the world, the impact of similar knowledge in the developing countries is now becoming evident. Starting at a later date, Japan mechanized its agriculture in 20 years while the same extent of progress in mechanization was accomplished in 50 years in the United States. Also, farmers in Punjab state of India adopted the new wheat varieties as fully in 5 years as the United States adopted hybrid corn in 20 years. Even more recent developments in mechanization and varieties should be adopted with the same or more rapidity under favorable and advanced communication systems. However, it is now apparent that if knowledge of more productive and profitable inputs is generated, is communicated broadly, and can be applied in a favorable economic climate, farmers everywhere will make rapid progress in applying them.

COST-FREE TECHNOLOGIES

Thus far in the green revolution new technologies largely have been nearly cost-free to the countries and governments which have realized the large recent benefits. The dwarf wheat varieties were largely financed and developed by the Rockefeller Foundation with some help from the Mexican government. For the new rice varieties, the major financial inputs have come from the Rockefeller Foundation, the Ford

4. Zvi Griliches. "Research Costs and Social Returns: Hybrid Corn and Related Innovations." *Journal of Political Economy* 40 (2): 393–405; T. W. Schultz. "Some Reflections on Agricultural Production, Output and Supply." *Journal of Farm Economics* 38 (3): 748–62; Charles Meiburg. "Non-farm Inputs as a Source of Agricultural Productivity." *Journal of Farm Economics* 44 (5): 1433–38; Earl O. Heady. *Agricultural Policy under Economic Development*. Ames: Iowa State University Press, 1962, pp. 594–627.

5. Yujiro Hayami. "Resource Endowments and Technological Change in Agriculture: U.S. and Japanese Experiences in Agricultural Productivity." *American Journal of Agricultural Economics* 51 (5): 1293–1303; Bruce Johnson. "Agriculture and Economic Development: The Relevance of Japanese Experience." *Food Research Institute Studies* 6 (3): 1–119; T. W. Schultz. *Transforming Traditional Agriculture*. New Haven: Yale University Press, 1964; Reed Hertford. *The Measured Success of Mexican Agricultural Production and Productivity*. U.S. Department of Agriculture, Economic Research Service, 1968.

Foundation, and other international agencies. Even the acquisition of seed supplies was financed by outside sources for some of the countries where the new wheat varieties have had initial success. Afghanistan, Pakistan, India, Iran, Lebanon, Nepal, and Turkey have made very large percentage increases in acreage of the new wheat varieties but have had practically no cost for their development. The same is true with respect to rice for India, the Philippines, South Vietnam, Indonesia, Singapore, and Burma, where adoption has been quite rapid. However, some of these countries did invest quite heavily in obtaining and distributing sufficient supplies of seed. Of equal importance was emphasis on price and supply policies which favored use of complementary inputs. For example, in the half-dozen years following the 1965–66 crop year, fertilizer use nearly tripled in India, Ceylon, Pakistan, and Turkey. It more than doubled in Indonesia, Iran, Lebanon, and South Vietnam.

The green revolution has emphasized the need for "package" approaches to agricultural improvement. Increases are much greater when the yield potentials are exploited through higher fertilization levels, pest control, and irrigation water. Studies have not been completed to determine how much of output increases are imputable to each of these inputs. However, experience indicates that both farmer adoption and yield success are much greater when other inputs are available in economically appropriate quantities. The monsoon rains alone guarantee great progress in yield for single wheat crops over much of Asia where complementary practices are used with the new varieties. However, even greater benefits can be derived where multiple-cropping systems are possible through water supplied from private wells or publicly developed surface water sources. New seed varieties, fertilization, and associated practices allow the developing countries to rise to a new plateau of production possibilities. A further and higher plateau is possible through improved use and management of water in multiple-cropping systems in those world regions of extended or year-round growing seasons. Multiple cropping allows a more intensive employment of both land and labor where climate allows a sufficiently large supply of energy over the year. In a rough geographic fashion, it is the less developed countries which possess climatic conditions favorable to multiple cropping.

Hence, the food supply potentials and plateaus attainable in the future will require larger indigenous efforts and investments by developing countries. The successful spread of an economic water supply, in combination with multiple cropping and other inputs applied with high-yielding varieties, will not be a cost-free technology adopted from other countries. Instead, domestic investments and institutions will be required to get it under way. Even for crop varieties, developing countries will need to invest more heavily in research for crops produced in regions now bypassed by the green revolution.

IMPLICATIONS TO PROJECTED FOOD SUPPLIES

The projections of this study do not incorporate recent variables of the green revolution. Hence, we believe that the fairly optimistic projections we make for the next 30 years are on the conservative side. Our results indicate that world food supplies can increase relative to demand over the next 30 years if ongoing trends are perpetuated and no explosive revolution in technology takes place. Further, our results indicate that if lower levels of population growth occurred over the next three decades, important progress could be made in improving human nutrition aggregatively for the world. The potential thrust posed by the green revolution only extends these optimistic possibilities.

However, neither the existence of these trends nor an initial start on the green revolution will bring about balances or equilibria in food supply and demand that guarantee improved nutrition and declining misery on the part of the world's subsistence population. Attainment of supply potentials and population checks will prevail in the future only if developing countries initiate and implement policies which are appropriate for their attainment. The remainder of this chapter deals with these policy needs and frameworks.

POLICY AND INSTITUTIONAL METHODS TO ACCOMPLISH PROJECTED FOOD SUPPLIES

The results of this study provide some optimism on world food supplies relative to demand over the next 30 years. The phenomena of new practices and their adoption represented by the initial stages of the green revolution while not easily incorporated as variables for our estimates provide further evidence that diets can even be improved over the next three decades. To assume that the processes involved in attaining greater food supplies would come about automatically in the developing countries and then do nothing about it would likely lead to disaster or underattainment of food supply levels that seem feasible. Beyond three decades, whether world diets remain at the levels which are feasible to be attained in that period depends on population policies the world over. Because of problems of the economics of birth control methods, the needed improvement in their technology, the attitudes drawn from previous generations of current childbearing people in many countries, and the role of children as definite income-generating resources in families of some less developed countries and religious and cultural attitudes, effective dampening of world population growth rates will require as long as 30 years. By 2000 it is important that these restraints and parameters to lower birthrates be effectively imposed. The projected food-producing potential, plus the prospects posed in the initial stages of new varieties and associated practices, does provide a "breathing spell" or time to implement the necessary policy elements for reduced population growth. But they must be implemented effectively in this 30-year period. Similarly, even if they are implemented, effective policies on the food supply side need to be put into practice to be sure that this breathing spell does last even as long as 20 years. Hence, in this section, we review the foundations in agricultural policy needed to provide the necessary degree of certainty on the supply side.

SEARCH FOR RESPONSE RESULTS

Recent progress, especially for wheat, in the adoption of new varieties and associated practices in southeastern United States indicate that cultivators with limited education do respond readily when there are profit prospects in doing so and when an appropriate economic climate prevails for these purposes. Of course, one need not rest on such recent examples in the passage of time. The development of North American agriculture in a short span of time by farmers who had limited education and little publicly supplied technical information is a hallmark example of accomplishments within an appropriate economic climate. The conditions of North America cannot be replicated in the developing countries (except for some regions *within* these countries where there is a supply of undeveloped land and a work force can migrate from other regions of the country) because they were based partly on a large expanse of unsettled land and a labor force which migrated both from other countries and other regions. Hence, "production" of this labor force seldom came out of the surplus income of the newly developing regions but was part of the economic surplus generated in other regions, and especially European countries. What was important was that markets and prices provided a favorable profit and economic climate for this rapid development of agriculture. The broad public policies of both Canada and the United States, through land settlement and emigration programs, kept the price of land and labor low. Agriculture responded accordingly at that time when the main inputs of farming were land and labor and capital had a small imputable share. Through land-grant provisions, both of these countries also furthered the development of transportation and associated marketing facilities and hurried the extension of railroads accordingly. These developmental policies were extended even further after the public domain or land supply was nearly exhausted. Extensions of the same general development policies, emphasizing an enlarged supply and low price of resources, were reflected in public research and its communication, public credit supplying mechanisms, and in the United States especially, the public development of irrigation land. Knowledge, capital, and water were all supplied to farmers at a low real price and farmers continued the rapid development of farming. With continuance of these public programs, the attainment of higher levels of education by farmers, and the participation of private industry in generating and communicating knowledge of new technologies (as farmers became highly capitalized and drew the majority of capital inputs from the industrial sector), farmer response accentuated into a crescendo of oversupply. Compensation policies had to be enacted accordingly to offset public developmental policies in causing glutted markets and low prices and income. With the compensation policies which provide high price supports and the developmental policies which generate more new technological inputs, agriculture in North America has maintained its growing food supply momentum.

Although the conditions of land and labor supply of early North America cannot be replicated in the developing countries, the processes of agricultural development in the early periods of Canada and the United States are important foundations upon which response policies can be built. They illustrate that under conditions of appropriate supplying and pricing of resources, with favorable price and market conditions for commodities, farmers respond readily in increasing output. Other characteristics of early American agriculture which facilitated this response were tenure conditions which assured the farmer that he would reap the rewards of his labor and other inputs. The land settlement scheme based on owner operation provided this framework, although it also can be provided for rented farms under appropriate share arrangements.[6] Still, we need not reach back so far in time to identify the variables and policies relevant for mounting and sustaining momentum in farmer adoption of new technologies and in rapidly augmenting food supply. The action of farmers in northern India, Pakistan, Afghanistan, Turkey, and Iran in the last half-dozen years is evidence of a similar nature.

The search for mysteries explaining increased farm productivity and innovation has been a major activity of economists and agriculturalists for over a decade. Why these mysteries exist is itself a mystery. Much knowledge is already at hand to explain how farm productivity is increased. The important ingredients are rather obvious; the factors to stress are evident. What is less obvious is how to overcome the political, cultural, and intellectual restraints which prevent nations from enacting appropriate policies and boosting agricultural productivity. These restraints, for example, include price relationships which discourage the substitution of capital for land and labor and marketing facilities which cause seasonal gluts, depressed prices, and uncertainty for farmers with primitive storage facilities. They discourage the supplying of new

---

6. For these conditions, see Earl O. Heady. "Optimal Sizes of Farms under Varying Tenure Forms, Including Renting, Ownership, State and Collective Structures." *American Journal of Agricultural Economics* 53(1):17–25. Earl O. Heady. *Economics of Agricultural Production and Resource Use.* New York: Prentice Hall, 1952, Ch. 17; Earl O. Heady and John L. Dillon. *Agricultural Production Functions.* Ames: Iowa State University Press, 1961, pp. 57–59.

knowledge, the improving of lease arrangements, and the improvement of the general environment within which agriculture functions. They prevent credit from getting into the hands of farmers who are short on assets and in low income positions.

Economic theory as well as practical experience indicates what factors must be applied in agriculture to boost productivity. Ample evidence exists to indicate that the same principles apply regardless of whether farmers are highly educated or not. Of course, in practice we must recognize that farmers lag in reacting to improved conditions—for example, to increased supplies and lower prices of inputs, to higher farm commodity prices, and to increased knowledge. It is too much to expect farmers to react instantaneously to changes in the economic variables or parameters or to simply increase the food output because officials wish them to. There are no examples of an overnight transformation of agriculture, although recent rapid advances for the technological package surrounding new wheat varieties are in this direction.

Even in the United States, sometimes taken as the hallmark of agricultural development, adjustment to improving production conditions has lagged. As mentioned previously, it took 25 years for agriculture to become mechanized after fairly efficient tractors and tractor-drawn farm equipment were developed. It took nearly 50 years after the creation of public research and educational facilities and a moderate increase in the supply of knowledge for U.S. agriculture to become highly oriented to science. It will be another 20 years before existing knowledge is rather fully exploited. All countries have some conditions which cause some farmers to react rather slowly to change. Most typically, lagged reaction in general is due to small demand outside agriculture for some productive factors used in farming; the supply of such factors used in agriculture is relatively fixed or highly inelastic. Thus such factors remain in farming use rather than being replaced by new capital technology or migrating to other economic sectors. New technology replaces them only when the supply conditions change or their productive life and services are depleted.

No new scientific breakthrough is required to explain the conditions which cause farmers to use more and different resources and to increase farm productivity. The knowledge is already at hand; there are many practical examples of success. If enough farmers are encouraged to react, productivity in a given country or area can be increased. Increasing the productivity of agriculture means using more capital; it means substituting one form of capital for another or for land or labor; it means increasing total farm output. These changes are encouraged only if certain conditions exist: the prices of productive resources and farm products, and the farmer's tenure situation must be favorable; and the farmer must be promised sufficient payoff from a long-run investment in his enterprise.

If support prices for U.S. farmers are increased, more resources are put into agriculture and output increases. If knowledge and satisfactory price relationships are provided for Japanese farmers, more fertilizer is used and more money is invested in tractors. Provide a favorable price outlook and Greek cultivators convert from cereals to long-term investments in citrus groves. Provide an adequate investment horizon and degree of certainty and private Polish farmers invest in orchards and buildings in the midst of a socialized economy. Provide packing facilities with a market and Ethiopians sell off cows which they have long hoarded. Provide a supply of resources and adequate price incentives and selected villages of Indian cultivators move toward Japanese-type farming technology or toward rapid adoption of Mexican wheat varieties. At every point over the world where sufficient data are available, it has been found that farmers respond to changing farm product prices and farm input prices. The mysteries of agricultural development are small indeed. More mysterious and complex are the "outside" policy, planning, political, and cultural processes which prevent the changes needed to increase farm productivity.

NORTH AMERICAN RECIPE

If one wanted to find the most efficient plan for increasing agricultural productivity, he would look to the United States and Canada. However, the way North American farms are organized now is in part determined by the stage of development and the surrounding economy. Thus North American farms cannot be used entirely as a model for agriculture in other countries. These highly industrialized economies provide productive new capital inputs at prices which favor their substitution for both labor and land. Hence, mechanical technology is favored over labor technology, and economies of scale encourage the operation of fewer but larger farms. Public programs have been highly consistent with economic theory which specifies what is needed to boost farm productivity. Neither the United States nor Canada is noted for central planning. Yet the long-term public programs for American agriculture have been the most consistent and successful in the entire world, including socialist countries where the crux of life is government planning. The planning to boost productivity and modernize agriculture, especially in the United States, was often unwitting; the public did not always know that the instruments used were highly adapted to agricultural progress. Several instruments conducive to a flow of resources into agriculture and a greater output of farm products were emphasized.

First, in the United States, for example, a large supply of productive resources was provided for farmers and resource prices were kept low. More land was acquired and given to farmers or sold at low prices.

Farmers responded to this incentive by applying more and more labor and land to farming. When there was no additional land to be farmed, the nation turned in other directions to increase the supplies of farm inputs and to reduce the price of these inputs. Through public research and educational facilities more technical knowledge was generated and put into use. This complemented the new capital inputs, thus making possible more productive capital technology. Public facilities were created which increased the supply and lowered the price of capital and credit and encouraged much greater use of those resources. Bureau of Reclamation and Soil Conservation Service programs reduced the cost of irrigation, fertilizer, lime, and similar specific capital items. The lower prices encouraged a spurt in the use of these inputs or technologies and helped increase farm output. Other government programs raised and stabilized farm commodity prices and thus encouraged farmers to use more resources and new capital technologies. In addition, farm tenure systems, though not ideal, stabilized cost and return relationships, promising farmers a profit. Of course, farm productivity did not climb at the same rate throughout the United States in the last half-century. The increases generally varied in proportion to incentives provided through resource and capital prices, available knowledge, tenure restraints, relation of farm costs and returns, and farm commodity prices.

So a recipe for increasing farm productivity is available, if it is still sought. Lower the prices and increase the availability of productive resources, including new techniques and their knowledge. Increase and stabilize farm commodity prices. Blend in a farming system that takes into consideration the marginal productivity and prices of inputs, as well as farm commodity prices, in determining what bundles of inputs to use. Supply a tenure system which retains the productivity of resources as expressed through biological processes and market relationships. Extend capital availability at reasonable economic prices to all strata of farmers. This mixture can be brought to a boil in a container of commercial farming, and not successfully in a purely subsistence environment outside the market economy.

Here we need a word of caution. The recipe cannot always be completed immediately. Delay will depend on how many of the ingredients are used, also on the extent to which a very few specific cultural factors exist. For example, a new state of mind must be created; cultivators who previously were oriented to producing to subsist in the year ahead must be induced to produce for the market. Families must be acquainted with the principles of managing credit once it has been put into their hands. This recipe has been tested and proved successful over many parts of the world—so much so that it is doubtful that anyone will ever come up with a better one. Hence, to achieve agricultural development, the urgency is to create the conditions implied above. There is no mystery relative to what the process induces.

Agriculture has failed to respond as hoped or expected in many countries. But the reasons are often obvious. Too frequently in the past, priority was not given to agricultural development. The leap to a modern industrial economy, including steel mills and international airlines operated at deficits, was too frequently given precedence over farm improvement. Just as frequently, prices in underdeveloped countries have not encouraged the use of new and more capital resources such as fertilizer, insecticides, and improved seed varieties. Input prices have been kept too high and output prices have been kept too low. Capital has not been put into the hands of subsistence farmers so they can produce for the market economy. Sometimes taxes have been placed on inputs or exports, as a negative substitute for other fiscal policies.

In planning industrial development, too little early attention has been given to inputs such as fertilizer for agriculture; the importance of providing such inputs in the quantities and at the prices which favor their use in agriculture has been overlooked. In numerous countries the price of fertilizer is still high relative to crops. Frequently, countries have neither a supply of such inputs nor the facilities to move and store them. Moreover, trade policies of underdeveloped countries often have rejected the idea of importing such needed inputs at low prices; instead, they have favored development of their own industries and used their foreign exchange for that purpose.

For a number of reasons farm product prices also have been kept too low. The emphasis frequently has been on low prices for the consumer. While this policy may be needed, a better policy would be producer prices which favor growth in output but subsidize consumers at lower prices—somewhat along the lines of recent British price policy. In a few cases, export taxes serve as a major source of government revenue and deprive farmers of the portion of the world market price they would otherwise receive. Ethiopia, where the bulk of exports is represented by farm commodities, is an example. In Saudi Arabia, subsidies on imports for consumer benefit have driven down domestic farm prices and dampened development. In some countries, the acceptance of foreign loans or aid funds tied to the farm commodity imports has acted in a manner to dampen prices and lower the payoff for improving the domestic agriculture. Whether or not these aid programs are detrimental depends on how well imported supplies are insulated from the market of domestic producers.

The argument is not against adequate and cheap food for consumers. Rather it is against low farm prices and low incentive for farmers. It is not against industrialization; it is against an inadequte supply to farmers of industrial inputs which represent modern technology and a high payoff in terms of farm output. It is against systems that fail to generate and supply knowledge and provide productive inputs to farmers. In backward agricultures, the inputs necessary in the mix typically are complementary. It is rather futile to establish an extension education machine when there is no new or adapted research knowledge to go with it. Yet this has been done in many countries. It is unproductive to supply credit when fertilizer, insecticides, and improved seeds are not physically available, or to supply fertilizer when adapted crop varieties are lacking. It likewise is futile to supply credit which can get into the hands of only the richer farmers, while those who need it must go without and sacrifice potential benefits of development. It takes no new theories or mysterious explanations of the agricultural development process to know that at some level these resources complement each other, that when the supply of one is limited the productivity of others is adversely affected. Even for new technological knowledge, its supply certainly will not be augmented if investment in public facilities for research is too small; political institutions prevent effective, long-run investigations; and well-trained manpower is not kept on hand to produce it. United States agricultural policy causes higher commodity prices and output at home and a tendency toward lower commodity prices and output abroad. A shift to the export of subsidized or low-cost resources would tend to discourage production in the United States and encourage it in other countries. In general, less developed countries need to turn more in the direction of the input sectors and knowledge supplies which are so highly advanced and low in cost in Western countries. Over the next decade this certainly needs to have priority over home development of industry for numerous countries. There is, of course, a relation between the development of nonfarm sectors which fabricate farm inputs at low prices and the economic progress of agriculture. The price to farmers of these inputs can be low only if the agriculture of the country provides a sufficiently large market for capital items. At the outset, in many countries, the small market for capital inputs can be supplied initially at lower prices from foreign sources. Development of local industrial sources on a sufficient scale can come at a later time when agriculture has advanced in technology and capitalization to merit such local industrial development.

PRIORITIES IN THE FARMING SECTOR AND DISTRIBUTIVE EFFECTS

There are other priorities in modernizing a country's agriculture. Situations can be outlined which help specify when agriculture should be given priority over industry in development and vice versa.

Case I—Farm output is low; diets are miserable; hunger prevails. Both the agricultural and industrial sectors are characterized by labor unemployment or underemployment, and export possibilities are unfavorable for farm products. At this stage, priority needs to be given to improving agriculture, but not all facets of agriculture. Crop and related biological innovations should be given precedence. The capital items required are improved seeds, fertilizer, insecticides—and irrigation, where it is not costly and has a high short-run payoff. These capital inputs serve as substitutes for land. Emphasis should not be given mechanization and labor substitutes. Investment might be made in boosting production of staples, the demand for which is about constant regardless of consumer income. Little priority should be given to investment in producing livestock or other products in which consumption varies considerably according to people's income level and consumption is chiefly by high income people. Many nations fit this category, particularly those which face another decade or two of increases in farm labor and population because industrialization cannot proceed rapidly enough to keep pace with the birthrate. More countries would fall in this category if the world food crisis ever moved to the intensity now being projected by numerous people.

In general, these types of innovation are labor neutral or labor using. They also can be scale neutral and their use and income benefit from them need not depend on farm size. The higher yields from new varieties, fertilizers, plant protection, and water use and management can require more labor in agricultures of low mechanization. In fact, there is considerable indication of a spreading of benefits to the unlanded workers of northern India and Pakistan where seasonal wages increased considerably with larger harvests. Engineering innovations which largely speed harvest and land preparation for increased cropping intensity or multiple cropping also can have main effects of boosting output and augmenting the demand for labor in agriculture.

Case II—Food supplies are adequate for the basic items of diet. Agricultural labor is highly underemployed. Labor even may be migrating to other countries. Here, no priority should be given to any aspect of agricultural development, except as the returns from investing more in farming equals that to be obtained from investing in nonfarm sectors —and where export potentials exist. Otherwise, priority should be given to industrial development to create jobs for the unemployed on farms and in towns. Examples of this category include regions such as southern Italy. With high priority on farm innovations, leading farmers who innovate rapidly can gain even though domestic prices decline. However, small farmers who have few resources for improvement are faced with reduced income and no nonfarm employment opportunities.

Case III—Food supplies are adequate, a high level of employment exists in nonagricultural sectors, incomes are relatively low in agriculture, and underemployment abounds in agriculture. Industrial development should be further emphasized to expand nonfarm employment opportunities for those migrating from farms. However, some special emphasis could be given even here to the development of agriculture. As capital becomes relatively cheaper than labor, it would substitute for labor. Production of livestock and other products for which demand increases as consumer incomes rise should be encouraged. Too, a most important investment may be in helping human resources move out of agriculture. There are many examples of countries which fit into this category in various degrees. Even included here are broad reaches of American farming. Also included are Southern Europe, Eastern Europe, and a majority of countries in Western Europe.

Other categories could be presented. But enough have been suggested to indicate that there is no universal rule nor are there any conditions which can specify all priorities for industrial development over agriculture or vice versa. In a very few cases, no priorities should exist for agriculture. In others, the urgency is to employ resources and exports which stimulate the output of food from basic plant food sources or export crops. At other stages of development, the urgency steps up to livestock and mechanization. The important thing is to "get at" the development of agriculture. There are few, if any, good examples where a nation has invested too much in agricultural development relative to industrialization.

True, there have been large mistakes in agricultural investment. Examples are extension services without research knowledge to communicate, fertilizer plants without distribution and storage facilities, machines without spare parts, infeasible irrigation projects, and mechanization in countries where relative prices of capital and labor specify a labor technology to be most appropriate. But overinvestment in sensible agricultural development as a whole is hard to find. True, what is needed is balanced development. If we have sufficient knowledge of the production and supply possibilities of the major economic sector, of the resource demand situation peculiar to each industry, of consumer demand and what people in society desire, we can specify just what is needed for balanced development. But in the absence of this information the possibility of error certainly is in the direction of investing too little in agriculture in developing countries. Perhaps the United States provides the single clear-cut example of overinvestment drawing forth too productive an agriculture. Other countries, in a food supply pinch, would prefer this error to their own error of underinvestment in agriculture.

PRIORITIES IN OTHER NATIONS

In the less developed countries emphasis should be on applied research, except where fundamental knowledge is lacking for conditions unique to the country. Fundamental knowledge is the least-cost import and requires only a good set of journals and translators. On the other hand, good applied research, which adapts modern technologies to the conditions of the country, typically is lacking. Often developing countries overemphasize fundamental research and underemphasize applied research because of the preferences of the returning graduate student. After completing his study in the United States, the returning graduate student finds greater status in speaking to his colleagues abroad through the scientific journals than in developing applied technologies for his country.

We have already mentioned that distorted investment occurs when elaborate extension or advisory services are established in the absence of knowledge to extend. Again, balance is required in investing in these two activities. When a country's agriculture is backward, the government must give priority to research. This is true because the major inputs of agriculture are land and labor. Capital represents too small a proportion of the total to provide an adequate market for capital inputs. Hence, agricultural private concerns do little research. When agriculture is highly developed, the major inputs are capital items from nonfarm sources. Thus industry turns heavily to agricultural research as a means of further expanding the demand for new farm technologies. These new technologies are chemicals in the form of fertilizers and seeds, steel in the form of machines, etc. At this stage, the public need not place such high priority on agricultural research, since the momentum will be carried forward by the private sector.

At low levels of development, priority should be on crop biology. Emphasis of farm engineering needs to come at a later stage of economic development, when the relative prices of labor and capital encourage mechanization. Similarly, high priority should be given to improving the livestock sector through research only at higher stages of development and per capita income.

FARM SCALE

Increasing the size of farming units also should be directly related to economic growth. Massive units such as the collective farms of Eastern Europe are consistent neither with the stage of economic development in these countries, the economies of scale which can be realized, nor the relative prices of productive inputs. Large-scale units should have no particular priority at low stages of economic development. Certain economies may possibly be realized where there are fewer and larger farms to which knowledge is communicated. Aside from this, however, the most profitable scale of operations is directly related to the stage of economic growth in the country and to the relative prices of inputs. At low stages of development, when capital is in short supply and has a high price relative to labor, the best resource mix in farming is a large amount of labor and small amount of capital. Under a labor technology, costs cannot be cut by increasing the size of the farm. Most of the cost economies from using modest capital items are largely exhausted as soon as the bullock team, horse, or camel which provides the power are fully employed. High stages of development, where capital prices are relatively lower than labor prices, call for a resource mix made up largely of capital. Economies can then be realized by operating larger farms. As mentioned previously, there is no basic reason for emphasis on mechanization in developing agricultures. From the standpoint of both capacity and food supply, the sustained emphasis needs to be on innovations and practices that are biological in nature, largely serve as land substitutes, and are scale-neutral.

OTHER ASPECTS OF PRIORITY

In establishing priorities for the development of agriculture, several other factors must be considered. If we wished most rapid development and could neglect human welfare, we would concentrate efforts on commercial farmers and forget subsistence farmers. In some countries, of course, the majority of farmers fall in the latter category. Their scale of operations must be extended so they produce beyond family requirements. Their product must enter the market so that they are influenced by prices and related stimuli. A generation may be required before many traditional subsistence farmers are converted to the "market state of mind." But their sons may react much more rapidly.

We are moving into a stage of development when the problems of equity are urgent. Perhaps some countries were nearly correct in their early emphasis on commercial farmers, to the neglect of small and low income cultivators, in getting the initial stages of speeded development under way. Now, however, the world in general requests equity in the speed of developmental benefits. It makes little sense in human welfare to increase food production and population in tandem, so that in 30 years there are only more people under misery at the same old subsistence level of many countries. More people with the same or more misery cannot be proved as progress or an advance in aggregative consumer welfare. Neither can welfare or community utility advance be guaranteed in programs which only aid the richer farmers to gain more in supplying a greater quantity of food at a lower real price to the richer consumers of the country. Left side by side are the smaller cultivators who can only produce the same amount at a lower price because they lack the capital for the new seeds and fertilizer; or because they must sell during a larger seasonal supply and depressed prices and, because of lack of storage, buy back for their own food supplies at a later time when prices have recovered. The distribution of benefits has not been entirely equitable under the green revolution of countries such as India and Pakistan. The large gain goes to those farmers who have the funds to buy the high-yielding seeds and the fertilizer and insecticides to attain higher interactions in productivity gains. These inputs are further augmented for those farmers who can sink a tube well and get an adequate supply of water in the dry period. But to make a tube well, or even use of a Persian wheel, profitable enough one must have a large enough land area. Too, he must own his land or have sufficient certainty of tenure. Small or landless farm families fall outside this realm of gain from technological revolutions in developing countries. If the emphasis turned to include general mechanization, their loss would be even greater as economies of scale cause land to be bid away from them and as machines are substituted for the labor they perform on larger neighboring farms.

Similarly, the supply of consumer goods which trickles out to remote villages often needs to be materially increased—and the price of such goods reduced. If such goods are lacking or too expensive, the villagers' incentive to sell their farm commodities on the market is reduced. Typically, modern consumer goods are much more expensive in the villages than in the cities. Still the impetus of these consumer goods supplies is illustrated by the transistor radios, bicycles, and similar items which show up even in remote places of India, Thailand, Peru, and other countries.

If increasing the food supply were to take precedence over all else, governments could operate their own commercial farms, which would serve as an example for the rank and file of cultivators. Such farms could help "sell" improved farming practices, for persons with little education tend to act on the basis of quantitative evidence rather than on the basis of deductive or theoretical evidence. Similarly, the government might "hire" a large number of representative farmers to follow prescribed farm plans and adopt improved technology. If the development posed really has a payoff, it should permit cultivators to increase their income as well as hire a managerial supervisor to lead a group of farms. If the development suggested will not cover these two costs, undoubtedly (a) it has too low a payoff to be considered in a developing country or (b) inputs are priced too high and output too low. Finally, for a rapid spurt in output, the possibility of giving franchises to experienced foreign farmers could be considered where land is not too limited. A franchise of perhaps 5 or 10 years would assure these foreign farmers sufficient payoff under an appropriate set of prices. Thus the food supply could be increased while farming know-how was being spread to native cultivators. A good example of the latter possibility is in Ethiopia, where Dutch farmers have developed highly efficient sugar operations, or in the farming franchises being proposed in Saudi Arabia. While the technology used may not be adapted to large numbers of the country's farmers, the operations have been successful in rapidly increasing the output of one commodity. Other countries have sufficient land to justify borrowing this technique. A short-run franchise could prevent concern over the threat of colonialism.

Yet important equity considerations are involved in these steps to speed development. Where they might be enacted, or where a very large number of low income families already prevail, a large investment in public works might be initiated to provide an income flow to these families. Perhaps little can be done toward true progress in equity and improving welfare of the masses in countries such as India without a large-scale public works program for housing, transport, market facilities, elementary schools, etc.

INTERCOUNTRY DIFFERENCES

Food production has moved ahead much more rapidly than population and food demand in highly developed countries. Accordingly, the import-export pattern in food has been reversed between highly developed and underdeveloped countries. Prior to World War II, there was one general trade pattern: Western Europe was the only importing region, and the rest of the world exported to it. There were six grain-exporting areas: North America, about 5 million tons; Latin America, about 9 million tons; Eastern Europe, 5 million tons; and each of three other regions—Asia, Africa, and Oceania (New Zealand and Australia)—small quantities. Western Europe has maintained its position as an importer. To meet greater population needs it buys about the same amount of grain as in the immediate prewar period. But mammoth changes have taken place in trade among other world regions. Only North America and Oceania remain major exporters. Asia and Africa have become net importers along with Latin America and Eastern Europe. Prior to World War II, the net annual flow of grains from the less developed regions was about 11 million tons. Now, annual shipments of grain from developed countries to less developed countries are about 25 million tons.

In the advanced countries, knowledge and technology have been able to hold birthrates in check and to increase food output faster than population growth. Moreover, investments in capital processes and technical knowledge may give rise to large food supplies from nonagricultural resources before limited farm resources restrain output and boost the real price of food in these countries. The pressing short-run problem, a span of the next three decades, is in the less developed countries. Most but not all of these countries are only recently independent nations able to determine their own national policies. The balancing of the demand and supply of food is one of the major problems that most of them must solve in the next decade or so. Any threatening world food crisis, with population growth rates thrust sharply above food growth rates, is perhaps three decades away. But for individual countries in the above category, it could be only a decade or so away. Balancing food needs and production involves investment both in modern farming and in population management.

Actually, the problem is not a simple one of balance. Food output and consumption will be balanced in three decades even if it means twice as many people subsisting on a miserable 2,000 calories per day. The basic problem is more to manage food supplies and populations so they are balanced at levels allowing adequate diets and human welfare. Investments of both types are required. Investment to increase the knowledge and improve the technology of birth control is no less important, and certainly pays a much higher return in the long run than investment in expanding the food supply.

How does agriculture fit into this complex in highly developed countries? Cannot the abundance of food and the potential of greater output in these countries be channeled to the food-deficient countries, thus warding off the crisis and even helping to lift the level of human well-being? This would be a simple solution—if it were possible. And it would satisfy the sincere humanitarian interests and intentions of many individuals, groups, organizations, and nations. But it is unrealistic as the major answer to the world's pending food crisis.

To be certain, the agricultural resources of the United States and other developed nations can have an important and significant role. But it is not in providing the increased food needed for an uncontrolled increase in the world's population over the next half-century. It is obvious that world population cannot go forward forever unchecked. Present rates of increase would eventually absorb all the untapped food-producing potential of both developed and undeveloped countries. Then, when the world could boost food production no more, there would be even greater masses of people to starve or live in hunger and misery. Human suffering would be multiplied and the negative effect could well be greater than if excess food stocks were withheld as a check against population growth. Ethical questions even arise as to whether societies should provide more and improved health and medical services to decrease mortality rates without parallel investments to increase the food supply for the greater number of persons who are thus kept on hand to consume.

Blind increases in production in the advanced countries, to be converted to food aid for less developed nations, will not solve the world's population problem. To an extent, they can even discourage improvement of local agriculture and growth of food supplies in less developed countries. Availability of such food can lessen the urgency and lower the motivation to increase productivity in food-short, developing countries. Perhaps food aid in the last decade and a few initial green revolution successes have diverted too much attention and investment from the more fundamental long-run problems of birth control and population management in less developed countries.

To produce food in advanced countries and send it as gifts or deferred loans is not the long-run answer to the world population and food problems. This approach only postpones the day of reckoning. Apart from political considerations, investment for purposes of world food supply should be made to increase agricultural productivity where it is most economic and where it returns the greatest payoff. In the short run, the payoff often will be greater in countries such as the United States with a highly developed agriculture and an underutilized capacity. This is true because of the educational and organizational restraints involved in short-term improvement of agriculture in most less developed countries. But over the long run, the payoff is almost certain to be greater in improvement in agriculture in developing countries which are endowed with favorable resources but tardy in technological development. These countries are using resources or inputs at low levels; thus the production response from applying additional units of an input should be much greater than in developed nations where resource combinations are more nearly ideal. Of course, some developed nations have a clear long-run comparative advantage in food production, and in some less developed countries industry has clear advantages over agriculture. In such countries further developments should follow these lines, with trade stimulated by appropriate international, commercial, and fiscal policies.

OTHER RESOURCES, GROUPS, AND INSTITUTIONS

Our discussion thus far has been in terms of commercial farmers who have access to resources. Under favorable economic policies which can readily be provided by governments, they obviously do respond if resource supplies and prices, commodity prices, and tenure and credit conditions are favorable. The green revolution can be extended a considerable distance before the potentials of these food suppliers are exhausted. But once they have been, the subsequent steps in development are not so easy.

The most rapid advances in new wheat varieties have been in areas where water is available and farmers have the resources to use it. The majority of the world and its farmers do not have access to supplemental or main-line irrigation water. Whole regions, the more arid regions of India and the Middle East for example, can gain little from higher-yielding varieties and fertilizer because moisture is too limited. In some cases, water can be made available at reasonable public cost. But in other cases, advantage of modern technologies cannot be realized without mammoth public irrigation developments. The cost/benefit ratio of many of these would be less than that realized through investment in appropriate industries and trade.

The proportion of farmers now participating in the green revolution and its benefits is much smaller than the proportion left out of its reaches because of small units, lack of capital, subsistence orientation, and related causes. To move these farmers into rapid innovation and supplying of markets is not easy. Hence, there are complex and difficult problems involved in extending the green revolution on an ever-growing basis, and the second stage will certainly be more difficult than the first.[7] Complexities also will continue to arise through market and third-country effects. Since only about 5 percent of world rice production moves into the world market, only slight increases in domestic production thrown into world trade for balance-of-payments reasons can have devastating effects, such as those already falling on Thailand as a historic rice exporter.

Further extension of the green revolution will run into growing problems of capacities in transport, milling, processing, and storage facilities. It will be faced with more commodities which are produced in surplus while advance in others is little because of historically molded and inflexible price support structures. We cite these and other complexities of sustained advances in agricultural productivity to indicate that the path upward is also attended by the so-termed second- and third-generation problems of agricultural development. But again these, as with attaining response from commercial farmers at the outset, are problems which can be solved with appropriate policies. Again the mystery is not whether or how they can be solved, but why policies might not be initiated soon enough to minimize their impact. In the total stretch of policies involved, including not only those of response and equity in the initial stage of development but also those for population and the later generations of agricultural transformation, the need is for the development of planning—administrative and institutions generally, which can foresee the problem and create a program for its solution before negative outcomes give rise to crises.

POPULATION EMPHASIS

Investments in agricultural development and institutions cannot ward off long-run world food crises stemming from population growth. Only investments in knowledge and the technology of birth control can do so. It is the only long-run solution to the problem of population versus food supply. Moreover, compared with agricultural development,

---

7. We cite the equity and other second- and third-generation problems of agriculture to indicate that they are important and will need policy emphasis if successful momentum in food supplies is to be attained and maintained. Otherwise, our purpose in this volume is in quite a different direction. Several recent studies detail the upcoming problems of employment, marketing facilities, and related phenomena and include: Walter P. Falcon. "The Green Revolution: Generations of Problems." *American Journal of Agricultural Economics* 52 (5): 698–710; Bruce Johnson. "Consequences of Rapid Population Growth: Unemployment and Underemployment." *Proceedings, Conference on Technological Change and Population Growth*, California Institute of Technology. Pasadena, 1970; V. M. Dandekar. "Overpopulation and the World and the 'Asian Drama.'" *Ceres* 2 (1969): 52–55.

such an investment will return much more in bringing population and future food requirements into a realistic and humanitarian balance. But it, too, is an investment which does not have immediate payoff. Effective population control programs require considerable time to be effective. They will provide the appropriate payoff only with sufficient time and effort to bring knowledge to less literate parents, to overcome fears and superstitions, and to provide birth control methods that are certain and cheap.

The projections of this study and recent experience in speeding the technological tempo in selected countries indicate the ability of the world to increase food production while improving somewhat the level of diets for a larger world population. The thrust toward a greater population already prevails in the large majority of the world's population now under 25 years of age and the slow progress being made in population planning and policies.

However, 30 years is a considerable span of time and could allow developing countries to initiate and implement effective population policies. We have emphasized that the mystery on the food supply side is not that of the necessary variables and policy to expand output; the mystery more nearly is one of why these policies and changes are not put into effect. This dilemma applies equally or more so for population and its associated effect on food demand. To be certain, more effective birth control technologies need to be developed. They must be simple to use, relatively effective, be of very low cost, and overcome cultural restraints. Still, technological know-how already exists if there were sincere interest and concern in attaining sufficiently reduced population growth rates in less developed countries. Pills, contraceptives, sterilization, and abortion have been applied effectively either in entire countries or in particular income groups of them to indicate that success is attainable. With these devices, technological means for restraining population growth prevails. To be certain, the failure to control it over the next 30 years will not be due to the absence of technology. The technology for reducing population growth rates is more nearly available than the technology to guarantee food supplies large enough to meet unchecked birthrates over the next half-century. The process can be aided by improved technologies and work must go forward on the side effects of existing and prospective birth-reduction devices. Yet the side effects to whole populations in not attaining a wide use of existing technologies promises to be more important than the side effects in using them. The potential number of persons doomed for subsistence diets, misery, and even death from inadequate food supplies over the long run poses a more serious societal side effect for many countries than do the side effects of existing control devices.

Hence, while our results suggest that the world is not faced with calamity in the next 30 years, this will be true only if the politicians and public administrators of selected developing countries do enact agricultural development and trade policies which hurry to guarantee adequate food supplies. Over the longer run, however, praise or blame for these same politicians and administrators will rest on their actions in initiating and implementing appropriate population policies. The lack of adequate birth control technologies is not a sufficient excuse for nonattainment. Needed immediately and on a much more intensive basis are much larger and more effective communication or propaganda programs to bring sufficient awareness to all of the population; larger public investments to provide the staffs, personnel, and administrative facilities to accomplish the task; effective economic incentives either in the cost of the techniques or in the return for their application; and actual sincerity and concern for future generations to stir the present generation of public officials into action. Of course, the ultimate goal is economic growth and per capita incomes at levels which cause families to exert their own initiative. Perhaps one threshold level is attained when the level of affluence of children causes them to draw on family income more heavily as consumers than they contribute to it as resources. But the world can hardly wait for this threshold level to be attained in all countries. The politicians and officials of these countries must speed the attainment of this threshold level by prior and effective public population policies. Whether the citizens of their countries live in misery and at levels of food subsistence in a half-century will depend on the actions they take in the next two decades. Leaders of developed countries can give encouragement and some financial assistance, but success or failure depends on the leaders and citizenry of developing countries during this and the next decade. The technologies for population management are already at hand, even if they can be improved upon in subsequent years. That they can be effectively used is illustrated in many countries and population strata. These statements refer not especially to the next three decades, given the results of our projections and the initial evidence of the green revolution, but to the generations of people that live thereafter. Perhaps the statues of current world leaders, politicians, and administrators should not be built by this generation of humanity but by those generations which prevail under happiness or misery, depending on decisions and actions of this generation, at the end of the next half-century. Man has imagination to tackle complex problems and provide their solutions. The world awaits exercise of this imagination with respect to future population size and welfare.

TRADE NEEDS AND POSSIBILITIES

Even with full success on both the food supply and population control sides by the majority of developing nations, we hardly expect that each would attempt self-sufficiency in farm commodities. Given the extent and distribution of natural resources over the world, great differentials in comparative advantage prevail among agricultural commodities and between food and industrial products. While there are some useful national reasons for placing a degree of restraint on imports and exports, the large possibility for the world to be well fed is through trade. Hence, the remainder of this volume is devoted to application of a trade model under certain food supply and demand projections presented previously.

The model developed and applied is a rather modest one, but important is the fact that it is the only one which has been applied thus far to food and fertilizer trade. With the foundations which it provides, improved models can be developed and applied in greater and more sophisticated detail. A start has now been made.

PART THREE

Model of Trade Patterns

Framework and Model of International Trade in Cereals, Fertilizer, and Phosphate Rock

PREVIOUS CHAPTERS of this book dealt with the prospective world food situation and the projection of production potential and demand or requirements under numerous assumptions of land availability, population growth, economic development, and per capita income growth. The possibilities suggested in comparison of certain of these alternatives can, of course, be attained only under increased trade and reduction to barriers in trade. Hence, Part 3 of this book deals with potential trade patterns, including indication of the origin and destination, respectively, of exports and imports of cereals, including rice, in future time periods. Cereals were selected for consideration in this section since they tend to reflect the major international dimension of food trade and are indicative of the evolving dimension of the world food situation. Since attainment of food production possibilities may be realized only with appropriate inputs, the analysis has been extended to *include* fertilizers and phosphate rock. We use different assumptions of short-run and long-run fertilizer production to determine how various levels of domestic fertilizer production may affect imports and exports for the 96 countries included in the analysis of previous chapters. Also, in a more summary and less detailed analysis, Mainland China is added to the countries to determine how its entry into the world food markets might affect total trade and the pattern of imports and exports among the various countries and regions of the world.

This chapter presents the framework and models within which the analysis of potential future trade was made. Subsequent chapters provide the data used in the analysis, and the results and an interpretation of them in terms of trade potentials and land use in the various countries included in the study. Only certain of the levels of population and per capita income growth included in the previous production and demand projections are used in the trade analysis. It is realized, of course, that ideal data are not at hand to determine how the conditions and patterns of world trade in food and fertilizers can be best organized in solving the world food problem. Many of the restraints to this process are of economic and political forms which are not readily removed. We

257

do not deal with these facets of the problem, but only with trade, production, and land use potentials if trade restraints are largely removed.[1] With the data available it is not possible to prescribe exactly how future world food problems might best be solved simultaneously with an optimal allocation, through trade, of food and fertilizer resources. However, with the production-consumption data generated in this study and certain assumptions about the conditions of trade, it is possible to evaluate the level and cost of regional imports of selected commodities required under various competitive positions of major export countries in relation to trade possibilities and the quantitative distribution of the resulting exports. This information has potential importance in appraisal, within the limitations of the data, of present agricultural policies of both exporting and importing countries: importers might thus evaluate the cost of quantities imported in relation to domestic alternatives in food supplies and development programs. Exporters could consider ways of improving their competitive position and/or reallocating resources.

NATURE AND OBJECTIVES OF TRADE ANALYSIS

One objective of the following analysis is to develop a model suitable for analyzing implications of projected quantities of (a) world trade in cereal grains, fertilizers, and phosphate rock and (b) regional use of existing and potential production capacity, given the prior determination of regional import requirements and potential exports for cereal commodities.[2] As a simultaneous objective the model has been applied in determining the levels and patterns of interregional trade in cereal grains, fertilizers, and phosphate rock in 1975, 1985, and 2000 to indicate the corresponding extent to which productive capacities for food and fertilizer might be used in future time periods.

We also estimate, for each of the commodities contained in the analysis, regional and world total import costs and fertilizer plant capacity investment expenditures to attain certain world food demand levels. Certain adaptations of the model are used in determining the implication of (a) the U.S. requirements that at least one-half of its total export sales under government-sponsored programs move in domestic flag vessels (with rates approximately double those of other ves-

---

1. It is assumed that exports are free to move in international trade at their 1963–65 average real export price and that the actual trade patterns are determined in accordance with minimization of export price plus transportation cost.

2. Cereals were divided into three categories: wheat and rye, rice, and other grains. Only the three major types of fertilizers—nitrogen, phosphate, and potash—were included in the analysis. Mainland China, North Vietnam, and North Korea, as well as some small countries with low population and production levels, were excluded because of inadequate data. For similar reasons, Mainland China was included in only one of the analyses, and even then, only in a summary fashion.

sels), (b) short-run fertilizer and phosphate rock plant capacity constraints of the world, and (c) alternate assumptions with respect to the proportion of nitrogen, phosphate, and potash in projected increases in world fertilizer use. Of course, an overall purpose of the analysis is to evaluate the implications of the results specified in the trade analysis relative to outlook and world food and agricultural development programs. In completing the analysis from data generated in previous chapters, and a general paucity of data types that would be preferred, it was necessary to estimate the following basic quantities: (a) the regional levels of fertilizer, phosphate rock, and sulfur use associated with the projected production of cereals, (b) per unit export supply price (f.o.b.) by exporting region for each potential export commodity included in the analysis, and (c) the cost of transporting each potential export commodity from the producing region to and within each importing region.

To keep the study manageable, only cereals have been included in this aspect of the study. Cereals consumed directly and indirectly form the major part of the world food diet and constitute nearly three-fourths of the world's harvested crop area. They provide well over half the direct supply of food energy. In addition, they provide a large part of the remainder indirectly in the form of livestock products. When measured in both tonnage and calories, grains heavily dominate world trade in foodstuffs. Cereal crop data also are more complete and of better quality than those for other crops.

For this second phase of the study, a linear programming model was used with the 96 nations grouped into producing-consuming regions. A set of one or more "representative" ocean ports was specified in each region for purposes of the trade analysis. Future regional requirements for nitrogen, phosphate, and potash fertilizers and for phosphate rock were estimated from the data on production potentials and future food requirements. This analysis also required estimation of fertilizer capacity in 1975 and regional supplies of potash and phosphate rock reserves. It also was necessary to estimate regional export prices, interregional ocean transportation costs, and intraregional rail transportation rates for each commodity included in the cereals analysis. Assumed accordingly for the analysis were (a) that the interregional and intraregional activities required to increase fertilizer plant capacity, to produce fertilizer commodities, and to produce and export all commodities could be specified as linear functions and (b) that conditions of "free" trade existed in the international markets for cereals, fertilizers, and fertilizer raw material.

MODEL USED

The model was constructed to minimize the cost of obtaining and, in the case of fertilizer, expanding the capacity to produce the commodities required to satisfy projected requirements for cereals, nitrogen, phosphate and potash fertilizers, and phosphate rock.[3] The constraints included in the model were (a) projected regional import requirements or export capacity for the cereal classes—wheat and rye, rice, and other grains; (b) associated regional nitrogen, phosphate, and potash fertilizer use levels; (c) potash and phosphate rock requirements associated with fertilizer production; (d) 1975 regional nitrogen, phosphate, and potash fertilizer and phosphate rock plant capacities; and (e) regional potash and phosphate rock reserves. The trade, interregional movements, or the imports for countries unable to meet their own requirements were measured as requirements in excess of production in the region.

To fulfill the study objectives subject to the specified constraints, the following activities were defined in the model, and activities were constructed to incorporate the following conditions: (a) simultaneous production and export of each cereal grain class from each region for which a regional excess is projected to each region with a corresponding production deficit, (b) wheat-for-rice substitution allowing production and export of wheat from regions with excess capacity to regions with rice deficits remaining after satisfaction of minimum rice requirements and utilization of all projected world rice capacity, (c) simultaneous production and export of nitrogen and phosphate fertilizer from each region with plant capacity in excess of domestic requirements to each region with capacity inadequate to meet their corresponding requirements in 1975, (d) simultaneous production and export of potash fertilizer and phosphate rock from regions with sufficient raw material reserves to each region with inadequate internal capacity in 1975 or inadequate potash reserves to support production in later years, (e) production of potash fertilizers and/or phosphate rock to meet domestic requirements in each region with reserves of potash and/or phosphate rock, (f) nitrogen and phosphate fertilizer production in each region up to the 1975 plant capacity and up to quantity demanded in each region for 1985 and 2000, (g) a fertilizer plant capacity increasing activity for production of nitrogen and phosphate fertilizer in each region, (h) phosphate rock and potash fertilizer plant capacity expansion in each region with the required reserves, and (i) land diversion for regions where land is not utilized in production for domestic consumption or exports (related to slack, or disposal variables permitting grain exports to be less than the potential quantity available).

The cereals considered available for export are those supplies in excess of domestic requirements. In subsequent discussion the term *excess requirements* will be used to denote food requirements exceeding domestic production potential of a region.

3. Potash fertilizer production was assumed possible only in countries with potash reserves since the shipment of the raw potash is impractical.

MATHEMATICAL NATURE OF THE MODEL

The objective function of the programming model is

$$\text{Min } Z = \sum_{i=1}^{38} \sum_{j=1}^{n_i} \sum_{k=1}^{38} \sum_{l=1}^{m_k} \sum_{g=0}^{7} C_{ijklg} X_{ijklg} + \sum_{k=1}^{38} \sum_{g=4}^{7} C_{ig}{}^D F_{ig}$$

$$+ \sum_{i=1}^{38} \sum_{g=4}^{7} C_{ig}{}^E E_{ig} \tag{12.1}$$

where variables have the following meaning:

Z: total cost of producing and transporting the cereals, fertilizers, and phosphate rock required to satisfy regional excess requirements, plus the investment cost required to produce the implied additional fertilizer and phosphate rock plant capacities;

C_{ijklg}: cost of producing the gth commodity in the ith region and transporting it from the jth port to the kth region through its lth port;

X_{ijklg}: quantity of commodity g produced in the ith region and exported from the jth port to the kth region through the lth port;

$C_{ig}{}^D$: per unit cost of producing commodity g in region i;

F_{ig}: quantity of commodity g produced in region i for domestic consumption;

$C_{ig}{}^E$: investment cost of adding one unit of capacity to produce commodity g in region i;

E_{ig}: total expansion of capacity to produce commodity g in region i.
The constraints used in the programming model are as follows:

$$\sum_{i=1}^{38} \sum_{j=1}^{n_i} a_{ijg} X_{ijklg} \geqq R_{klg} \quad \text{for } k \text{ and } l \text{ and for } g = 0,1,2,3 \tag{12.2}$$

$$\sum_{i=1}^{38} \sum_{j=1}^{n_i} \sum_{l=1}^{m_k} a_{ijg} X_{ijklg} + d_{kg} F_{kg} \geqq \sum_{l=1}^{m_k} R_{klg} \quad \text{for all } k \text{ and } l$$

and for $g = 4,5,6.7$ \hfill (12.3)

$$\sum_{k=1}^{38} \sum_{l=1}^{m_k} \bar{a}_{ij0}(X_{ijkl0} + X_{ijkl3}) \leqq A_{ij0} \quad \text{for all } i \text{ and for } j = 1,2,3, \ldots, n$$

\hfill (12.4)

$$\sum_{k=1}^{38} \sum_{l=1}^{m_k} \bar{a}_{ijg} X_{ijklg} \leqq A_{ijg} \quad \text{for all } i, \text{ for } j = 1,2,3, \ldots, n \text{ and for } g = 1,2$$

\hfill (12.5)

$$\sum_{j=1}^{n_i} \sum_{k=1}^{38} \sum_{l=1}^{m_k} \overline{a}_{ijg}X_{ijklg} + \overline{d}_{ig}F_{ig} - e_{ig}E_{ig} \leqq \sum_{j=1}^{n_i} A_{ijg} \text{ for all } i$$

and for $g = 4,5,6,7$ $\hspace{6cm}$ (12.6)

$$\overline{d}_{i6}F_{i6} \leqq \sum_{j=1}^{n_i} A_{ij9} \text{ for all } i \hspace{4cm} (12.7)$$

$$\overline{d}_{i7}F_{i7} \leqq \sum_{j=1}^{n_i} A_{ij8} \text{ for all } i \hspace{4cm} (12.8)$$

Definition of terms used and not previously explained are:

R_{klg}: excess requirement for commodity g associated with port l in consuming region k where $0 \leqq g \leqq 3$ and the total requirements for commodity g associated with port l in consuming region k where $3 < g \leqq 7$;

A_{ijg}: the projected surplus of commodity g associated with port j in producing region i where $0 \leqq g \leqq 2$, the total available capacity to produce commodity g associated with port j in region i where $4 \leqq g \leqq 7$, and the total estimated reserve of the commodity g associated with port j in region i where $8 \leqq g \leqq 9$;

a_{ijg}: the quantity of commodity g produced in region i and exported from port j to any specified region k and port l per unit of activity X_{ijklg};

$\overline{a}_{ijg}$: the quantity of commodity g consumed per unit of activity X_{ijklg};

d_{kg}: the quantity of commodity g produced for satisfaction of requirements per unit of activity F_{kg};

$\overline{d}_{ig}$: the quantity of domestic capacity required per unit of activity F_{ig};

e_{ig}: the capacity increase generated for each unit of activity E_{ig};

g: the subscript indicating the commodity involved. The commodity associated with each numeric value is 0—wheat and rye, 1—rice, 2—other grains, 3—wheat as a rice substitute, 4—nitrogen, 5—phosphate, 6—potash, 7—phosphate rock, 8—phosphate rock reserves, and 9—potash reserves;

m_k: number of ports in importing region k;

n_i: number of ports in exporting region i.

Conventional conditions to guarantee only feasible solutions are:

$X_{ijklg} \geqq 0$ $\hspace{0.3cm}$ (12.9) $\hspace{1.5cm}$ $F_{ig} \geqq 0$ $\hspace{0.3cm}$ (12.10) $\hspace{1.5cm}$ $E_{ig} \geqq 0$ $\hspace{0.3cm}$ (12.11)

for $i = 1,2, \ldots , 38, j = 1,2, \ldots , n_j, k = 1,2, \ldots , 38, l = 1,2, \ldots , m_k$, and $g = 0,1, \ldots , 7$.

REGIONAL GROUPING OF COUNTRIES

The 96 countries were grouped into world regions using three levels of aggregation. Countries were aggregated first into geographic regions according to common ocean ports selected to minimize differences in marine distance between a particular region and all other regions. Trade area aggregates then were specified in accordance with normal trade patterns. Finally a two-way developmental grouping was used; countries with average per capita income of $800 or less were classified as developing countries; those with higher per capita income levels were classed as developed.[4] The countries included in the study and their grouping according to the above criteria are indicated in Table 12.1.

The port or ports for receipt and dispatch of cargo in each of the 38 regions of Table 12.1 were specified. Multiple ports were indicated for Canada, the United States, India, and Australia, due to their spatial characteristics. The proportion of regional excess capacity and/or requirements, respectively, allocated to each port within a region was based on historic export or production patterns and the distribution of regional population. Figure 12.1 indicates the 38 regions delineated and the ports (by name) specified therein.

PROJECTED CEREAL GRAIN REQUIREMENTS

Projections of food production potentials and food consumption requirements are those from previous chapters.[5] For purposes of the analysis which follows a "most probable" consumption-requirements set was selected from the combinations possible. The set selected was that of medium population growth and low income growth as defined in previous chapters. One modification was made in the assumed income set. Where a country's low income projection indicated a declining per capita income for that country, a constant per capita trend was substituted. This modification guarantees at least maintenance of base period consumption levels for all commodities. Selection of population and income levels used in the programming analysis is based on the following considerations. The FAO population study suggests that the medium population projections are most probable for the future, based on trends, policies, and the possibilities now apparent. High income nations have low income elasticities of demand for staple foods such as cereals. Low income nations typically have not had high income growth

---

4. The classification criteria for the aggregates were modified in a few instances to prevent problems of cross-classification.

5. The 1975 projections not previously introduced are based on the methods employed for projections presented in early chapters (and were generated in the process of making the 1985 and 2000 estimates of previous chapters).

TABLE 12.1. Countries Included in the Study and Their Group Delineations

Country	Geographic Region and Identification Number[a]	Trade Area	Development Class[b]
United States	1—United States	United States	DD
Canada	2—Canada	Canada	DD
Mexico	3—Mexico		
Cuba Dominican Republic Haiti Jamaica	4—Caribbean		
British Honduras Costa Rica El Salvador Guatemala Honduras Nicaragua Panama	5—Central America		
Trinidad and Tobago Colombia Venezuela	6—Northern South America	Latin America	DG
Brazil	7—Brazil		
Bolivia Ecuador Peru	8—Western South America		
Argentina Uruguay Chile Paraguay	9—Southern South America		
Belgium and Luxembourg West Germany France Netherlands Italy	10—European Economic Community	European Economic Community	DD
Ireland United Kingdom	11—Ireland-United Kingdom		
Denmark Norway Sweden Finland	12—Scandinavia	Other West Europe	DD
Spain Portugal	13—Spain-Portugal		
Austria Switzerland	14—Austria-Switzerland		

[a] The region numbers indicated are used for regional identification in application of the model.
[b] DD represents developed and DG represents developing countries.

TABLE 12.1   (cont.)

Country	Geographic Region and Identification Number[a]	Trade Area	Development Class[b]
East Germany Poland Czechoslovakia	15—Northern East Europe		
Yugoslavia	16—Yugoslavia	East Europe	DD
Hungary Rumania Bulgaria	17—Other East Europe		
USSR	18—USSR	USSR	DD
Greece Turkey	19—Greece-Turkey		
United Arab Republic	20—United Arab Republic		
Iran Iraq	21—Iran-Iraq	Middle East	DG
Israel Jordan Lebanon Syria Cyprus	22—Other Middle East		
Morocco Algeria Tunisia Libya	23—Northern Africa		
Ghana Guinea Ivory Coast Liberia Nigeria Senegal Sierra Leone Togo	24—Western Africa	Africa	DG
Angola Cameroun Congo (Kinshasa), Rwanda and Burundi	25—West Central Africa		
Ethiopia Sudan	26—Ethiopia-Sudan		
Kenya Tanganyika Uganda Malagasy Republic Malawi, Rhodesia, and Zambia	27—East Central Africa		

TABLE 12.1   (cont.)

Country	Geographic Region and Identification Number[a]	Trade Area	Development Class[b]
Republic of South Africa	28—Republic of South Africa		
Australia	29—Australia	South Africa-Oceania	DD
New Zealand	30—New Zealand		
India Ceylon	31—India		
Pakistan	32—Pakistan	Pakistan-India	DG
Burma	33—Burma		
Cambodia Thailand South Vietnam	34—Other Far East		
South Korea	35—South Korea	Other East Asia	DG
Federation of Malaya Indonesia	36—Malaya-Indonesia		
Philippines China (Taiwan)	37—Philippines-China (T)		
Japan	38—Japan	Japan	DD

rates. The assumption that agricultural production and income growth are closely correlated has been used in other production projections for developing countries.[6] Deviations of income from trends are assumed to be partially offset by corresponding deviations in production. The "most probable" production projections used are those resulting under the low land bounds discussed in earlier chapters.

EXCESS COMMODITY REQUIREMENTS

Excess requirements for each class of commodity studied were developed directly as the difference between projected production and requirements. Aggregations of excess requirement were first computed by country, then aggregated into the regional groups discussed earlier.

The analysis assumes that wheat can be substituted (in the long

---

6. United Nations, Food and Agriculture Organization. *Agricultural Commodities: Projections to 1975 and 1985.* Rome, Italy, 1962, paragraph 88, page xxx; and M. E. Abel and A. S. Rojko. *World Food Situation Prospects for World Grain Production, Consumption and Trade.* U.S. Department of Agriculture, ERS-Foreign 35, 1967, Appendix A.

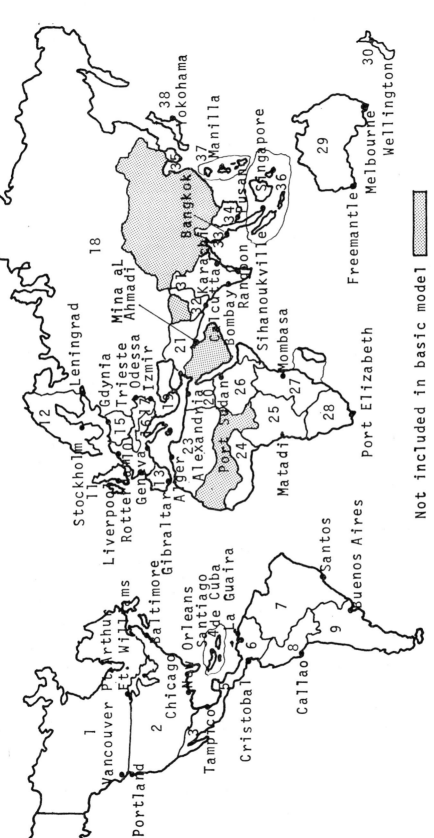

Fig. 12.1. Geographic regions as numbered in Table 12.1 and ports within each region.

Not included in basic model

run) for rice where rice supplies are inadequate to meet consumption requirements. Substitution is constrained as follows: (a) substitution of wheat for rice is possible only after all available rice supplies are exhausted; (b) no substitution is possible in developed regions; (c) where substitution is required, rice imports by developing importers (countries designated *DG* in Table 12.1) are distributed in accordance with regional minimum rice requirements; (d) the minimum rice requirement in each developing region with rice deficits is (1) the level of domestic production if that level provides sufficient rice to allow consumption at or above 1960 per capita consumption or (2) the average maximum per capita availability for all developing region importers if domestic production is less than 1960 per capita consumption; and (e) wheat substitutes for rice on a one-for-one poundage basis.

FERTILIZER PRODUCTION AND USE

The intensification of agricultural production and the utilization of new high-yielding crop varieties are highly synonymous with increased fertilizer use. Fertilizers represent the largest single intensive capital input in the package of practices normally used in programs to increase technological improvement. Fertilizer use levels and their regional distribution have direct implications for current and future distribution of fertilizer plant capacity, as well as for competition over fertilizer raw material deposits. Short-run planning in the fertilizer industry may be affected substantially by increased information on the extent and duration of current overcapacity, especially in developed countries.

FERTILIZER PLANT CAPACITIES AND FERTILIZER RAW MATERIAL RESERVES

The current levels of fertilizer and phosphate rock production capacity were obtained from unpublished data of the Tennessee Valley Authority for 1975. Regional potential to produce potash and phosphate rock was constrained by regional potash and phosphate rock reserves. Phosphate rock reserve data were taken from the World Survey of Phosphate Deposits.[7] Reserve data on potash deposits were taken from the World Survey of Potash.[8] The actual plant capacity and material reserve data by trade area and development class and the 96 countries' totals are presented in Chapter 13. Corresponding data for lower levels of aggregation are presented in the Appendix.

---

7. The British Sulphur Corporation, Ltd. *A World Survey of Phosphate Deposits,* 2nd ed. London, 1964.

8. The British Sulphur Corporation, Ltd. *A World Survey of Potash.* London, 1966.

Country estimates of fertilizer use for each of the 96 nations were developed, employing one of several methods, including the use of simple grain to fertilizer ratios, projection of past trends, and use of rates of growth in fertilizer consumption, all in conjunction with nutrient-mix assumptions based on judgment of soil fertility experts.

Extensive investigation of alternatives preceded selection of methodology to use in estimation of future fertilizer use. A simple ratio estimate was one method considered. However, its implicit assumption of a universally equal fertilizer response causes the method to be discarded when other alternatives could be applied. The data for a sample of representative countries were analyzed to evaluate available alternatives. Data available on grain yield and fertilizer use per hectare for those countries were used to test alternative projection methods, including (a) utilization of functional fertilizer use-grain yield relationships or simple production functions, (b) the use of time trends, and (c) the use of simple fertilizer use growth rates.

The FAO production function approach was tested with 1961–63 FAO data for the United States and India.[9]

The functional relationship used was:[10]

$$Y = 769.23 + 0.9526X + 134.12\sqrt{X} \qquad (12.12)$$

where Y is grain yield and X is fertilizer used in kilograms per hectare. In applying this function on a per acre basis and multiplying by the number of acres, estimates of fertilizer use and production were grossly overestimated for the United States and grossly underestimated for India (in comparison with actual use in the 1961–63 period). The form of the function results in a very high fertilizer response at low levels and low response at high levels. Hence, the function was judged inappropriate as a basis for projections. Country production functions then were estimated with time series data covering the period 1953–65 for Japan, India, Mexico, Spain, Czechoslovakia, United States, Yugoslavia, and France. The variables were grain yield (Y), time (T), and fertilizer use (X).[11] The functions fitted were:

---

9. M. S. Williams and J. W. Couston. *Crop Production Levels and Fertilizer Use.* United Nations, Food and Agriculture Organization. Rome, Italy, 1962.

10. For discussion of other estimates developed by Council of Scientific Advisors to President Johnson using this type of function, see Organization for Economic Cooperation and Development. *Supply and Demand Prospects for Chemical Fertilizer in Developing Counties.* Mimeographed preliminary report. Paris, France, 1967.

11. Grain yield was expressed as the sum of wheat, rye, barley, oats, corn, and rice divided by the corresponding area, and fertilizer per hectare was measured as total fetilizer use divided by grain area. Grain yield was expressed as 100 kg/ha and fertilizer as kg/grain ha.

$$Y = a + bX + cX^2 \qquad (12.13) \qquad Y = a + bX \qquad\qquad\qquad (12.14)$$
$$Y = a + bX + c\sqrt{X} \qquad (12.15) \qquad Y = a + bT + cX + d\sqrt{X} \qquad (12.16)$$
$$Y = a + bX + cX^{-1} \qquad (12.17)$$

Only the linear form gave results which were consistently satisfactory, in comparison with actual fertilizer use and output. While other forms often gave statistically acceptable estimates of parameters, they posed conceptual problems and yielded unreasonable predictions.[12] While we prefer statistically acceptable nonlinear forms to represent the "pure" fertilizer-yield relationship, other factors necessitate use of the linear form for the approximation of fertilizer requirements. It implies that when fertilizer use is used as a proxy variable to represent the response of crop yield to the level of all production inputs, as well as to reflect changing technology, the relationship is linear. This possibility is not unrealistic for the time series data and the changing technology associated with time.[13]

Time functions of fertilizer use also were estimated. These simple functions were estimated after prior examination of more logical fertilizer demand functions incorporating variables representing commodity prices, fertilizer prices, technology, and various other relevant variables. While demand functions of this type could be estimated successfully for a few highly developed countries, data were too inadequate to allow their application to most of the world. Hence, as a further foundation for determining fertilizer use projection methods, the following simple fertilizer use-time functions were estimated from time series data:

$$X = a + bT \qquad\qquad (12.18) \qquad X = a + b \ln (T) \qquad (12.19)$$
$$X = a + bT + c \ln (T) \qquad (12.20)$$

where X is total use of fertilizer and T is time by year. Comparisons of estimated fertilizer use and actual use indicated that the time functions selected from equations 12.18 to 12.20 in each case served as a better basis for projections than most other methods employed, with yield expressed as a linear function of the fertilizer proxy variable and fertilizer use as a linear function of time (and conversely, for specified future yields, fertilizer requirements could be derived similarly).

---

12. For example, the quadratic form for the United States had an r^2 of 0.88 and a small standard error. But the function's maximum yield of 31.4 (100 kg/ha) was specified for 156 kilograms of fertilizer per grain hectare. The actual 1965 grain yield of 30.0 (100 kg/ha) was achieved with 163.2 kilograms of fertilizer per hectare.

13. For other discussions of the potential applicability of linear forms see R. K. Perrin. "Analysis and Prediction of Crop Yields for Agricultural Policy Purposes." Unpublished Ph.D. thesis. Ames: Iowa State University Library, 1968. L. V. Mayer. "An Analysis of Future Resource Supplies, Resource Utilization, Domestic and Export Demand and Structural Change in the Agricultural Economy in 1980." Unpublished Ph.D. thesis. Ames: Iowa State University Library, 1967.

Finally, data available from FAO fertilizer trials and demonstrations were considered. These data consisted of estimated fertilizer use and yields by crop and regions within a limited number of countries in Africa, the Near East, and northern Latin America. While available for a limited number of countries, review of the information indicated that it could be aggregated into a form with potential for use in fertilizer projections on the basis of rates of change in fertilizer consumption.

Based on our investigation and study of alternatives, the following estimation methods were selected. Where area and fertilizer use data were available, time trends of fertilizer use were developed from the functional forms expressed in equations 12.18 to 12.20. The trend equation best reflecting fertilizer use-time relationships was selected for each country. Equations of this type were used for 64 of the 96 individual countries. Those 64 countries account for some 95 percent of total projected cereal production in 1985. In the remaining cases where data on total fertilizer consumption in recent years were available but cropland area information was inadequate, regional compound fertilizer growth rates were used. In some others, growth rates used were derived from FAO[14] and OECD[15] data which expressed as percent per annum were for Africa—7.9, Asia—13.5, and Latin America—10.5. Due to the explosive nature of high compound growth rates, these were used only through 1975. Consumption was assumed to double between then and 2000 for the four countries to which these estimates were applied. For a few countries, the only available information was that of total grain production. In these cases, ratio estimates were developed. The ratio used was 1 kg of fertilizer per 10 kg of grain produced.

SULFUR AND PHOSPHATE ROCK REQUIREMENTS

Regional requirements for sulfur and phosphate rock (raw materials required for phosphate fertilizer production) were also necessary for the analysis. A fixed requirement of each per unit of phosphate fertilizer produced was determined and the total requirement was specified as the product of the per unit requirement and total regional phosphate fertilizer production. The requirements were three tons of phosphate rock and two tons of sulfur per ton of P_2O_5 produced.

14. United Nations, Food and Agriculture Organization. Freedom from Hunger Campaign, "Fertilizer Program Consolidated Review of Trial and Demonstration Results, 1961/62–1963/64." Unpublished mimeographed report. Rome, Italy, 1967.

United Nations, Food and Agriculture Organization. Freedom from Hunger Campaign," Fertilizer Program Review of Trial and Demonstration Results, 1964–65." Unpublished mimeographed report. Rome, Italy, 1967.

15. Organization for Economic Cooperation and Development. *Supply and Demand Prospects for Chemical Fertilizer in Developing Countries.* Mimeographed preliminary report. Paris, France, 1967.

THE FERTILIZER NUTRIENT MIX

General agreement on the proportions of nitrogen (N), phosphate (P_2O_5), and potash (K_2O) which will be used in the future does not exist. Commenting on two ratios which have been used in projections, OECD stated:[16]

> Consultations with a number of individual experts in the fertilizer industry suggest that the ratio 1:1:1 may be underemphasizing the role that nitrogen will have to play, while the ratio 2:2:1 may be underemphasizing the role of potash and perhaps also phosphorus.

Nor do past trends in the nutrient mix reveal any convergence toward one general mix. After examining several alternatives and previous projections, a 2:1:1 ratio was used in allocation of projected increases in fertilizer use among nutrients beyond 1975. However, with strong evidence against choosing a particular ratio, projections under an alternative proportions assumption were also developed and used for comparative analysis. The one selected was a 2:1:1 ratio for developing country increases between 1975 and 1985 and a 1:1:1 ratio for developed country increases beyond 1975 and developing country increases beyond 1985. In all instances, assumed proportions were applied only to increments in fertilizer use with base year consumption being in those proportions actually consumed. The 2:1:1 ratio is termed assumption A and the alternate 1:1:1 ratio for developed countries beyond 1975 and all countries beyond 1985 is termed assumption B.

Fertilizer use projections developed by the methodology and assumptions outlined are presented later. Also fertilizer and fertilizer raw material plant capacities and fertilizer raw material reserves are reported by trade area, development class, and as 96-country totals in Chapter 13.

TRANSPORT COSTS WITHIN AND BETWEEN REGIONS

Transportation costs were required for ocean transport of cereals, fertilizers, and phosphate rock. Inland transport costs were also required for cereals. An initial attempt was made to develop them as a historic series. However, data on ocean and inland transportation rates within and between regions for many commodities were unavailable, and available data series contained marked and irregular fluctuations over time. Consequently, techniques based on available transportation research were developed and used to determine transportation rates. Ocean transportation rates were based primarily on research reported in "Maritime Transportation of Unitized Cargo: A Comparative Analysis of Bulk

16. Ibid., p. 14.

Break and Unit Load Systems."[17] In adapting the basic data, the following assumptions were made: (a) Fertilizer, cereals, and phosphate rock have a cargo storage factor of 50 cubic feet per ton. (b) Commodities moving in trade are transported in vessels with a bale cubic capacity equivalent to 15,000 tons. (c) A full load is 13,500 tons and 90 percent of a ship's bale cubic capacity is utilized. (d) Vessel speed is 14 knots per hour. (e) Labor costs on foreign flag vessels are 53 percent of those on U.S. flag vessels. (f) Vessels obtain 60 percent of a full load in ports where they discharge grain cargo. (g) Costs of discharging a vessel, plus profit for the voyage, are 10 percent of vessel ownership, at-sea, and port expenses. (h) Days in port for loading and discharging are 5 and 14, respectively, for cereals and phosphate rock and 7 and 18, respectively, for fertilizers. (i) Only one port call is made per voyage. (j) Vessel construction cost equals $2,475,000.[18]

These and associated assumptions lead to specification of a vessel with 750,000 feet of bale cubic capacity and a cubic number of 15.4. The vessel's normal shaft horsepower was set at 4,300 horsepower.[19] Given the derived vessel specifications and adapting the methodology accordingly, costs of owning and operating the vessel were developed as follows: (a) Vessel ownership expenses per voyage day: amortization, $530; crew wages, $620; insurance, $180; maintenance and repair, $220; stores and supplies, $90; subsistence, $85; in-port fuel, $170; and miscellaneous, $20. (b) At-sea expenses per sea day: at-sea fuel, $350. (c) Port expense: per call, $525; and per port per day, $130.

Given those conditions and quantities, the per ton cost of shipping a commodity was expressed as:

$$\text{Cost per ton} = [\text{Vessel ownership expense (voyage days)} + \text{At-sea expenses (sea days)} + \text{Cost per port call (port calls)} + \text{In-port expenses (days in port)}]\ 1.5/13,500.[20]$$

Days spent at sea were calculated as distance between ports[21] divided by

17. Maritime Cargo Transportation Conference. "Maritime Transportation of Unitized Cargo: A Comparative Analysis of Bulk-Break and Unit Load Systems." National Academy of Sciences-National Research Council, Washington, D.C., 1959.

18. Based on Ocean Freighting Research, Ltd. *World Freights 1965: A Comprehensive Review of the Global Shipping Scene.* London: The British Sulphur Corporation, Ltd., 1966.

19. Maritime Cargo Transportation Conference. "Maritime Transportation of Unitized Cargo: A Comparative Analysis of Bulk-Break and Unit Load Systems." National Academy of Sciences-National Research Council, Washington, D.C., 1959.

20. The figure 1.5 in the relationship is an adjustment for profit and nonfull return from ports where grain cargo is discharged.

21. Distance and canal passage data were obtained from Waterman Steamship Corporation. *Marine Distance and Speed Tables.* Mobile, Alabama, 1941.

distance per day. Substitution of the appropriate estimates in the relationships and simplification of the resulting form yields the relation:

Total cost per ton $= 4.15 + 0.25167$ (at-sea days) for cereal crop and phosphate rock transportation costs and
Total cost per ton $= 5.74 + 0.25167$ (at-sea days) for fertilizers.

Where one or more canal passages were required on a voyage, 23 cents per ton per passage were added (the equivalent of an additional day in port). Commodities exported from Great Lakes ports of North America had $1.60 per ton added, based on both Hutchinson's[22] estimate that the total voyage time between Chicago and Antwerp is twice that between Montreal and Antwerp and his calculated cost of 23 cents per ton per day of delay. These data indicate a 7-day delay associated with passage of the St. Lawrence seaway at a cost of 23 cents per ton per day or a total cost of approximately $1.60 per ton. Discharging in Antwerp-Rotterdam was assumed completed in 7 days less than in other ports, and cost was reduced $1.60 per ton accordingly.

The regional inland transportation costs were based on information contained in the Directory of World Railways[23] under these assumptions: (a) expenses associated with freight movement are the same proportion of total expenses as freight cars are of total rail cars used, and (b) the commodities hauled are average freight commodities. Per ton freight rates were developed through use of the following relationship:

Cost per ton $=$ (number of freight cars/number of rail cars) (total expenses/total volume of freight carried).

Given the resulting country rates by commodity, regional rates weighted by country imports within regions were developed.

FERTILIZER PLANT INVESTMENT COSTS

Fertilizer investment costs were not available on a refined basis because of current excess production capacity and the tight security in which fertilizer manufacturers hold their cost data. Therefore, fertilizer production and plant investment costs were based entirely on the aggregate costs available from OECD.[24] Fertilizer costs (f.o.b.) are their average production costs minus a $10.00 adjustment for transportation,

22. T. Q. Hutchinson. *Heavy Grain Exports in Voyage Chartered Ships: Rates and Volume.* U.S. Department of Agriculture, ERS—Marketing Research Report 812, 1968.
23. H. Sampson, ed. *World Railways 1961–62*, 7th ed. London: Sampson Low's "World Railways" Ltd., 1962.
24. Organization for Economic Cooperation and Development. *Supply and Demand Prospects for Chemical Fertilizer in Developing Countries.* Mimeographed preliminary report. Paris, France, 1967.

since costs specified by OECD include insurance and freight.[25] Since phosphate rock is a component of the phosphate fertilizer production cost estimates, production cost is assumed to be zero for it. Investment costs required to increase capacity to produce nitrogen, phosphate, and potash fertilizers are from OECD.[26] The only variation used in their level among regions is a 1.4[27] adjustment factor applied to developing countries.

MODEL ANALYSES CONDUCTED

Seven solutions were derived from the programming model outlined previously: the first three were for the years 1975, 1985, and 2000 and the most probable set of parameters for land quantity, income, and population. The most probable parameter set used included the low land level, the low income level, and the medium population projection of earlier chapters. A fourth model solution was computed to evaluate the implications of the high land level or bounds. The solution is for the 2000 year and involved adjusting only the constraint level for land. Three more solutions were then computed. One was designed to measure the implications of Mainland China's exclusion from the general analysis. Another, through removal of capital cost requirements associated with increasing potash and phosphate rock production capacity, considers the competitiveness of phosphate rock and potash reserves in respect to their spatial distribution.

The final solution was developed to evaluate U.S. requirements that at least one-half of all commodities sold under government programs must be transported in domestic flag ships. It allows freight rates for shipment of the entire volume of U.S. commodities to developing regions to be reduced to foreign vessel rates.

---

25. Recent changes which indicate lower potash prices in future years required a further $10.00 reduction in potash fertilizer prices.

26. Organization for Economic Cooperation and Development. *Supply and Demand Prospects for Chemical Fertilizer in Developing Countries.* Mimeographed preliminary report. Paris, France, 1967, Table 3, p. 83.

27. Ibid., Tables 5 and 22.

CHAPTER 13

The Data Used in the Model

THIS CHAPTER SUMMARIZES the basic data developed for use in the world trade programming model. It has been derived by the methods explained earlier. The material, summarized by trade area to facilitate presentation and interpretation, indicates the quantities of cereals produced in each of the trade area groupings under the high and low land bounds or levels explained in earlier chapters. These production levels then are compared with consumption requirements of the trade areas to indicate the excess capacity or the deficit[1] to be met through optimized international trade as suggested by the programming model. The resulting surpluses and deficits are also presented. The consumption requirements are those representing variables of low income and medium population growth rates.[2] As reviewed in earlier chapters, these demand conditions, with projected yield and supply growth at levels discussed previously and under alternative assumptions of land availability, allow a potential world surplus of production over consumption. Other levels of demand variables, for example high income and high population growth rates, would pose a demand increase greater than the supply increase generated for some projection periods under the assumptions of the study.

Trade area data are also provided on land availability, fertilizer requirements, and cost of fertilizer plant capacity for various groupings of countries. The trade patterns generated by the linear programming model then are presented in Chapters 14 through 16.

The data presented in this chapter describe inputs for the programming model applied in the analysis of trade potentials under the assumptions of the model. As mentioned previously in this chapter, the data presented generally are on a trade area basis. Corresponding data for the 38 world regions used in the programming model appear in Chapters 14 and 15 and in the Appendix.

---

1. Regional deficits specified as commodity imports are presented in Table 14.1.
2. The actual population projections and income projection indices used for each region are presented in the Appendix.

PROJECTED CEREAL GRAIN PRODUCTION AND REQUIREMENTS
BY DEVELOPMENT CLASS AND TRADE AREA

Tables 13.1, 13.2, and 13.3 include projected levels of grain production and requirements of each cereal class for the years 1975, 1985, and 2000, respectively, for the 13 specified trade areas. Data also are presented for the two development classes and the 96 nations' total.

The total requirement quantities include grain used directly for food, for industrial uses, and livestock feed. Also, they incorporate allowances for seed and waste associated with the projected requirements regardless of whether or not it is all produced domestically. Feed includes grain used for domestic livestock and livestock products imported. In other words, grain for seed and waste, as well as for imported livestock products, is included in requirements of the importing country (since indirectly these are requirements for the imports).

HIGH LAND BOUNDS

The special soils study explained in earlier chapters indicated that two alternative upper bounds on cropland expansion were specified for 33 countries. However, in the programming analysis for 1975, 1985, and 2000, cropped area at the lower bounds did not prove to be a restraint in 24 of these 33 countries. The alternative of a higher land bound then had no effect on production in these 24 countries. In the other 9 countries, production was restrained even at the high bound for one or more of the three projection years. The 9 countries for which even the high land bound was restrictive were Iran, Israel, Jordan, Syria, India, Pakistan, Thailand, South Vietnam, and the Philippines. The data in Table 13.4 summarize for these 9 countries the effect of the higher land bound on cereal grain production and fertilizer requirements for the three projection years. Under the 1985 projections, wheat and rye production would be 1,071 million tons greater under the high land bounds, rice would be 2,213 million tons greater, and other grains would be 412 million tons greater. The differences in grain production are still greater under projections for the year 2000. These computations suggest quite large flexibility and capacity in meeting total food demand within the 9 countries. This is true not even considering some of the more exciting and romantic hopes expressed for discrete technological leaps representing the green revolution. However, the picture is different under alternative assumptions such as the low level of cropped area or land availability and high growth rates in per capita incomes and population in the developing countries. It is the requirement quantities summarized in Tables 13.1, 13.2, and 13.3 which serve as the demand restraints to be fulfilled in the linear programming model while the production possibilities summarized in these tables

TABLE 13.1. Requirements by Production and Deficit or Surplus of Cereal Grains for 1975; 96-Nation Total, Developing and Developed Nations and Trade Areas (1,000 metric tons)

Area	Domestic Requirements	Domestic Production		Requirements Less Production	
		Low Land Bounds	High Land Bounds	Low Land Bounds	High Land Bounds
96-Nation Total					
Wheat, rye	281,863	311,761	311,995	−29,898	−30,132
Rice	132,504	136,523	137,787	−4,019	−5,283
Other grain	416,954	464,024	469,139	−47,070	−52,185
Developing Nations					
Wheat, rye	83,179	57,522	57,756	25,657	25,423
Rice	117,672	117,806	119,070	−134	−1,398
Other grain	117,593	110,421	110,531	7,172	7,062
Developed Nations					
Wheat, rye	198,684	254,239		−55,555	
Rice	14,832	18,717		−3,885	
Other grain	299,361	353,603		−54,242	
United States					
Wheat, rye	18,435	61,245		−42,810	
Rice	619	2,703		−2,084	
Other grain	126,819	190,563		−63,744	
Canada					
Wheat, rye	4,509	20,312		−15,803	
Rice	45	0		45	
Other grain	12,873	12,188		685	
Latin America					
Wheat, rye	22,538	12,578		9,960	
Rice	7,619	7,919		−300	
Other grain	34,270	36,736		−2,466	
EEC Countries					
Wheat, rye	35,977	38,631		−2,654	
Rice	615	590		25	
Other grain	37,929	34,780		3,149	
Other West Europe					
Wheat, rye	20,746	14,290		6,456	
Rice	624	444		180	
Other grain	31,217	27,554		3,663	
East Europe					
Wheat, rye	40,021	35,209		4,812	
Rice	530	104		426	
Other grain	36,360	36,018		342	
USSR					
Wheat, rye	68,728	69,178		−450	
Rice	869	278		591	
Other grain	41,442	41,977		−535	
Middle East					
Wheat, rye	25,801	19,289	19,523	6,512	6,278
Rice	2,298	2,569	2,569	−271	−271
Other grain	16,629	10,626	10,736	6,003	5,893
Africa					
Wheat, rye	6,843	3,462		3,381	
Rice	3,138	2,741		397	
Other grain	28,833	25,521		3,312	
South Africa-Oceania					
Wheat, rye	4,854	13,963		−9,109	
Rice	104	154		−50	
Other grain	6,704	9,618		−2,914	

TABLE 13.1 (cont.)

Area	Domestic Requirements	Domestic Production Low Land Bounds	Domestic Production High Land Bounds	Requirements Less Production Low Land Bounds	Requirements Less Production High Land Bounds
India-Pakistan					
Wheat, rye	25,492	21,752	21,752	3,740	3,740
Rice	62,533	61,535	61,535	998	998
Other grain	30,215	27,651	27,651	2,564	2,564
Other East Asia					
Wheat, rye	2,505	441	441	2,064	2,064
Rice	42,084	43,042	44,306	—958	—2,222
Other grain	7,646	9,887	9,887	—2,241	—2,241
Japan					
Wheat, rye	5,414	1,410		4,004	
Rice	11,426	14,444		—3,018	
Other grain	6,017	905		5,112	

TABLE 13.2. Requirements by Production and Deficit or Surplus of Cereal Grains for 1985; 96-Nation Total, Developing and Developed Nations, and Trade Areas (1,000 metric tons)

Area	Domestic Requirements	Domestic Production Low Land Bounds	Domestic Production High Land Bounds	Requirements Less Production Low Land Bounds	Requirements Less Production High Land Bounds
96-Nation Total					
Wheat, rye	321,652	354,718	355,339	—33,066	—33,687
Rice	163,137	163,468	165,681	—331	—2,544
Other grain	494,936	569,783	570,195	—74,847	—75,259
Developing Nations					
Wheat, rye	105,092	66,482	67,103	38,610	37,989
Rice	147,073	141,500	143,713	5,573	3,360
Other grain	150,256	128,576	128,988	21,680	21,268
Developed Nations					
Wheat, rye	216,560	288,236		—71,676	
Rice	16,064	21,968		—5,904	
Other grain	344,680	441,207		—96,527	
United States					
Wheat, rye	21,242	83,286		—62,044	
Rice	724	3,198		—2,474	
Other grain	147,604	243,867		—96,263	
Canada					
Wheat, rye	5,457	22,607		—17,150	
Rice	55	0		55	
Other grain	15,613	13,445		2,168	
Latin America					
Wheat, rye	28,773	13,753		15,020	
Rice	9,953	10,150		—197	
Other grain	44,916	44,071		845	
EEC Countries					
Wheat, rye	37,724	44,152		—6,428	
Rice	660	650		10	
Other grain	42,237	46,128		—3,891	
Other West Europe					
Wheat, rye	21,095	14,862		6,233	
Rice	673	469		204	
Other grain	33,597	34,872		—1,275	

TABLE 13.2 (cont.)

Area	Domestic Requirements	Domestic Production Low Land Bounds	High Land Bounds	Requirements Less Production Low Land Bounds	High Land Bounds
East Europe					
Wheat, rye	43,294	38,355		4,939	
Rice	582	107		475	
Other grain	40,809	41,097		−288	
USSR					
Wheat, rye	75,784	68,027		7,757	
Rice	1,031	317		714	
Other grain	49,265	51,716		−2,451	
Middle East					
Wheat, rye	32,943	21,012	21,633	11,931	11,310
Rice	2,990	3,115	3,186	−125	−196
Other grain	21,365	11,172	11,397	10,193	9,968
Africa					
Wheat, rye	9,013	3,551		5,462	
Rice	3,936	3,230		706	
Other grain	37,475	28,833		8,642	
South Africa-Oceania					
Wheat, rye	5,922	15,803		−9,881	
Rice	131	177		−46	
Other grain	8,606	9,956		−1,350	
India-Pakistan					
Wheat, rye	31,071	27,563	27,563	3,508	3,508
Rice	76,082	74,107	74,107	1,975	1,975
Other grain	36,345	31,586	31,586	4,759	4,759
Other East Asia					
Wheat, rye	3,292	603	603	2,689	2,689
Rice	54,112	50,898	53,040	3,214	1,072
Other grain	10,155	12,914	13,101	−2,759	−2,946
Japan					
Wheat, rye	6,042	1,144		4,898	
Rice	12,208	17,050		−4,842	
Other grain	6,949	126		6,823	

TABLE 13.3. Requirements by Production and Deficit or Surplus of Cereal Grains for 2000; 96-Nation Total, Developing and Developed Nations, and Trade Areas (1,000 metric tons)

Area	Domestic Requirements	Domestic Production Low Land Bounds	High Land Bounds	Requirements Less Production Low Land Bounds	High Land Bounds
96-Nation Total					
Wheat, rye	391,054	421,426	424,491	−30,372	−33,437
Rice	219,208	198,837	208,795	20,371	10,413
Other grain	630,170	740,440	743,089	−110,270	−112,919
Developing Nations					
Wheat, rye	145,378	78,692	81,757	66,686	63,621
Rice	201,751	171,974	181,932	29,777	19,819
Other grain	213,072	155,407	158,056	57,665	55,016
Developed Nations					
Wheat, rye	245,676	342,734		−97,058	
Rice	17,457	26,863		−9,406	
Other grain	417,098	585,033		−167,935	

TABLE 13.3   (cont.)

Area	Domestic Requirements	Domestic Production		Requirements Less Production	
		Low Land Bounds	High Land Bounds	Low Land Bounds	High Land Bounds
United States					
Wheat, rye	26,359	122,558		—96,199	
Rice	904	3,888		—2,984	
Other grain	183,809	334,790		—150,981	
Canada					
Wheat, rye	7,202	26,007		—18,805	
Rice	73	0		73	
Other grain	20,578	15,243		5,335	
Latin America					
Wheat, rye	41,021	15,524		25,497	
Rice	14,453	13,842		611	
Other grain	65,815	55,648		10,167	
EEC Countries					
Wheat, rye	40,435	50,507		—10,072	
Rice	727	729		—2	
Other grain	48,119	64,018		—15,899	
Other West Europe					
Wheat, rye	21,652	15,422		6,230	
Rice	740	498		242	
Other grain	36,820	45,289		—8,469	
East Europe					
Wheat, rye	47,594	42,691		4,903	
Rice	645	110		535	
Other grain	46,348	49,646		—3,298	
USSR					
Wheat, rye	87,848	66,804		21,044	
Rice	1,274	367		907	
Other grain	60,991	65,456		—4,465	
Middle East					
Wheat, rye	45,362	22,709	24,137	22,653	21,225
Rice	4,185	3,673	3,918	512	267
Other grain	29,389	11,689	12,185	17,700	17,204
Africa					
Wheat, rye	13,014	3,668		9,346	
Rice	5,637	3,938		1,699	
Other grain	55,934	33,581		22,353	
South Africa-Oceania					
Wheat, rye	7,878	18,234		—10,356	
Rice	187	210		—23	
Other grain	12,402	10,512		1,890	
India-Pakistan					
Wheat, rye	41,206	35,972	37,609	5,234	3,597
Rice	100,264	89,080	94,407	11,184	5,857
Other grain	46,794	36,344	38,109	10,450	8,685
Other East Asia					
Wheat, rye	4,775	819	819	3,956	3,956
Rice	77,212	61,441	65,827	15,771	11,385
Other grain	15,140	18,145	18,533	—3,005	—3,393
Japan					
Wheat, rye	6,708	511		6,197	
Rice	12,907	21,061		—8,154	
Other grain	8,031	79		7,952	

281

TABLE 13.4. Projected Cereal Grain Production Possibilities and Fertilizer Require-
ments under High and Low Land Availabilities for the Projection Years.
Nine Countries Where Cropland Expansion Was Bounded by the Low
Land Restraint (1,000 metric tons)

Production and Fertilizer Use Effects by Period	Total Effects under Each Land Assumption and the Difference		
	Land high	Land low	Difference
	1975		
Production—Wheat and rye	28,502	28,268	234
Rice	83,308	82,044	1,264
Other grain	34,914	34,804	110
Fertilizer use	2,489	2,473	16
	1985		
Production—Wheat and rye	35,781	34,710	1,071
Rice	99,600	97,387	2,213
Other grain	41,322	40,910	412
Fertilizer use	3,880	3,818	62
	2000		
Production—Wheat and rye	46,517	43,452	3,065
Rice	124,955	114,997	9,958
Other grain	52,363	49,714	2,649
Fertilizer use	6,326	5,850	476

and in Table 13.4 serve as the production restraints under assumptions
of low and high land bounds or cropped area. Of course, requirement
and production projections were first developed for the 96 individual
countries, then summed to the 38 world regions in developing the programming model. The requirement (demand) quantities and the production (given supply quantities) in the 38 regions serve as the restraints
in the model.

Since the situation with the high land availability provides binding
restraints in only a few of the countries, discussion of results under that
situation is restricted to results for the year 2000. These are presented
in Chapter 16. In other words, the summary of surpluses and deficits
in this and succeeding chapters generally relates to quantities under
the low land bound.

SUMMARY BY STAGE OF DEVELOPMENT AND TRADE AREA

The largest deficits for developing nations are in wheat and rye
(Tables 13.1, 13.2, and 13.3). The deficit increases from a 1960 level of
14.6 million tons to 66.7 million tons in 2000. The projected 2000 deficit
of other grains for developing nations is only slightly less, 57.7 million
tons. The developing nations have a small rice surplus in 1975 but
a deficit of 29.8 million tons is projected for 2000.

For developed areas, 2000 production of wheat-rye, rice, and other grains is projected to exceed domestic requirements by 39, 54, and 40 percent, respectively. The 1985 projected surpluses for developed countries are 71.7, 5.9, and 96.5 million tons of wheat-rye, rice, and other grains. While projected aggregate production suggests a supply great enough to meet food requirements as defined in this study, the total deficits in the developing nations indicate the need for widespread trade if demand restraints are to be met.

A review of the trade area data in Tables 13.1, 13.2, and 13.3 indicates that the United States and Canada will continue to have large exportable surpluses, or the potential for them, to the year 2000. Domestic production in the Latin American trade area is inadequate to meet rising needs. Projected wheat production is only 56 percent of domestic requirements for 1975. By the year 2000, the projected deficits are 25.5, 0.6, and 10.2 million tons, respectively, of wheat-rye, rice, and other grains and represent 24, 2, and 13 percent of the 96 nations' aggregate deficits. The EEC trade area has projected surpluses of wheat, rye, and other grains by 2000 and can be self-sufficient in rice production (in contrast to 1960 when total demand exceeded production in the trade area). While Other Western European countries are not transformed to a surplus position by 2000, production is adequate to stabilize wheat and rye deficits at 6.0 to 6.5 million tons and the rice deficit at 0.2 million tons.[3] Production of other grains rises from 69 percent of domestic requirements to 123 percent of requirements by 2000. For the Eastern European trade area, deficits and surpluses are projected to remain quite stable for the next 30 years. The level of wheat deficits varies from 4.8 to 4.9 million tons and that for rice from 427,000 to 534,000 tons. For the USSR the quantity of wheat and rye produced remains quite stable over the projection period but the relative importance of other grains increases markedly. In 1960 other grains accounted for 31.6 percent of the USSR total grain production. They are projected at 49 percent of 2000 total grain production.

Projections made for the Middle East trade area are typical of those for other trade areas made up of developing countries. It has a small initial surplus of rice while a deficit is projected for later years. While deficits for other cereals are small, they increase over time. Production of wheat-rye, rice, and other grain was 82, 106, and 89 percent, respectively, of domestic requirements in 1960. Projected 2000 production as a percent of requirements is only 50 percent for wheat, 88 percent for rice, and 40 percent for other grains.

African domestic requirements for grain increase 48.3 million tons while domestic production rises only 16.8 million tons over the period

---

3. As is true in the EEC, Eastern Europe, and the USSR, rice is neither a major grain crop nor a major component of domestic cereal consumption in Other Western European countries.

from 1960 to 2000. However, the South Africa-Oceania area continues as one with exportable surpluses to the year 2000. In the nearby trade area made up of India and Pakistan, domestic rice production in 1985 is projected to be large enough to cover 97 percent of requirements; the area's rice deficit is 27 percent of the gross deficit for the 96 countries included in this study. Relative deficits under the methods and assumptions of the study are projected to decline only slightly by 2000.

Projected trends for Other East Asia indicate growing wheat deficits, growing surpluses of other grain, and a shift from a surplus to a deficit situation for rice by 2000. The 15.8 million tons deficit for rice is 50 percent of the 96-nation total in 2000. Finally, Japan is projected to have a growing deficit of wheat and other grains but a large surplus of rice by 2000.

If average per capita availability of cereals is not to decline to the year 2000, a 110 percent increase in cereal production over the 1960 level is required for the 96 nations. The population increase for the developing nations is 2 billion people or 156 percent during the period. It is only 0.5 billion or 49 percent for the developed nations. Consequently, substantially higher increases in domestic availability of cereals is required in developing nations to maintain 1960 levels of consumption. In 1960, 43 percent of the 96-nation total population lived in the developed areas; by 2000 it will be only 30 percent under the set of population projections used. The effects of income changes are also largest in developing areas where income elasticities are highest.[4]

The projected surpluses of wheat and other grains and the 2000 rice deficit (Table 13.3) reflect the corresponding trends in production and requirements for each grain class. Production of wheat and rye is projected to increase approximately 83 percent from 1960 to 2000 while requirements rise only 68 percent. Production of other grains increases 134 percent while requirements rise 101 percent. In contrast, rice production rises by only 107 percent while requirements increase 133 percent. The 30 million tons of surplus wheat, under conforming changes in dietary habits, could readily meet the 20 million ton rice deficit.

FERTILIZER USE, PLANT CAPACITY, AND PRODUCTION

Fertilizer and nutrient use by trade area, development class, and the 96-nation total are presented in Tables 13.5, 13.6, 13.7, 13.8, 13.9, and 13.10 for the base periods, 1975, 1985, and 2000. Fourteen percent of 1975 projected fertilizer consumption for the 96 nations is in developing nations, an amount only 3 percent above that of the base

---

4. These considerations were important in the level of per capita incomes selected for the programming analysis. Realization of this prompted us to present detailed supply-demand projections using the high income assumption on the first phase of this study. This enables the reader to become explicitly aware of the difference in results produced under the high income versus low income assumption.

TABLE 13.5. Fertilizer and Nutrient Use in the Base Period by Trade Area, Development Class, and 96-Nation Total (million tons)

| Area | Fertilizer Use | | Nutrient Use[a] | | | | | |
| | Total | Percent[b] | Total | | | Percent[c] | | |
			N	P	K	N	P	K
96-Nation Total	36,852	100	13,555	13,334	9,963	37	36	27
Developed Nations	32,904	89	11,444	12,036	9,424	35	37	29
Developing Nations	3,948	11	2,111	1,298	539	53	33	14
United States	9,651	26	3,933	3,256	2,462	41	34	26
Canada	453	1	124	220	109	27	49	24
Latin America	1,247	3	534	470	243	43	38	19
EEC	7,480	20	2,363	2,581	2,536	32	35	34
Other West Europe	4,284	12	1,547	1,510	1,227	36	35	29
East Europe	3,951	11	1,238	1,411	1,302	31	36	33
USSR	3,898	11	1,396	1,453	1,049	36	37	27
Middle East	705	2	430	244	31	61	35	4
Africa	315	1	108	149	58	34	47	18
S. Africa-Oceania	1,400	4	124	1,113	163	9	79	12
India-Pakistan	776	2	542	151	83	70	19	11
Other East Asia	905	2	497	284	124	55	31	14
Japan	1,787	5	719	492	576	40	28	32

SOURCE: United Nations, Food and Agriculture Organization. *Fertilizer: An Annual Review of World Production, Consumption and Trade, 1965.* Rome, Italy, 1967.
NOTE: 1963–65 average except where lack of data required use of an earlier group of years.
[a] N—nitrogen, P—phosphate (P_2O_5), K—potash (K_2O).
[b] Percent of 96-nation total.
[c] Percent of area total.

TABLE 13.6. Projected Fertilizer and Nutrient Use under Assumptions A and B for 1975 by Trade Area, Development Class, and 96-Nation Total (million tons)

| Area | Fertilizer Use | | Nutrient Use[a] | | | | | |
| | Total | Percent[b] | Total | | | Percent[c] | | |
			N	P	K	N	P	K
96-Nation Total	60,982	100	23,023	22,204	15,755	38	36	26
Developed Nations	52,279	86	18,385	19,280	14,614	35	37	28
Developing Nations	8,703	14	4,638	2,924	1,141	53	34	13
United States	16,843	28	6,865	5,681	4,297	41	34	26
Canada	832	1	227	404	201	27	49	24
Latin America	2,951	5	1,375	1,056	520	47	36	18
EEC	10,392	17	3,284	3,585	3,523	32	35	34
Other West Europe	5,726	9	2,068	2,030	1,628	36	35	28
East Europe	7,427	12	2,435	2,800	2,192	33	38	30
USSR	6,105	10	2,187	2,275	1,643	36	37	27
Middle East	1,283	2	787	439	57	61	34	4
Africa	1,256	2	481	591	184	38	47	15
S. Africa-Oceania	2,303	4	253	1,775	275	11	77	12
India-Pakistan	1,505	2	1,052	292	161	70	19	11
Other East Asia	1,708	3	943	546	219	55	32	13
Japan	2,651	4	1,066	730	855	40	28	32

NOTE: Differences between assumptions A and B occur only in respect to the 1985 and 2000 projections.
[a] N—nitrogen, P—phosphate (P_2O_5), K—potash (K_2O).
[b] Percent of 96-nation total.
[c] Percent of area total.

TABLE 13.7. Projected Fertilizer and Nutrient Use under Assumption A for 1985 by Trade Area, Development Class, and 96-Nation Total (million tons)

Area	Fertilizer Use Total	Fertilizer Use Percent[b]	Nutrient Use[a] Total N	Nutrient Use[a] Total P	Nutrient Use[a] Total K	Nutrient Use[a] Percent[c] N	Nutrient Use[a] Percent[c] P	Nutrient Use[a] Percent[c] K
96-Nation Total	86,749	100	35,905	28,647	22,197	41	33	26
Developed Nations	72,251	83	28,371	24,274	19,606	39	34	27
Developing Nations	14,498	17	7,534	4,373	2,591	52	30	18
United States	25,378	29	11,133	7,815	6,430	44	31	25
Canada	1,190	1	406	494	290	34	41	24
Latin America	5,473	6	2,634	1,687	1,152	48	31	21
EEC	12,977	15	4,576	4,232	4,169	35	33	32
Other West Europe	7,297	8	2,853	2,423	2,021	39	33	28
East Europe	10,426	12	3,935	3,549	2,942	38	34	28
USSR	8,279	10	3,274	2,818	2,187	40	34	26
Middle East	1,864	2	1,077	585	202	58	31	11
Africa	2,167	2	937	819	411	43	38	19
S. Africa-Oceania	3,407	4	805	2,051	551	24	60	16
India-Pakistan	2,375	3	1,487	509	379	63	21	16
Other East Asia	2,619	3	1,399	773	447	53	30	17
Japan	3,297	4	1,389	892	1,016	42	27	31

[a] N—nitrogen, P—phosphate (P_2O_5), K—potash (K_2O).
[b] Percent of 96-nation total.
[c] Percent of area total.

TABLE 13.8. Projected Fertilizer and Nutrient Use under Assumption B for 1985 by Trade Area, Development Class, and 96-Nation Total (million tons)

Area	Fertilizer Use Total	Fertilizer Use Percent[b]	Nutrient Use[a] Total N	Nutrient Use[a] Total P	Nutrient Use[a] Total K	Nutrient Use[a] Percent[c] N	Nutrient Use[a] Percent[c] P	Nutrient Use[a] Percent[c] K
96-Nation Total	86,749	100	32,577	30,310	23,862	38	35	28
Developed Nations	72,251	83	25,043	25,937	21,271	35	36	29
Developing Nations	14,498	17	7,534	4,373	2,591	52	30	18
United States	25,378	29	9,710	8,526	7,142	38	34	28
Canada	1,191	1	347	524	320	29	44	27
Latin America	5,473	6	2,634	1,687	1,152	48	31	21
EEC	12,978	15	4,146	4,447	4,385	32	34	34
Other West Europe	7,296	8	2,591	2,554	2,151	36	35	29
East Europe	10,426	12	3,435	3,799	3,192	33	36	31
USSR	8,280	10	2,912	3,000	2,368	35	36	29
Middle East	1,864	2	1,077	585	202	58	31	11
Africa	2,167	2	937	819	411	43	38	19
S. Africa-Oceania	3,406	4	621	2,142	643	18	63	19
India-Pakistan	2,375	3	1,487	509	379	63	21	16
Other East Asia	2,619	3	1,399	773	447	53	30	17
Japan	3,296	4	1,281	945	1,070	39	29	32

[a] N—nitrogen, P—phosphate (P_2O_5), K—potash (K_2O).
[b] Percent of 96-nation total.
[c] Percent of area total.

TABLE 13.9. Projected Fertilizer and Nutrient Use under Assumption A for 2000 by Trade Area, Development Class, and 96-Nation Total (million tons)

Area	Fertilizer Use Total	Fertilizer Use Percent[b]	Nutrient Use[a] Total N	Nutrient Use[a] Total P	Nutrient Use[a] Total K	Percent[c] N	Percent[c] P	Percent[c] K
96-Nation Total	130,475	100	57,769	39,578	33,128	44	30	25
Developed Nations	106,543	82	45,516	32,846	28,181	43	31	26
Developing Nations	23,932	18	12,253	6,732	4,947	51	28	21
United States	42,120	32	19,504	12,000	10,616	46	28	25
Canada	1,798	1	710	646	442	39	36	25
Latin America	9,669	7	4,733	2,736	2,200	49	28	23
EEC	16,503	13	6,339	5,113	5,051	38	31	31
Other West Europe	9,680	7	4,044	3,019	2,617	42	31	27
East Europe	15,073	12	6,259	4,711	4,103	42	31	27
USSR	11,542	9	4,905	3,634	3,003	43	31	26
Middle East	2,682	2	1,487	789	406	55	29	15
Africa	3,891	3	1,799	1,250	842	46	32	22
S. Africa-Oceania	5,360	4	1,781	2,539	1,040	33	47	19
India-Pakistan	3,765	3	2,182	857	726	58	23	19
Other East Asia	3,925	3	2,052	1,100	773	52	28	20
Japan	4,467	3	1,974	1,184	1,309	44	27	29

[a] N—nitrogen, P—phosphate (P_2O_5), K—potash (K_2O).
[b] Percent of 96-nation total.
[c] Percent of area total.

TABLE 13.10. Projected Fertilizer and Nutrient Use under Assumption B for 2000 by Trade Area, Development Class, and 96-Nation Total (million tons)

Area	Fertilizer Use Total	Fertilizer Use Percent[b]	Nutrient Use[a] Total N	Nutrient Use[a] Total P	Nutrient Use[a] Total K	Percent[c] N	Percent[c] P	Percent[c] K
96-Nation Total	130,477	100	47,221	44,854	38,402	36	34	29
Developed Nations	106,546	82	36,473	37,369	32,704	34	35	31
Developing Nations	23,931	18	10,748	7,485	5,698	45	31	24
United States	42,120	32	15,291	14,107	12,722	36	33	30
Canada	1,798	1	549	726	523	31	40	29
Latin America	9,667	7	4,032	3,086	2,549	42	32	26
EEC	16,503	13	5,321	5,622	5,560	32	34	34
Other West Europe	9,682	7	3,386	3,349	2,947	35	35	30
East Europe	15,075	12	4,984	5,349	4,742	35	35	30
USSR	11,542	9	3,999	4,087	3,456	35	35	30
Middle East	2,683	2	1,351	858	474	50	32	18
Africa	3,891	3	1,512	1,393	986	39	36	25
S. Africa-Oceania	5,357	4	1,271	2,793	1,293	24	52	24
India-Pakistan	3,765	3	1,950	973	842	52	26	22
Other East Asia	3,925	3	1,903	1,175	847	48	30	22
Japan	4,469	3	1,672	1,336	1,461	37	30	33

[a] N—nitrogen, P—phosphate (P_2O_5), K—potash (K_2O).
[b] Percent of 96-nation total.
[c] Percent of area total.

period. However, the major share of the 96-nation total fertilizer use is still in developed areas. Projected fertilizer use rises to 86.7 million tons in 1985, an amount considerably larger than Parker's 1980 estimate of 70 million tons.[5] The projected 1985 use of 14.5 million tons in the developing nations is substantially less than the 40 million tons estimated by the President's Science Advisory Committee.[6] The committee estimate is not directly comparable since it relates production to projected population growth. Comparison of the three major plant nutrients (nitrogen, phosphate, and potash) under fertilizer use assumptions A and B indicates the degree to which the projections vary under alternate nutrient ratios. Data in Tables 13.7 and 13.8 for 1985 indicate that under assumption A the 96-nation total nitrogen consumption is 3.3 million tons higher than under assumption B, while phosphate and potash consumption are each approximately 1.7 million tons lower.

FERTILIZER MATERIAL RESERVES, PLANT CAPACITY ESTIMATES, AND 1965 PRODUCTION LEVELS

The location of potash and phosphate rock reserves is important in the analysis of this part of the study. The manufacture of potash fertilizers and of phosphate rock is restricted to locations where reserves are present and are constrained by these reserve levels.

Vast reserves of both potash and phosphate rock exist and no shortage of either is likely to occur through the year 2000. Estimated reserve levels are presented by trade area, development class, and for the 96-nation total in Table 13.11. To specify constraints for the programming model, reserve levels were divided by 25 under the assumption that production facilities developed would be restricted to capacities which would not exhaust a particular deposit before 2000. The geographic region reserve levels corresponding to those in Table 13.11 are presented in Table 15.6.

Table 13.12 presents the 1975 nitrogen, phosphate, and potash fertilizer plant capacity estimates,[7] the 1965 corresponding levels of nutrient production, and the percent each production level is of estimated 1975 capacity. Large excess fertilizer plant capacity is indicated. The percent of capacity utilized in 1965 is understated because of its evaluation relative to 1975 capacities. However, 1965 world fertilizer consumption

5. F. W. Parker. "Fertilizer and Economic Development: Fertilizer Technology and Use." *Soil Science Society of America 1962 Short Courses Proc.*, 1963, pp. 1–21.

6. The 40 million tons includes an estimated 5 million tons consumed by Mainland China.

7. The estimates presented in Table 13.12 are trade area estimates. The corresponding geographic region estimates are presented in the Appendix.

TABLE 13.11. Reserves of Potash and Phosphate Rock for Trade Areas, Development Classes, and the 96-Nation Total (million tons)

Country	Potash	Phosphate Rock
96-Nation Total	49,720.8	18,923.6
Developed Nations	47,533.4	10,802.7
Developing Nations	2,187.4	8,120.9
United States	400.0	4,747.0
Canada	6,400.0	. . .
Latin America	42.4	2,697.4
EEC	10,306.4	. . .
Other West Europe	490.0	. . .
East Europe	12,700.0	. . .
USSR	17,237.0	5,907.0
Middle East	2,000.0	1,007.5
Africa	145.0	4,416.0
S. Africa-Oceania	. . .	148.7
India-Pakistan	. . .	. . .
Other East Asia	. . .	. . .
Japan	. . .	. . .

SOURCES: The British Sulphur Corporation, Ltd. *A World Survey of Phosphate Deposits*, 2nd ed. London, 1964; *A World Survey of Potash*. London, 1966.
NOTE: Million tons of K_2O and of phosphate rock.

of nitrogen, phosphate, and potash was 15.3, 13.6, and 11 million tons, respectively.[8] Each of the corresponding 1965 nutrient production levels contained in Table 13.12 is larger than these consumption levels. Hence, if fertilizer consumption were to rise to levels required for full utilization of the estimated 1975 capacity, increases of 80 to 100 percent in world fertilizer consumption would be required.[9]

COMMODITY ACQUISITION PRICES, FERTILIZER PLANT INVESTMENT COSTS, AND TRANSPORTATION RATES

In solutions of the programming model to specify the least-cost acquisition costs of cereal grains by importing nations, and for attaining the discrete world demand or requirements specified at the various time periods, it was necessary to estimate the costs of grain f.o.b. at each port in exporting countries and the transportation costs from each port of one country to all ports of other countries.

Values used as f.o.b. costs of acquiring cereal grains to be exported from each region (the price in exporting regions) are presented in Table 13.13. We then assume these prices as the f.o.b. real prices for the future period (although we are aware that various nations may cause relative acquisition prices to change through domestic policies). The

8. United Nations, Food and Agriculture Organization. *Fertilizer: An Annual Review of World Production, Consumption and Trade, 1966*. Rome, Italy, 1968.
9. Maximum production is generally assumed to be 80 percent of capacity in developing countries and 90 percent of capacity in developed countries.

TABLE 13.12. 1965 Trade Area, Development Class, and 96-Nation Total Capacities for and Production of Nitrogen, Phosphate, and Potash Fertilizers and Phosphate Rock; and 1965 Production as Percent of 1975 Capacities (1,000 million tons)

Area	Nitrogen			Phosphate			Potash			Phosphate Rock		
	1975 Capacity	1965 Production	%[a]	1975 Capacity	1965 Production	%[a]	1975 Capacity	1965 Production	%[a]	1975 Capacity	1965 Production	%[a]
96-Nation Total	48,191	17,029	35	32,642	13,366	41	25,501	12,112	47	97,172	52,166	54
Developed Nations	40,582	15,453	38	28,550	12,825	45	24,321	11,856	49	66,322	36,150	55
Developing Nations	7,609	1,576	21	4,092	541	13	1,180	256	22	30,850	16,016	52
United States	11,314	4,465	39	7,243	3,652	50	3,250	2,516	77	37,572	25,895	69
Canada	1,276	341	27	942	339	36	6,408	1,067	17			
Latin America	2,453	571	23	945	182	19	220	24	11	650	586	90
EEC	8,142	4,048	50	4,278	2,936	69	4,461	4,208	94		47	
Other West Europe	3,836	1,663	43	2,092	1,346	65	302	314	104			
East Europe	6,035	1,343	22	2,914	1,209	41	2,400	1,857	77		89	
USSR	6,000	2,099	35	7,428	1,407	19	7,500	1,894	25	25,000	6,200	25
Middle East	1,113	249	22	1,166	164	14	360	232	64	7,200	1,420	20
Africa	316	149	47	574	19	3	600			23,000	14,010	61
S. Africa-Oceania	921	100	11	2,370	1,321	56				3,750	3,919	105
India-Pakistan	2,750	328	12	1,101	132	12						
Other East Asia	967	279	29	306	44	14						
Japan	3,058	1,394	46	1,283	597	47						

SOURCE: United Nations, Food and Agriculture Organization. Fertilizer: An Annual Review of World Production Consumption and Trade, 1966. Rome, Italy, 1968; J. R. Douglas, Jr., and E. A. Harre. TVA, Muscle Shoals, Alabama. Fertilizer Plant Capacity Estimates. Private communication, 1968.

[a] Percent 1965 production is of 1975 estimated capacity.

TABLE 13.13. Regional 1963–65 Average Cereal Grain Prices for Regions with Export Potential (dollars per metric ton)

Region	Cereal Class		
	Wheat and Rye[a]	Rice (milled)	Other Grain[b]
United States	$64.10	$154.10	$54.45
Canada	67.85	...	...
Northern South America	...	141.35	...
Mexico	61.80	136.20	57.15[c]
Brazil	...	130.25	...
Southern South America	59.95	...	51.10
EEC	64.45	...	76.20
Scandinavia	...	...	71.45
Spain-Portugal	...	141.40	57.65[c]
Austria-Switzerland	62.80	...	...
Northern E. Europe	...	...	77.15
Yugoslavia	74.70	...	...
Other East Europe	74.70	...	81.75
USSR	70.35	...	66.30
UAR	...	137.87	...
Iran-Iraq	...	163.60	...
East Central Africa	...	161.33	...
S. Africa	...	...	51.40[d]
Australia	58.55	137.85	57.30
New Zealand	...	...	63.75[d]
India	63.85	124.40	...
Burma	...	106.75	55.00[d]
Other Far East	...	112.60	56.40
Malaya-Indonesia	...	...	55.00[d]
Japan	...	124.40	...

SOURCE: United Nations, Food and Agriculture Organization. *Trade Yearbook, 1966.* Rome, Italy, 1967.

[a] Wheat price.

[b] Average corn equivalent prices weighted by the 1985 proportions of regional surplus corn and barley in the corn plus barley surplus for that region.

[c] Corn price.

[d] Weighted average world price.

corresponding fertilizer prices used in the analysis were nitrogen, $87; phosphate, $63; and potash, $30.[10]

To determine whether fertilizer deficit countries should expand plant capacity or increase imports, it was necessary to estimate investment costs of adding fertilizer capacity in developed and developing countries. The same investment costs were used for each projection year. The fertilizer plant investment costs per ton of productive capacity used to determine per unit investment costs in the analysis are presented in Table 13.14. The data presented in the table are costs per unit of

10. Based on Organization for Economic Cooperation and Development. *Supply and Demand Prospects for Chemical Fertilizer in Developing Countries.* Mimeographed preliminary report. Paris, France, 1967. The fertilizer prices are for materials with nutrient concentrations of 46 percent for nitrogen, 46 percent for phosphate, and 60 percent for potash fertilizer.

TABLE 13.14. Investment Cost Associated with Expansion of Fertilizer Plant Capacity in Developing and Developed Countries (dollars per metric ton of productive capacity)

Nutrient Produced by the Plant	Investment Costs	
	Developed Countries	Developing Countries
Nitrogen	$286	$400
Phosphate	125	175
Potash	71	100
Phosphate Rock	57	80

SOURCE: Organization for Economic Cooperation and Development. *Supply and Demand Prospects for Chemical Fertilizer in Developing Countries.* Mimeographed preliminary report. Paris, 1967.

NOTE: Productive capacity is normally assumed to be 80 percent of actual capacity in developing countries and 90 percent of actual capacity in developed countries.

productive capacity added. The OECD report from which they were taken suggests they are minimum cost estimates and imprecise. Given that and the fact that trade in nitrogen and phosphorus is not allowed beyond 1975, the nitrogen and phosphorus plant cost data were used as cost per unit of actual rather than productive capacity added. In the case of potash and phosphate rock, for which trade is allowed in all projection years, it was assumed that plants in both developed and developing countries operate at 80 percent efficiency and that costs of productive capacity additions for both are $71 for potash and $57 for phosphate rock.[11] The 1.4 factor difference between costs presented in the table for developed and developing countries is assumed to reflect the higher cost associated with lack of manufacturing facilities, skilled craftsmen, and professional personnel in developing countries.

TRANSPORTATION COSTS

Rates for specific shipment routes can be derived simply by applying the appropriate formula presented earlier, after determining the interport distance. Calculated rates for a sample of potential shipping situations are included in Table 13.15. They provide an indication of the general level of ocean shipping rates for cereal grain on selected routes. The corresponding historic rates presented with the calculated rates are an indication of the validity of transportation cost estimation equations used in the analysis.

Inland transportation rates for individual countries are presented in the Appendix. Regional rates were developed as country rates

---

11. To include a 40 percent increase in costs for developing countries and assume a 10 percent lower efficiency for them would preclude their participation in international trade. Given the assumption made, information is gained in respect to their competitiveness in terms of geographic location.

TABLE 13.15. A Sample of Foreign Flag Ocean Rates, Calculated Using the Method Developed for Use in This Study, and Comparable Recent Rates for Foreign Flag Vessels (dollars per metric ton)

Voyage			Ocean Rates			
			Calculated[a]	Actual Rates Selected from Available Data for 1965 and 1966		
Origin	Destination			A	B	C
Quebec	Liverpool		$ 6.30	$ 6.56	$ 7.60	$ 6.15
Baltimore	Liverpool		6.67	6.41	8.02	6.25
New Orleans	Liverpool		7.80	7.78	. . .	. . .
New Orleans	Bombay		11.32	12.19	13.04	. . .
New Orleans	Calcutta		12.58	13.43	. . .	. . .
New Orleans	Yokohama		11.07	10.85	. . .	. . .
New Orleans	Rio de Janeiro		8.05	8.50	. . .	. . .
Baltimore	Bombay		10.57	. . .	. . .	. . .
Portland	Bombay		11.45	12.02	. . .	. . .
Portland	Calcutta		10.82	11.48	. . .	. . .
Portland	Yokohama		7.42	7.83	7.51	. . .
Freemantle	Liverpool		11.32	. . .	12.22	11.20
Capetown	Yokohama		8.80	. . .	8.06	. . .
New Orleans	Naples		8.30	. . .	. . .	8.30
Baltimore	Trieste		7.95	. . .	. . .	8.50
Portland	Liverpool		10.57	. . .	. . .	7.97
Buenos Aires	Liverpool		8.93	. . .	. . .	13.30
Sydney	Liverpool		12.84	. . .	. . .	12.25

SOURCES: Ocean Freighting Research, Ltd. *World Freights, 1965: A Comprehensive Review of the Global Shipping Scene.* London: The British Sulphur Corporation, Ltd., 1966; T. Q. Hutchinson. *Heavy Grain Exports in Voyage Chartered Ships: Rates and Volume.* U.S. Department of Agriculture, ERS-Marketing Research Report 812, 1968; U.S. Department of Agriculture. *Agricultural Statistics,* 1949 through 1965, 1950 through 1966.

[a] To obtain the calculated rate for U.S. flag vessels, multiply the foreign flag rate by two.

weighted by the proportion of a particular commodity imported by each country in the region. Rates weighted by country imports within regions were developed for 1975, 1985, and 2000. They are presented in Table 13.16. Since variation in weighted rates over time was small, the 1985 rates were used for all time periods.

1965 PATTERNS AND LEVELS OF TRADE

The 1965 pattern of trade in cereals is summarized by the 13 trade areas in this section as background for the trade patterns projected to be optimal, in terms of the specific linear programming model, at future points in time. The 1965 patterns of trade in fertilizers and rock phosphate also are summarized. (Since 1965 was the most recent year for which reasonably complete data were available, data for that year were used.)

TABLE 13.16. Regional Inland Transportation Rates Developed for Use in the Analysis (dollars per metric ton)

Region	Inland Transportation Rates[a]		
	Wheat and Rye	Rice (milled)	Other Grain
United States	$ ...	$ ...	$ ...
Canada	...	8.44	8.44
Mexico	...	5.37	5.37
Caribbean	3.91	4.05	3.79
Central America	4.64	4.31	4.18
Northern South America	2.30	3.60	2.34
Brazil	2.49	...	2.49
Eastern South America	4.71	5.62	4.61
Southern South America	8.01	7.41	4.78
EEC	6.19	4.69	8.01
Ireland-United Kingdom	6.97	6.14	5.96
Scandinavia	5.38	6.56	4.17
Spain-Portugal	7.40	7.76	5.46
Austria-Switzerland	5.33	5.48	5.35
Northern E. Europe	5.21	5.21	5.21
Yugoslavia	...	7.87	7.87
Other East Europe	7.87	7.87	7.87
USSR	5.60	5.60	...
Greece-Turkey	6.56	7.36	7.30
United Arab Republic	4.13	...	4.13
Iran-Iraq	9.18	6.62	8.29
Other Middle East	4.75	5.61	3.96
Northern Africa	4.13	4.13	4.13
Western Africa	7.85	7.02	10.19
West Central Africa	5.13	4.91	4.96
Ethiopia-Sudan	9.58	...	10.07
East Central Africa	6.95	8.46	8.36
South Africa	3.40	3.40	3.40
Australia	...	...	...
New Zealand	7.94	7.94	...
India	3.42	3.42	4.50
Pakistan	5.08	5.08	5.08
Burma	6.23	...	6.23
Other Far East	4.00	7.14	2.68
South Korea	2.21	2.21	2.21
Malaya-Indonesia	5.38	5.27	5.56
Philippines-China (T)	5.29	5.16	5.14
Japan	5.18		5.19

[a] Where no imports of a particular cereal class are required by a region no rate is specified.

TRADE IN CEREAL GRAINS

The volume of world trade in cereal grains has increased constantly in postwar years, the annual rate of increase exceeding 6 percent. Feed grain trade has averaged nearly an 8 percent annual increase while food grain trade has increased by more than 5 percent annually. Corn trade alone has approached a 14 percent annual increase.

The United States and Canada are the largest exporters of food grain and account for about 60 percent of world food grain exports. The largest exporters of feed grains are the United States, Latin America, and EEC countries. Together they account for about 75 percent of world feed grain exports. Rice exports are greatest in Southeast Asia, accounting for 55 percent of world rice exports and over a third of total regional export earnings. The major grain importers are Japan, EEC countries, the European Free Trade Association (EFTA), Eastern Europe, Latin America, North Africa, and South Asia. The USSR and Communist Asia have become major importing regions since 1960. These nine major importing regions accounted for over 85 percent of world grain imports in 1964. Western Europe (EEC and EFTA) alone has accounted for about a third of world grain imports, a fifth of world wheat imports, 7 percent of world rice imports, and over 60 percent of imports of all other cereals. Trade statistics for wheat and rye, rice, and other grains are presented in Tables 13.17, 13.18, and 13.19 for 1965 by the trade areas of this study.

1965 IMPORT-EXPORT LEVELS FOR NITROGEN, PHOSPHATE, AND POTASH FERTILIZERS AND PHOSPHATE ROCK

Information on patterns of trade in fertilizers and phosphate rock is not generally available and data on import-export levels are incomplete. To conduct a separate study of fertilizer trade statistics in the detail required to gain data on past interregional trade in fertilizer was not considered feasible. Tables 13.20 and 13.21 contain net import-export statistics for fertilizer and phosphate rock trade in 1965. Since data were unavailable on trade in phosphate rock, the import-export statistics were developed as the difference between domestic rock consumption and production. FAO indicates that 20 percent of the total world supply of phosphate rock is used for technical purposes.[12] If the indicated phosphate rock export levels are reduced by this percentage, the resulting import-export levels are comparable.

Developing countries are the main importers of nitrogen and phosphate fertilizers, and imports are in most cases the fertilizer exports of developed regions. The presence or absence of reserves of potash and phosphate rock largely determine whether a region imports or exports these materials.

SULFUR AND PHOSPHATE ROCK REQUIREMENTS

The following summary on requirements indicates the projected level of phosphate rock requirements relative to known phosphate rock reserves. The data demonstrate clearly the vastness of phosphate rock reserves relative to projected fertilizer use or demand. Some concern

12. United Nations, Food and Agriculture Organization. *Fertilizer: An Annual Review of World Production, Consumption and Trade, 1966.* Rome, Italy, 1968.

TABLE 13.17. 1965 Net Import-Export Levels and Patterns of Trade among Trade Areas for Wheat and Rye (1,000 metric tons)

Importing Area	Exporting Areas (as numbered on the left)													Total Imports
	1	2	3	4	5	6	7	8	9	10	11	12	13	
1 United States		110												110
2 Canada														
3 Latin America	2,373	874					449							3,696
4 EEC	831	1,469	1,327							1				3,628
5 Other W. Europe	540	2,512	862	762			15			707				5,398
6 E. Europe	1,418	1,758	393	1,233			1,001							5,803
7 USSR	46	970		99						787				1,902
8 Middle East	2,433	38	29	704			60			400				3,664
9 Africa	989	88	4	831			1			61				1,974
10 S. Africa-Oceania	1	79	5											85
11 India-Pakistan	7,673	275		172			9			739				8,868
12 Other E. Asia	1,224	251		61						435				1,971
13 Japan	1,656	1,429	2	27						443				3,557
Total Exports	19,184	9,853	2,622	3,889			1,535			3,573				40,656

source: U.S. Department of Agriculture. *The World Grain Trade 1963–64/1964–65 and 10-Year Summary.* U.S. Department of Agriculture, ERS-Foreign 180, 1966.

TABLE 13.18. 1965 Net Import-Export Levels and Patterns of Trade among Trade Areas for Rice (1,000 metric tons of milled rice)

Importing Area	Exporting Areas (as numbered on the left) 1	2	3	4	5	6	7	8	9	10	11	12	13	Total Imports
1 United States														52
2 Canada	30		17							4				197
3 Latin America	144			1								53		178
4 EEC	52		31						3			92		240
5 Other W. Europe	77		32	56			1		19	5	4	46		260
6 E. Europe	10		51	13	60				42			84		202
7 USSR									64		20	118		316
8 Middle East	117		1	2	3						76	117		620
9 Africa	165		96								28	331		76
10 S. Africa-Oceania	71											5		1,182
11 India-Pakistan	220											962		321
12 Other E. Asia	321											494		817
13 Japan	290				33									
Total Exports	1,497		228	72	96		1		128	9	128	2,302		4,461

SOURCE: U.S. Department of Agriculture. World Trade in Selected Agricultural Commodities, 1951–65, vol. 2. U.S. Department of Agriculture, ERS-Foreign 45, 1968.

TABLE 13.19. 1965 Net Import-Export Levels and Patterns of Trade among Trade Areas for Other Grains (1,000 metric tons of corn equivalent)

Importing Area	Exporting Areas (as numbered on the left)													Total Imports
	1	2	3	4	5	6	7	8	9	10	11	12	13	
1 United States														
2 Canada	300													300
3 Latin America	349	34					290			3				676
4 EEC	7,743	154	3,786				184			426				12,292
5 Other W. Europe	3,682	207	470	1,342			119			388				6,208
6 E. Europe	193		66	246	15		865		1	30				1,416
7 USSR														
8 Middle East	884	25	11	21	1							5		947
9 Africa	122			33	1					37				193
10 S. Africa-Oceania	1													1
11 India-Pakistan	187									1				188
12 Other E. Asia	219	12		4						45				280
13 Japan	3,097	151	297							390		729		4,664
Total Exports	16,777	583	4,630	1,646	17		1,458		1	1,319		734		27,165

SOURCE: U.S. Department of Agriculture. *The World Grain Trade 1963–64/1964–65 and 10-Year Summary*. U.S. Department of Agriculture, ERS-Foreign 180, 1966.

TABLE 13.20. Trade Area Levels of Net Trade in Nitrogen and Phosphate Fertilizers, 1965 (1,000 metric tons of nutrient)

| Area | Net Exports and Imports by Type of Fertilizer | | | |
| | Nitrogen | | Phosphate | |
	Exports	Imports	Exports	Imports
United States		70	304	
Canada	156		28	
Latin America		163		80
EEC	1,324		419	
Other W. Europe		50		257
E. Europe		121		103
USSR	55		42	
Middle East		293		98
Africa		60		18
S. Africa-Oceania		29		9
India-Pakistan		303		26
Other E. Asia		212		194
Japan	631		66	
96-Nation Total	2,166	1,301	859	785

SOURCE: United Nations, Food and Agriculture Organization. *Fertilizer: An Annual Review of World Production, Consumption and Trade, 1966.* Rome, Italy, 1968.

TABLE 13.21. Trade Area Levels of Net Trade in Potash and Phosphate Rock, 1965 (1,000 metric tons of nutrient)

| Area | Net Exports and Imports by Type of Material | | | |
| | Potash (K_2O) | | Phosphate Rock (material) | |
	Exports	Imports	Exports	Imports
United States		236	15,943	
Canada	837			660
Latin America		139		824
EEC	1,163			7,696
Other W. Europe		1,163		4,530
E. Europe	257			4,144
USSR	314		1,841	
Middle East	177		688	
Africa		20	13,563	
S. Africa-Oceania		164	580	
India-Pakistan		89		453
Other E. Asia		92		852
Japan		583		1,476
96-Nation Total	2,748	2,486	32,615	20,635

SOURCE: United Nations, Food and Agriculture Organization. *Fertilizer: An Annual Review of World Production, Consumption and Trade, 1966.* Rome, Italy, 1968.

TABLE 13.22. Projected Phosphate Rock Requirements for 1975, 1985, and 2000 and Corresponding 1960 Calculated Consumption (1,000 metric tons)

Area	Level of Requirements for Specified Years			
	1960	1975	1985	2000
96-Nation Total	40,002	66,612	85,941	118,734
Developed Nations	36,108	57,840	72,822	98,538
Developing Nations	3,894	8,772	13,119	20,196
United States	9,768	17,043	23,445	36,000
Canada	660	1,212	1,482	1,938
Latin America	1,410	3,168	5,061	8,208
EEC	7,743	10,755	12,696	15,339
Other West Europe	4,530	6,090	7,269	9,057
East Europe	4,233	8,400	10,647	14,133
USSR	4,359	6,825	8,454	10,902
Middle East	732	1,317	1,755	2,367
Africa	447	1,773	2,457	3,750
S. Africa-Oceania	3,339	5,325	6,153	7,617
India-Pakistan	453	876	1,527	2,571
Other East Asia	852	1,638	2,319	3,300
Japan	1,476	2,190	2,676	3,552

currently prevails on whether available quantities of sulfur will be adequate to meet future needs. The requirement estimates for sulfur provide an indication of future sulfur requirements relative to present consumption.

The sulfur and phosphate rock statistics in Tables 13.22 and 13.23 were estimated from 1960 actual and 1975, 1985, and 2000 projected phosphate fertilizer use. The phosphate rock and sulfur requirements

TABLE 13.23. Projected Sulfur Requirement for 1975, 1985, and 2000 and Corresponding 1960 Calculated Consumption (1,000 metric tons)

Area	Level of Requirements for Specified Years			
	1960	1975	1985	2000
96-Nation Total	26,668	44,408	57,294	79,156
Developed Nations	24,072	38,560	48,548	65,692
Developing Nations	2,596	5,848	8,746	13,464
United States	6,512	11,362	15,630	24,000
Canada	440	808	988	1,292
Latin America	940	2,112	3,374	5,472
EEC	5,162	7,170	8,464	10,226
Other West Europe	3,020	4,060	4,846	6,038
East Europe	2,822	5,600	7,098	9,422
USSR	2,906	4,550	5,636	7,268
Middle East	488	878	1,170	1,578
Africa	298	1,182	1,638	2,500
S. Africa-Oceania	2,226	3,550	4,102	5,078
India-Pakistan	302	584	1,018	1,714
Other East Asia	568	1,092	1,546	2,200
Japan	984	1,460	1,748	2,368

were calculated as multiples of P_2O_5 requirements. The multipliers used were three tons of phosphate rock and two tons of sulfur per ton of P_2O_5. The phosphate rock multiplier was based on the assumption that phosphate rock is on average 33 percent P_2O_5. The sulfur multiplier is based on data of other studies.[13] These studies estimate that in 1971 total phosphate fertilizer production would be approximately 50 percent complex fertilizer, 28 percent normal superphosphate, 17 percent concentrated superphosphate, and 5 percent basic slag. Assuming the respective percentage of P_2O_5 in these fertilizers to be 46, 20, 46, and 16 percent and that their respective sulfur requirements per ton of P_2O_5 produced are 0.96, 0.60, 0.67, and zero, the sulfur requirement per ton of P_2O_5 produced is two tons.

Table 13.22 indicates the 2000 projected phosphate rock requirements to be 18.7 million tons. This quantity is only 0.6 percent of estimated phosphate rock reserves. Therefore, with annual consumption of phosphate at the 2000 projected level, known reserves would be adequate for more than 140 years. Phosphate rock reserves thus are not an important constraint on phosphate fertilizer use and production. Actual trade levels, distribution of reserves, and plant capacity for production of phosphate rock are presented in Chapter 15.

Table 13.23 includes projected requirements for sulfur in 1975, 1985, and 2000, as well as 1960 consumption, estimated in the manner outlined. McCune and Harre estimate 1971 total fertilizer and industrial use requirements for sulfur at 35 to 35.9 million metric tons.[14] Their estimate suggests the projections in Table 13.23 may be high. The 1975 estimated requirement is 12 million tons larger than the 1969 capacity of 32.23 million tons estimated by McCune and Harre. Hence, for the levels of fertilizer use projected in this study, considerable expansion of sulfur production capacity will be required.

13. D. L. McCune and E. A. Harre. "Trends and Prospects of World Fertilizer Production Capacity as Related to Future Needs." Unpublished paper presented at International Symposium on Industrial Development, Athens, Greece, Nov.-Dec., 1967. Mimeographed. Muscle Shoals, Alabama. TVA Division of Agricultural Development, 1967.

T. P. Hignett, "Outlook for the Phosphate Industry: A National and International View." Unpublished paper presented at Industrial Seminar of the Western Phosphate Region, Butte, Montana, Oct., 1966. Mimeographed Muscle Shoals, Alabama. TVA Division of Agricultural Development, 1966.

14. D. L. McCune and E. A. Harre. "Trend and Prospects of World Fertilizer Production Capacity as Related to Future Needs." Unpublished paper presented at International Symposium on Industrial Development. Athens, Greece, Nov.-Dec., 1967. Mimeographed. Muscle Shoals, Alabama. TVA Division of Agricultural Development, 1967.

CHAPTER 14

<hr />

Levels and Patterns of International Trade in Cereal Grains and Rice and Their Implication for Cropland Use and Foreign Capital Requirements

ALTERNATE MEANS of satisfying world food needs include redistribution of production and/or international trade in commodities produced. The analysis conducted here was directed toward determination of future regional production based on past trends and the nature and magnitude of redistribution international trade required to meet projected needs to the extent possible. Projected production and requirement levels have been presented. In this chapter, levels and patterns of international trade in cereal grains and rice required in 1975, 1985, and 2000 are presented and discussed.

Several features dominate the import-export data presented in Figures 14.1 to 14.3[1] and in Table 14.1. One is the preponderance of a very small number of developed nations in the export trade throughout the projection period. The United States, Canada, Australia, Japan, and the EEC export 82.1, 84.9, and 89.9 percent of the grains moving in international trade in 1975, 1985, and 2000, respectively. Another is the consistency with which imports (commodity deficits) by developing countries rise. Almost without exception, deficits increase substantially for all three commodity classes between each projection period.

The trading patterns presented in Tables 14.2 and 14.5 indicate the levels and patterns of trade in wheat and rye, rice, and other grains in 1975, 1985, and 2000. They are those patterns which result, given commodities "free" to move in trade at the 1963–65 export supply price in a manner which minimizes the commodity plus transportation costs of satisfying regional import requirements.

TRADE IN WHEAT AND RYE

A feature of Table 14.2 which becomes immediately apparent is complete removal of Canada from the international wheat and rye market. The price of Canadian wheat ($67.85 per ton), given Canada's

<hr />

1. A tabular presentation of the information contained in Figures 14.1 to 14.3 is contained in the Appendix.

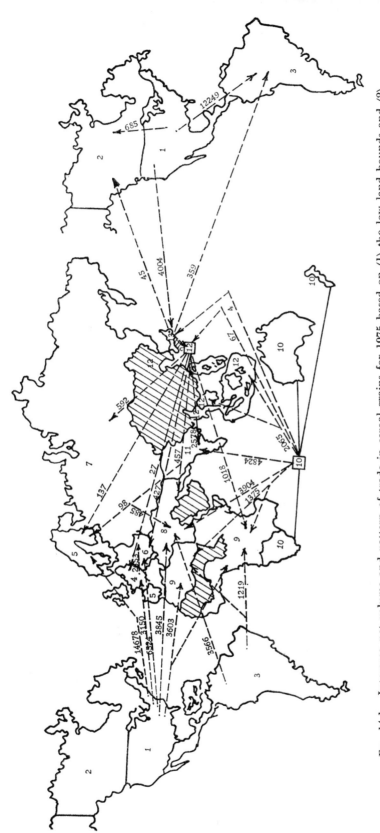

FIG. 14.1. Interarea net volume and pattern of trade in cereal grains for 1975 based on (1) the low land bounds and (2) fertilizer use assumption A (1,000 metric tons). Volume of cereal grain trade is the sum of projected trade in wheat and rye, rice, and other grains.

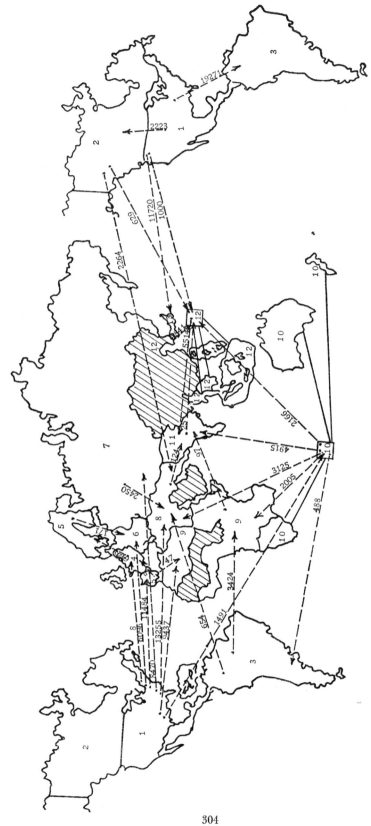

Fig. 14.2. Interarea net volume and pattern of trade in cereal grains for 1985 based on (1) the low land bounds and (2) fertilizer use assumption A (1,000 metric tons). Volume of cereal grain trade is the sum of projected trade in wheat and rye, rice, and other grains.

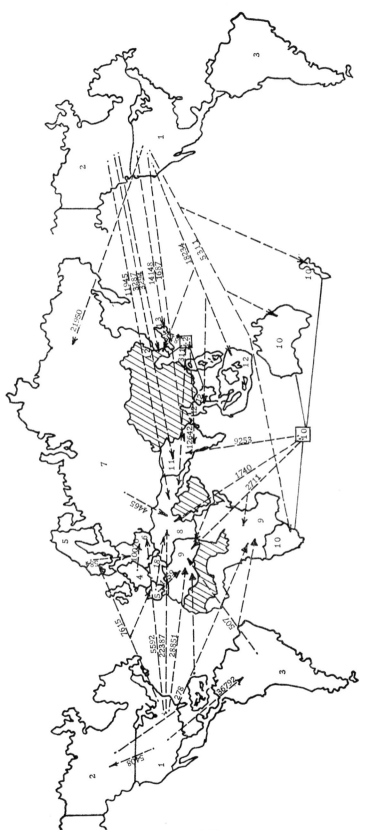

Fig. 14.3. Interarea net volume and pattern of trade in cereal grains for 2000 based on (1) the low land bounds and (2) fertilizer use assumption A (1,000 metric tons). Volume of cereal grain trade is the sum of projected trade in wheat and rye, wheat substituted for rice, rice, and other grains.

TABLE 14.1. Projected Regional Import-Export Levels for Cereal Grains and Rice in 1975, 1985, and 2000 (1,000 metric tons)

Region	Year	Exports Wheat and Rye	Rice	Other Grains	Wheat as a Rice Substitute	Imports Wheat and Rye	Rice	Other Grains	Wheat as a Rice Substitute
1 United States	1975	31,983		22,066					
	1985	48,485	2,171	33,809					
	2000	83,007	2,983	71,146	10,836.7				
2 Canada	1975						45	685	
	1985	2,893					55	2,168	
	2000	12,969			5,835		73	5,335	
3 Mexico	1975	554						494	
	1985	654	37	867					
	2000	252		882					
4 Caribbean	1975					652	266	220	
	1985					786	346	409	
	2000					1,027	346	777	10
5 Central America	1975					403	28	511	
	1985					545	79	1,109	
	2000					814	86.4	2,337	175
6 Northern South America	1975					5,077		610	
	1985		164			7,007		1,215	
	2000		238			11,305		2,626	118.6
7 Brazil	1975		37			3,896		810	
	1985		641			5,322		2,430	
	2000		386			8,043		6,540	
8 Western South America	1975					1,123	65	388	
	1985					1,531	162	773	
	2000					2,339	180.1	1,611	207.9

TABLE 14.1 (cont.)

Region	Year	Exports				Imports			
		Wheat and Rye	Rice	Other Grains	Wheat as a Rice Substitute	Wheat and Rye	Rice	Other Grains	Wheat as a Rice Substitute
9 Southern South America	1975	631		4,137			31		
	1985			4,206		488	63		
	2000			4,215		2,221	117		
10 European Economic Community	1975	2,653					27	3,150	
	1985	6,426					13		
	2000	10,070							
11 Ireland-United Kingdom	1975					4,898	121	6,657	
	1985					4,537	133	3,841	
	2000					4,707	152	371	
12 Scandinavia	1975					980	42		
	1985					1,539	46		
	2000					2,052	51		
13 Spain-Portugal	1975					454		422	
	1985		52			330		60	
	2000		45	1,082		57			
14 Austria-Switzerland	1975					126	72	1,141	
	1985	171					78	920	
	2000	584					85	139	
15 Northern East Europe	1975					8,068	331		
	1985					10,178	360		
	2000					12,526	395		

TABLE 14.1 (cont.)

Region	Year	Exports				Imports			
		Wheat and Rye	Rice	Other Grains	Wheat as a Rice Substitute	Wheat and Rye	Rice	Other Grains	Wheat as a Rice Substitute
16 Yugoslavia	1975						48	1,109	
	1985						54	2,040	
	2000						62	3,186	
17 Other East Europe	1975						48		
	1985						61		
	2000						77		
18 USSR	1975	450		535			592		
	1985			2,450		7,756	714		
	2000			4,465		21,043	907		
19 Greece-Turkey	1975					1,598	78	3,082	
	1985					3,908	131	4,995	
	2000					8,183	143.6	8,001	73.4
20 United Arab Republic	1975					2,462		1,235	
	1985		540			3,606		2,361	
	2000		361			5,828		4,499	
21 Iran-Iraq	1975					842		925	
	1985					1,932	59	1,785	
	2000					4,387	311	3,550	
22 Other Middle East	1975					1,608	166	761	
	1985					2,487	226	1,054	
	2000					4,254	274.4	1,651	72.6
23 Northern Africa	1975					2,312	33	855	
	1985					4,021	47	1,997	
	2000					7,089	51.7	4,041	24.3

TABLE 14.1 (cont.)

Region	Year	Exports Wheat and Rye	Exports Rice	Exports Other Grains	Exports Wheat as a Rice Substitute	Imports Wheat and Rye	Imports Rice	Imports Other Grains	Imports Wheat as a Rice Substitute
24 Western Africa	1975					436	410	856	
	1985					573	665	2,441	
	2000					883	902.7	7,364	485.3
25 West Central Africa	1975					148	68	326	
	1985					176	101	725	
	2000					246	129.4	1,685	62.6
26 Ethiopia-Sudan	1975					114		544	
	1985					158		1,180	
	2000					278		2,777	
27 East Central Africa	1975					373		740	
	1985		97			536		2,303	
	2000					902		6,490	44
28 S. Africa	1975					200	67		
	1985			606		392	88	853	
	2000					973	133	3,977	
29 Australia	1975	9,431		2,254					
	1985	10,430	137	2,137					
	2000	9,994	162	1,996	1,559.3				
30 New Zealand	1975					123			
	1985					158			
	2000					228			
31 India	1975		218			720		2,046	
	1985		71					3,850	
	2000	900					2,839.6	8,763	727.4

TABLE 14.1 (cont.)

Region	Year	Exports — Wheat and Rye	Exports — Rice	Exports — Other Grains	Exports — Wheat as a Rice Substitute	Imports — Wheat and Rye	Imports — Rice	Imports — Other Grains	Imports — Wheat as a Rice Substitute
32 Pakistan	1975					3,020	1,216	520	
	1985					4,408	2,049	909	
	2000					7,390	1,387.1	1,687	4,074.9
33 Burma	1975		3,064	15		33			
	1985		3,724			25		2	
	2000		4,536			11		42	
34 Other Far East	1975		1,070	2,506		189			
	1985			4,463		234	197		
	2000			8,416		309	630.4		
35 South Korea	1975					496	704	574	
	1985					629	1,222	1,000	
	2000					884	655	1,879	4,026.6
36 Malaya-Indonesia	1975			733		416	1,450		
	1985			677		541	2,701		
	2000			43		808	2,463.2		1,850
37 Philippines-China (T)	1975					931	1,027	441	
	1985					1,259	2,820	1,381	
	2000					1,943	4,403.4	3,539	3,798.8
38 Japan	1975		2,519			4,004		5,111	
	1985		4,841			4,897		6,823	
	2000		8,152			6,196		7,952	2,479.6
96-Nation Totals	1975	45,702	6,908	33,719		45,702	6,908	33,719	
	1985	69,959	12,475	48,624		69,959	12,475	48,624	
	2000	116,876	16,863	91,313	18,231	116,876	16,863	91,313	18,231
Developing Countries	1975	1,185	4,389	8,258		26,849	5,542	15,444	
	1985	1,554	5,274	10,228		40,172	10,868	31,919	
	2000	252	5,521	12,674		69,094	14,921	70,353	18,231
Developed Countries	1975	44,517	2,519	25,461		18,853	1,397	18,275	
	1985	68,405	7,201	38,396		29,787	1,607	16,705	
	2000	116,624	11,342	78,639	18,231	47,782	1,942	20,960	18,231

TABLE 14.2. 1975, 1985, and 2000 Interarea Net Trade Statistics for Wheat and Rye, Assuming (1) the Low Land Bounds and (2) Fertilizer Use Assumption A (trade flows in 1,000 metric tons; value of imports, c.i.f., in million dollars)

Importing Area	Year	Exporting Areas (as numbered on the left)													Total Imports	Value of Imports
		1	2	3	4	5	6	7	8	9	10	11	12	13		
1 United States	**1975**															
	1985															
	2000															
2 Canada	1975															
	1985															
	2000															
3 Latin America	1975	10,520													10,520	807
	1985	15,191									488				15,679	1,205
	2000	25,497													25,497	1,976
4 EEC	1975															
	1985															
	2000															
5 Other W. Europe	1975	6,458													6,458	506
	1985	6,406													6,406	502
	2000	6,817													6,817	534
6 East Europe	1975	5,415			2,653										8 068	607
	1985	3,581			6,426	171									10,178	739
	2000	1,872			10,070	584									12,526	889
7 USSR	1975	7,756													7,756	611
	1985	21,043													21,043	1,662
	2000															
8 Middle East	1975	2,638									2,868				6,510	506
	1985	5,510	2,264	554				450			3,125	380			11,933	956
	2000	9,078	11,872	654							1,702				22,652	1,885
9 Africa	1975	2,748									635				3,383	267
	1985	4,594	278								870				5,464	431
	2000	8,168									902				9,348	743

311

TABLE 14.2 (cont.)

Importing Area		1	2	3	4	5	6	7	8	9	10	11	12	13	Total Imports	Value of Imports
								Exporting Areas (as numbered on the left)								
10 S. Africa Oceania	1975	200													200	16
	1985	550													550	43
	2000	1,201													1,201	94
11 India-Pakistan	1975										3,740				3,740	269
	1985										3,913				3,913	283
	2000										7,390				7,390	535
12 Other E. Asia	1975										2,065				2,065	144
	1985		629								2,034	25			2,688	194
	2000	3,136	819												3,955	323
13 Japan	1975	4,004													4,004	308
	1985	4,897													4,897	377
	2000	6,196													6,196	476
World Totals	1975	31,983	2,893	554	2,653	171		450			9,308				44,948	3,429
	1985	48,485	12,969	654	6,426	584					10,430	405			69,464	5,340
	2000	83,008			10,070						9,994				116,625	9,119

TABLE 14.3. 2000 Interarea Net Trade Statistics for Wheat Substituted for Rice, Assuming (1) the Low Land Bounds and (2) Fertilizer Use Assumption A (trade flows in 1,000 metric tons; value of imports, c.i.f., in million dollars)

Importing Area	Exporting Areas (as numbered on the left)													Total Imports	Value of Imports
	1	2	3	4	5	6	7	8	9	10	11	12	13		
1 United States															
2 Canada															
3 Latin America	512													512	39
4 EEC															
5 Other W. Europe															
6 East Europe															
7 USSR															
8 Middle East	73	73												146	12
9 Africa	572									44				616	51
10 S. Africa-Oceania															
11 India-Pakistan		3,287								1,515				4,802	385
12 Other E. Asia	9,680	2,475												12,155	999
13 Japan															
Total Exports	10,837	5,835								1,559				18,231	1,486

313

TABLE 14.4. 1975, 1985, and 2000 Interarea Net Trade Statistics for Rice, Assuming (1) the Low Land Bounds and (2) Fertilizer Use Assumption A (trade flows in 1,000 metric tons; value of imports, c.i.f., in million dollars)

Importing Area		Exporting Areas (as numbered on the left)													Total Imports	Value of Imports
		1	2	3	4	5	6	7	8	9	10	11	12	13		
1 United States	1975															
	1985															
	2000															
2 Canada	1975														45	45	7
	1985	55														55	9
	2000	73														73	12
3 Latin America	1975														359	359	50
	1985	574														574	96
	2000	613														613	102
4 EEC	1975													27		27	3
	1985	8				5										13	2
	2000																
5 Other W. Europe	1975											98	137		235	31
	1985	257														257	43
	2000	288														288	48
6 East Europe	1975												427		427	53
	1985	475														475	80
	2000	534														534	90
7 USSR	1975												592		592	73
	1985	714														714	120
	2000	907														907	153
8 Middle East	1975												244		244	30
	1985																
	2000					18					38		311		367	48
9 Africa	1975			37									474		511	64
	1985			766		47									813	119
	2000	436		507		27					115				1,085	172

314

TABLE 14.4 (cont.)

Importing Area		Exporting Areas (as numbered on the left)													Total Imports	Value of Imports
		1	2	3	4	5	6	7	8	9	10	11	12	13		
10 S. Africa-Oceania	1975												67	4	71	8
	1985	88													88	15
	2000	133													133	22
11 India-Pakistan	1975												1,096		1,096	129
	1985								124	97			1,757		1,978	243
	2000										2		4,225		4,227	491
12 Other E. Asia	1975													2,111	2,111	283
	1985										132			4,841	4,973	671
	2000													8,152	8,152	1,104
13 Japan	1975															
	1985															
	2000															
Total Exports	1975			37								98	3,064	2,519	5,718	731
	1985	2,171		766		52			124	97	132		1,757	4,841	9,940	1,399
	2000	2,983		507		45					155		4,536	8,152	16,378	2,242

315

TABLE 14.5. 1975, 1985, and 2000 Interarea Net Trade Statistics for Other Grains, Assuming (1) the Low Land Bounds and (2) Fertilizer Use Assumption A (trade flows in 1,000 metric tons; value of imports, c.i.f., in million dollars)

Importing Area		Exporting Areas (as numbered on the left)													Total Imports	Value of Imports
		1	2	3	4	5	6	7	8	9	10	11	12	13		
1 United States	1975															
	1985															
	2000															
2 Canada	1975	685													685	43
	1985	2,168													2,168	136
	2000	5,335													5,335	336
3 Latin America	1975	1,729													1,729	115
	1985	3,506													3,506	234
	2000	10,170													10,170	681
4 EEC	1975	3,150													3,150	214
	1985															
	2000															
5 Other W. Europe	1975	8,220													8,220	554
	1985	4,821													4,821	325
	2000	510													510	34
6 East Europe	1975	1,109													1,109	79
	1985	2,040													2,040	144
	2000	3,186													3,186	225
7 USSR	1975															
	1985															
	2000															
8 Middle East	1975	1,207		3,012				535			1,036				6,003	425
	1985	7,745						2,450							10,195	746
	2000	13,236						4,465					213		17,701	1,294
9 Africa	1975	855									740				3,321	229
	1985	4,853		1,182							1,185				8,646	627
	2000	19,675		2,658		1,032					1,650		544		22,357	1,685

316

TABLE 14.5 (cont.)

Importing Area		Exporting Areas (as numbered on the left)													Total Imports	Value of Imports
		1	2	3	4	5	6	7	8	9	10	11	12	13		
10 S. Africa-Oceania	1975															
	1985	853													853	58
	2000	3,977													3,977	270
11 India-Pakistan	1975										1,084		1,482		2,566	174
	1985										1,002		3,757		4,759	322
	2000	1,687									346		8,417		10,450	717
12 Other E. Asia	1975															
	1985	1,000													1,000	69
	2000	5,418													5,418	387
13 Japan	1975	5,111													5,111	344
	1985	6,823													6,823	459
	2000	7,952													7,952	535
Total	1975	22,066		4,194				535			2,860		2,239		31,894	2,177
	1985	33,809		2,658				2,450			2,137		3,757		44,811	3,118
	2000	71,146				1,032		4,465			1,996		8,417		87,056	6,164

317

location relative to wheat markets, is not competitive with the lower U.S. price even with substantially higher transportation rates for U.S. wheat exports to developing countries.

In 1965 net interarea trade in wheat and rye totaled 40.7 million tons. The most noteworthy aspect of the 1975 level is perhaps the small increase of 4.3 million tons or 10.5 percent over the decade. This result is partly due to absence of Mainland China from the model, the grouping of all wheat and rye together without regard to quality differences, and the simplifying assumptions underlying the model.

Projections for 1975 also show increasing dominance of the export market for wheat and rye by developed trade areas. Only slightly more than half a million tons are exported by developing areas. In 1975, 58 percent of total imports are imports to developing countries, compared with 50 percent in 1965. Over the same period, imports of developed areas fall 2 million tons while those of developing regions rise 6 million tons.

Some trade areas experience marked changes in the level of wheat and rye imports received or level of exports in 1975. Imports to Latin America rise some 7 million tons while those for India-Pakistan fall from a 1965 level of 8.9 million tons to 3.7 million tons. As mentioned earlier, Canada does not export any wheat and rye in 1975 under the conditions of the analysis. Exports of more competitive regions, namely, Australia and the United States, replace Canadian exports. Export increases of 6 million tons are projected for Australia. Those of the United States rise 12.8 million tons.

By 1985 a 96-nation total import-export level of 69.5 million tons is projected for wheat and rye. That projection is 28.8 million tons above the 1965 level and represents an annual export-import increase of 1.44 million tons. The largest import increases projected for the decade 1975–85 occur in the areas of Latin America, the USSR, and the Middle East where imports rise 5.2, 7.8, and 5.5 million tons, respectively. The United States exports 48.5 million of the 69.5 million tons exported in 1985. Other regions which export substantial quantities include Australia, 10.4 million tons; the EEC, 6.4 million tons; and Canada, 2.9 million tons. Canada's entry into the wheat and rye export market reflects exhaustion of export supplies in U.S. ports with locational advantages relative to Canadian ports.

In 2000, if past production trends and the most probable set of requirements prevail, trade in wheat and rye will reach 116.6 million tons; a level nearly triple that of 1965. In projections for the year 2000, 59 percent of total wheat and rye exports are received by developing countries and the entire 96-nation total export volume originates in developed areas. Imports in excess of 20 million tons are received by each of Latin America, the USSR, and the Middle East. In 1965 imports to all three areas were 9.3 million tons. Exports from the United States total 83 million tons and account for 71 percent of 2000 wheat and rye exports. By 2000, import requirements are such that Canada becomes a significant exporter, exporting some 13 million tons.

PROJECTED TRADE CONDITIONS FOR RICE

Excess requirements for rice and consequently international trade in rice are small relative to the commodity classes wheat and rye and other grain. In 1965, 4.5 million tons of rice were traded relative to a trade volume totaling 67.8 million tons for wheat and rye plus other grains. The corresponding trade levels for each projection period are 5.7 million tons to 76.8 million tons, 9.9 million tons to 114.3 million tons, and 16.4 million tons to 203.7 million tons in 1975, 1985, and 2000, respectively. As indicated in Table 14.1, the projections show Japan, Burma, and the United States as the major exporters. Other regions export rice but the quantities they export are small and are in general projected to decline.

The patterns of area trade for rice presented in Table 14.4 further illustrate the major rice export roles of Japan, Burma, and the United States. They also clearly show the competitiveness of the limited quantities of rice exports available from lesser exporters. In 1975 the United States exports no rice. However, in 1985 when supplies are not available from developing regions which formerly had small quantities to export, the United States becomes a rice exporter. In 2000 there is an overall rice deficit and all available supplies are exported prior to the substitution of wheat for rice. Tables 14.1 and 14.3 show the 18.2 million tons of wheat required as a substitute for rice with Table 14.3, in addition, illustrating the sources of wheat used as a rice substitute.

Certain dimensions of the rice situation projected for each projection period stand out. The share of total imports received by developing areas increases from a 1965 level of 59 percent to 76 percent in 1975. Of total regional imports, 4.4 million tons of the 6.9 million tons are received by Pakistan, South Korea, Malaya-Indonesia, and Philippines-China (T). Table 14.4 shows that all except 135,000 tons of rice imports come from the trade areas of Japan and Other East Asia. As mentioned earlier, no rice exports flow from the United States in 1975.

Total interarea rice trade rises 4.2 million tons between 1975 and 1985 as compared with 1.3 million tons in the preceding decade. The inability of producers in India and Pakistan and countries in Other Far East trade area to keep pace with rising requirements results in rising imports to those areas. In 1985 exports from regions in Other Far East are only 57 percent of their 1975 level and imports to regions in that trade area rise 136 percent. In 1985 U.S. rice exports rise from a 1975 level of zero to 2.2 million tons.

TABLE 14.6. 2000 Minimum Rice Requirements for Geographic
Regions Which Required Them (1,0000 metric tons)

Geographic Region	Minimum Requirement
Caribbean	346.0
Central America	86.4
Eastern South America	180.1
Southern South America	34.4
Greece-Turkey	143.6
Iran-Iraq	212.6
Other Middle East	274.4
Northern Africa	51.7
Western Africa	902.7
West Central Africa	129.4
India-Ceylon	688.7
Pakistan	321.6
Other Far East	630.4
South Korea	655.0
Malaya-Indonesia	1,104.8
Philippines-China (T)	4,403.4

In 2000 all available rice exports are utilized since an overall rice
deficit is projected. The result is interarea rice trade of 16.4 million
tons. The volume is distributed in accordance with the constraint that
all developed country import requirements be met and no developing
country's supply be allowed to fall below the 1960 per capita level of
availability. The resulting regional minimum rice requirements are
presented in Table 14.6. Given their satisfaction and distribution of
all available rice exports, 18.2 million tons of wheat were imported in
lieu of rice. Developing countries dominate the rice import market
with Africa, India-Pakistan, and Other East Asia, respectively, accounting for 1.1, 4.2, and 8.2 million of the 16.4 million tons imported.

TRADE IN OTHER GRAINS CONSIDERED

While one might hypothesize that developed regions with relatively
high income would have, relative to developing regions, proportionally
higher import demands for feed grains, the projections presented indicate the opposite. A high and increasing proportion of feed grain imports for each projection period are imports of developing regions.

Those regions which account for most of the 96-nation total exports
are the United States, Southern South America, the USSR, Australia,
and the Other Far East. Of these, the United States is by far the largest
exporter with exports of 22.1, 33.8, and 71.1 million tons in each of 1975,
1985, and 2000 relative to 96-nation total projected 1975, 1985, and 2000
regional exports of 33.7, 48.6, and 91.3 million tons.

As shown in the import data of Table 14.1, each of Greece-Turkey,

India, and Japan consistently accounts for some 10 percent of total imports in each projection period but no importing region imports a disproportionately large percentage of total projected imports. The increasing levels of other grain imports by Canada, the decline of those for Ireland-United Kingdom, and the increasing role of the USSR as an exporter of other grains are noteworthy.

In respect to other grains, certain dimensions of Table 14.5 are interesting. Other grain trade in 1975 of 31.9 million tons reflects a relatively small increase over the 1965 level of 27.2 million tons. However, some significant changes in the position of particular countries occur. For instance, 1965 net imports of the EEC totaled 12.3 million tons. In 1975 the area imports only 3.1 million tons. Other grain imports increase most for India-Pakistan, Africa, and the Middle East.

As in the case of wheat and rye, and rice, net trade in other grains is projected to rise much more between 1975 and 1985 than for the previous decade. Trade in other grains rises 12.9 million tons in the period 1975–85 while only a 4.7 million ton increase is projected for 1965–75. Increased exports from the United States account for 11.7 million tons or 91 percent of the total interarea increase.

The areas with the largest increases in imports are the Middle East and Africa. Their other grain imports rise 4.2 and 5.3 million tons, respectively. The decline of EEC other grain imports from 3.1 million tons in 1975 to zero in 1985 is the largest single reduction in imports.

Other grain trade among regions is projected to reach 87 million tons by 2000. That projection implies an average annual increase of 1.7 million tons over the 35-year period 1965–2000. Between 1951 and 1965 trade in other grains rose at an annual average rate of 1.6 million tons.[2] Since the 1951–65 rate is expressed in terms of unadjusted feed grain volume rather than corn equivalent units used in the analysis, the results are not directly comparable. However, the historic pattern does indicate a trend of similar order. Unlike the 1965 situation where developing areas imported 2.3 million tons or 8.4 percent of total other grain imports, developing areas import 57.3 million tons or 63 percent of the 2000 total. Exports from developed trade areas account for 78.6 million tons of total other grain exports in 2000. The four developing trade areas—Latin America, the Middle East, Africa, and India-Pakistan —import more than 10 million tons each. The largest importer, Africa, receives 22.4 million tons. The major exporter of other grains in 2000, the United States, exports 71.1 million tons. The only developing nations exporter, Other East Asia, provides exports of 8.4 million tons.

2. L. L. Blakeslee. "An Analysis of Projected World Food Production and Demand in 1970, 1985, and 2000." Unpublished Ph.D. thesis. Ames: Iowa State University Library, 1967.

CROPLAND AREA AND CEREAL GRAIN IMPORT COST IMPLICATIONS

Two specific dimensions which the international grain import-export analysis elucidates are (a) the extent to which projected cropland area production capacity is utilized to satisfy regional import requirements and (b) costs to be incurred by importing regions to meet their excess requirements.

Table 14.7 indicates (a) the cropland upper bounds under the low land assumption, (b) 1965 reported cropland area,[3] (c) cropland area projected for each projection period,[4] and (d) projected area not used.[5] Review of the statistics presented there indicates that over the projection period projected production increases keep pace with increased requirements through 1985 but by 2000 decreases in projected area not used suggest pending future deficits. Vast areas of potential cropland remain unused, especially in Latin America, to say nothing of Africa where it was not possible to develop or obtain either estimates of potential cropland or projected cropland area.

As indicated in Table 14.7, 1975 projected cropland areas are only slightly different from those of 1965 in the EEC, Other West Europe, East Europe, and the USSR. In all other areas, sizable increases in area are projected. Areas with considerable projected capacity not utilized are the United States, Canada, Other West Europe, and East Europe. In the United States, the area required is some 4 million hectares (10 million acres) below the 1965 cropland area. Due to her failure to export wheat in 1975, Canada's excess capacity is very large.

The 1985 analysis indicates that most of the projected land area not utilized will be in Canada and the United States. Excesses of 15.6 and 8.3 million hectares are projected for the United States and Canada.[6]

The total 1985 U.S. cropland requirement of 86.7 million hectares is only 6.3 million hectares more than that projected for 1975. Comparison of the 1985 area projections with the cropland upper bounds shows the United States, Latin America, and the Middle East to have considerable potential for cropland area expansion. In Latin America, where potential is greatest, over 200 million hectares of potential cropland remain unused.

The 2000 cropland area situation indicates a trend toward complete utilization of projected capacity. It, as do solutions for earlier periods, indicates the competitiveness of developing world exporters. They utilize their entire projected cropland area for domestic and export

---

3. Forage crop area, orchards, and pastures are the main arable areas not included.
4. Area projected, using procedures explained in Chapter 4.
5. The amount of projected area not required to meet world grain demand restraints in the programming model.
6. The projections are based on U.S. yields of wheat and rye = 38.8 bushels, rice = 5,788 pounds (paddy), and other grains = 96.4 bushels per acre and Canadian wheat and rye yields estimated at 25.4 bushels per acre.

TABLE 14.7. Cropland Area Statistics for 1975, 1985, and 2000 as Related to Production and Trade Projections under the Low Land Bound Assumption (1,000 hectares)

Trade Area	1965 Cropland Area	Cropland Area Upper Bound	Projected Cropland Area		Projected Cropland Area Not Used
1 United States	84,035	125,000	1975 1985 2000	95,048 102,331 115,746	14,627 15,584 10,335
2 Canada	19,925	22,000	1975 1985 2000	21,523 21,723 22,001	10,001 8,337 none
3 Latin America	82,913	319,221	1975 1985 2000	90,737 100,559 114,754	512 none none
4 EEC	25,362	27,850	1975 1985 2000	25,622 25,169 24,317	none 829 2,659
5 Other W. Europe	21,513	21,515	1975 1985 2000	20,174 20,011 19,951	1,494 1,868 2,206
6 East Europe	41,989	43,670	1975 1985 2000	41,992 41,728 41,634	1,744 2,600 3,629
7 USSR	146,546	145,000	1975 1985 2000	144,999 145,000 145,000	none none none
8 Middle East	29,870	35,175	1975 1985 2000	32,238 33,344 33,799	171 none none
9 Africa	n/a	n/a	1975 1985 2000	n/a n/a n/a	121 6 none

TABLE 14.7 (cont.)

Trade Area	1965 Cropland Area	Cropland Area Upper Bound	Projected Cropland Area		Projected Cropland Area Not Used
10 S. Africa-Oceania	10,342	12,220	1975 1985 2000	12,182 12,193 12,210	42 18 20
11 India-Pakistan	165,563	189,702	1975 1985 2000	172,645 183,303 188,928	none none none
12 Other E. Asia	45,488	89,965	1975 1985 2000	52,620 57,897 63,802	none none none
13 Japan	5,606	6,470	1975 1985 2000	5,582 5,242 5,197	119 none none

TABLE 14.8 Per Capita and Per Hectare Equivalent Costs of Developing Nation Grain Imports

Projection Period	Per Capita Equivalent	Per Hectare (Acre) Equivalent
1975	$1.90	$ 8.40 ($ 3.40)
1985	2.65	13.40 (5.40)
2000	4.20	27.35 (11.05)

production while production potential of some developed regions, mainly the United States and regions in Continental Europe, go unused. In the United States, based on yields of 3.23 tons (48 bushels) of wheat and rye and 8.31 tons (132 corn equivalent bushels) of other grains per hectare (acre), 10.3 (25.5) million hectares (acres) of idle land are forecast relative to projected cropland area totaling 115.7 (286.0) million hectares (acres).

The import levels (see Table 14.1) projected by region in this analysis imply massive foreign exchange requirements for many developing nation trade areas. The estimated magnitudes are presented in Tables 14.2 to 14.6. When import levels on the order of those projected and associated foreign capital requirements are considered, increased domestic production alternatives warrant the attention of developing nations with scarcities of foreign capital.

In the analyses conducted, estimates of the per capita and per hectare equivalent cost of grain imports were developed. Those estimates are presented in Table 14.8.

Per hectare equivalent costs of the magnitude specified in Table 14.8 indicate the need to investigate potential returns to investment of similar order in domestic agriculture. If increasing returns to investment in agriculture exist for a country or countries, and are large enough, borrowing to finance high early investment in agriculture may be expedient and desirable. However, a word of caution is required. To ignore the influence of institutional and cultural factors and the level of development in other sectors of a nation's economy on its ability to respond to production technology would be foolhardy and potentially disastrous.

PROJECTED GRAIN TRADE AND ASSOCIATED IMPLICATIONS— A SUMMARY STATEMENT

To provide an overview summary of the 96-nation grain trade in each projection period, Figures 14.1 to 14.3 were developed. They illustrate the total volume of grains (wheat and rye, rice, and other grain) which flows among the 13 trade areas in each projection period. Comparisons made among them give an immediate indication of the vast increases in volume of grain trade projected. The essence of the material presented in those figures and throughout this chapter is that (a) given the population, income, and grain production projections employed[7] and limited substitution of wheat for rice, grain production will be adequate through the year 2000; (b) the volume of international trade in grains can rise spectacularly; (c) the dependence of developing nations on grain imports from developed nations can increase dramatically; (d) potential for intensification of cropland area use, the development of new and the introduction of existing advanced technologies and development, and use of vast areas of unused potential cropland area will remain great, and (e) the high foreign capital requirements which projected grain imports dictate will provide strong incentive for developing nations to explore potential means of increasing domestic agricultural production and/or earning foreign currencies.

---

7. Namely, medium population growth, low but not declining per capita income growth, and grain production projected to increase according to past trends.

CHAPTER 15

Trade in Fertilizers and Phosphate Rock and the Implication for Regional Plant Capacity

THE FERTILIZER USE projection methodology is presented in Chapter 12. The projected levels of fertilizer use and the specification of 1975 fertilizer and phosphate rock plant capacities are presented in Chapter 13. Raw material reserves are specified in Table 13.12. These provide the basis for an analysis of (a) future trade in fertilizers and phosphate rock and (b) the utilization and expansion of fertilizer and phosphate rock plant capacities throughout the world. Such an international analysis of the fertilizer industry was conducted. The purpose of this chapter is to present the results and implications of that analysis for each of the 1975, 1985, and 2000 projection periods.

To facilitate the understanding and interpretation of results to be presented in the chapter, several aspects of the analysis, most or all of which have been previously discussed, bear repeating: (a) the assumption that in all except the 1975 projection period the model required that total nitrogen and phosphate fertilizer requirements be met through domestic production, (b) the assumption that potash and phosphate rock plant construction costs, which must be borne in order to increase capacity through the model for the 1975 and 1985 solution, are equal in all 38 regions and that all operate at 80 percent efficiency,[1] and (c) that plants on stream (in production) in 1975 and on which 1975 capacity is based become defunct before 2000. Assumption c removes any advantage afforded regions with existing capacity for the 1975 and 1985 analyses.

IMPORT-EXPORT LEVELS IN THE FERTILIZER INDUSTRY

Given the specified assumptions and the supporting data discussed above, trade and productive capacity analyses were conducted for the 38-region model which encompasses fertilizer demand of 96 nations. Figures 15.1 to 15.4 illustrate the projected patterns of fertilizer trade

---

1. The effect of this assumption, given that some countries have existing plant capacity, is to make importing of potash and phosphate rock the generally least cost means of satisfying excess requirements even for countries with reserves.

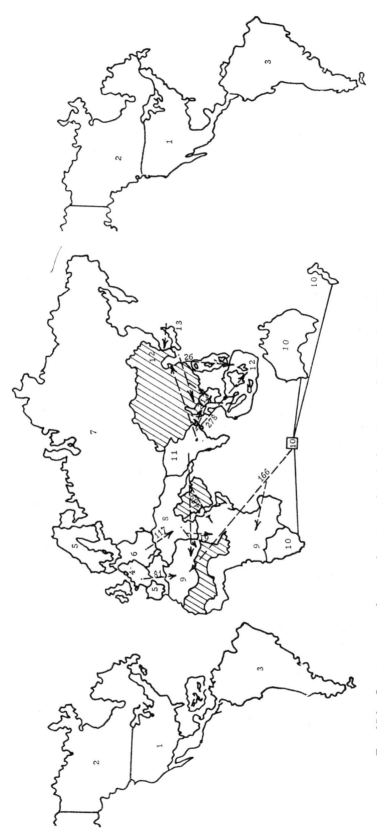

Fig. 15.1.   Interarea net volume and pattern of trade for nitrogen fertilizer for 1975, assuming (1) the low land bounds and (2) fertilizer use assumption A (1,000 metric tons of 46 percent material).

Fig. 15.2. Interarea net volume and pattern of trade for phosphate fertilizers for 1975, assuming (1) the low land bounds and (2) fertilizer use assumption A (1,000 metric tons of 46 percent material).

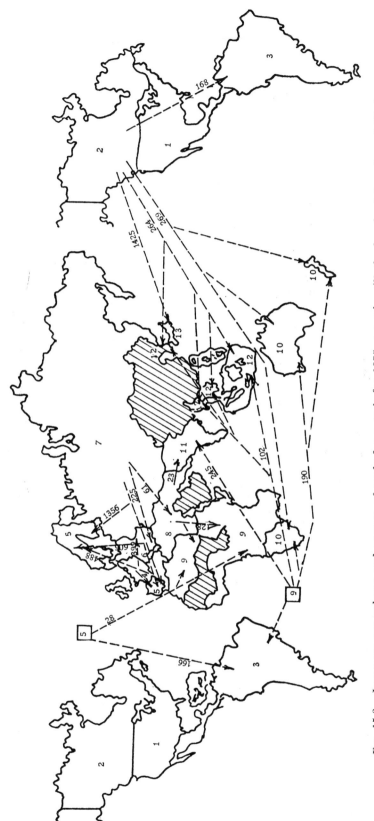

Fig. 15.3. Interarea net volume and pattern of trade for potash for 1975, assuming (1) the low land bounds and (2) fertilizer use assumption A (1,000 metric tons).

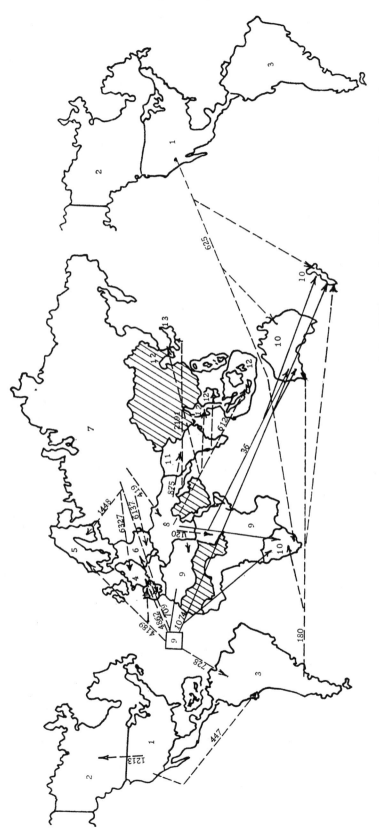

Fig. 15.4. Interarea net volume and pattern of trade for phosphate rock for 1975, assuming (1) the low land bounds and (2) fertilizer use assumption A (1,000 metric tons).

for 1975 projections. Import-export levels which result from the analysis for each projection period are presented in Table 15.1. The data presented mirror conditions in the fertilizer industry quite vividly. They indicate that (a) developed regions provide most of the nitrogen and phosphate fertilizer exports (b) while many developing regions import nitrogen and phosphate fertilizers, the quantities are generally small, and (c) the major portion of the large quantities of potash and phosphate rock moving in trade are imports of developed regions. That situation is a direct reflection of conditions in the industry, namely, one where developed regions, which use by far the major portion of fertilizers consumed, have large fertilizer plant capacities relative to domestic requirements.[2] Developing regions, which use very little fertilizer, have extremely limited domestic plant capacities even relative to domestic needs, and the location of world potash and phosphate rock reserves dictates possible locations for potash and phosphate rock production plants. Therefore, as shown in the table, developed regions with excess capacity export limited quantities of nitrogen and phosphate fertilizers to developing regions with limited requirements and even more limited domestic capacity, while regions with reserves and plant capacity for production of potash and phosphate rock export large quantities, the largest portion of which is imported by high fertilizer use regions, namely the developed ones. In respect to potash and phosphate rock production and trade in 2000, the results are somewhat modified since plants in use by 1975 have become obsolete and must be replaced.

PROJECTED INTERAREA TRADE IN FERTILIZERS AND PHOSPHATE ROCK

The fertilizer industry patterns of interarea trade are presented in Tables 15.2 to 15.5. The most significant feature of Tables 15.2 and 15.3, indicating patterns of trade in nitrogen and phosphorus, is the small magnitude of trade in those commodities. The absence of 1985 and 2000 trade statistics for them results from the forced self-sufficiency of each region in production of nitrogen and phosphorus for those projection periods.

The potash and phosphate rock trade projections indicate, as do the data in Table 15.1, the disproportionate quantity of imports received by developed areas. The abrupt shifts, in some cases quite large, in patterns and levels of trade which occur between projections for 1985 and 2000 are a result of the assumed obsolescence of present plant capacities by 2000.

In projections for 1975 no developed area imports nitrogen fertilizers and only two developing areas export them. The total trade vol-

---

2. This is, of course, not so in the case of potash and/or phosphate rock for regions with limited or no reserves of raw material.

TABLE 15.1. Regional Import-Export Levels for Fertilizers and Phosphate Rock, Assuming (1) the Low Land Bounds and (2) Fertilizer Use Assumption A (1,000 metric tons)

Region	Year	Commodity Exports				Commodity Imports			
		Nitrogen (46%) Material	Phosphorus (46%) Material	Potash (60%) Material	Phosphate Rock (33%) Material	Nitrogen (46%) Material	Phosphorus (46%) Material	Potash (60%) Material	Phosphate Rock (33%) Material
1 United States	1975		916		2,285				1,213
	1985			540	8,584				1,482
	2000				1,956				1,937
2 Canada	1975			2,126					
	1985			4,079					
	2000			2,919					
3 Mexico	1975		11		32			45	447
	1985				32			219	747
	2000				932			540	1,325
4 Caribbean	1975					285	273	183	23
	1985							309	627
	2000							504	978
5 Central America	1975					439	132	81	68
	1985							291	629
	2000							628	1,235
6 Northern S. America	1975	811					179	187	67
	1985							346	600
	2000				1,280			604	
7 Brazil	1975						331	232	338
	1985							346	1,001
	2000			1,227	1,924				879
8 Western S. America	1975			163	247	87	11		
	1985			149	205				
	2000			752	2,762				

TABLE 15.1 (cont.)

Region	Year	Commodity Exports				Commodity Imports			
		Nitrogen (46%) Material	Phosphorus (46%) Material	Potash (60%) Material	Phosphate Rock (33%) Material	Nitrogen (46%) Material	Phosphorus (46%) Material	Potash (60%) Material	Phosphate Rock (33%) Material
9 Southern S. America	1975			12		248			331
	1985							243	1,131
	2000							623	1,924
10 European Economic Community	1975								11,189
	1985		313	887					12,695
	2000			786					15,339
11 Ireland-United Kingdom	1975						16		2,232
	1985							1,154	2,830
	2000							1,474	3,663
12 Scandinavia	1975							845	1,448
	1985							948	1,633
	2000							1,095	1,897
13 Spain-Portugal	1975	81		194					1,600
	1985			58					1,845
	2000			975					2,228
14 Austria-Switzerland	1975						313	451	357
	1985							546	962
	2000							717	1,269
15 Northern E. Europe	1975			606			590		3,444
	1985			37					5,284
	2000			1,812					6,891
16 Yugoslavia	1975							399	709
	1985							553	985
	2000							786	1,405
17 Other East Europe	1975	117					535	225	2,693
	1985							751	4,379
	2000							1,562	5,839

TABLE 15.1 (cont.)

Region	Year	Commodity Exports				Commodity Imports			
		Nitrogen (46%) Material	Phosphorus (46%) Material	Potash (60%) Material	Phosphate Rock (33%) Material	Nitrogen (46%) Material	Phosphorus (46%) Material	Potash (60%) Material	Phosphate Rock (33%) Material
18 USSR	1975		1,140	1,641	14,330				
	1985			5,279	14,272				
	2000			1,955	29,967				
19 Greece-Turkey	1975	195	812					46	1,796
	1985							141	847
	2000							289	1,114
20 United Arab Republic	1975		29		2,846	312		5	
	1985				2,724			86	
	2000				2,403			194	
21 Iran-Iraq	1975	134	87					14	248
	1985							52	197
	2000							104	289
22 Other Middle East	1975		290	156	2,614	73			
	1985			482	3,013				
	2000			194	1,403				
23 Northern Africa	1975				9,818			128	
	1985		526		18,128			290	
	2000				7,161			595	
24 Western Africa	1975				1,743	81	56	92	
	1985				1,586			126	
	2000				879			206	
25 W. Central Africa	1975			446		17	4		17
	1985			439					
	2000			1,310					
26 Ethiopia-Sudan	1975			450		550	520		821
	1985			384					998
	2000			2,807					

335

TABLE 15.1 (cont.)

Region	Year	Commodity Exports				Commodity Imports			
		Nitrogen (46%) Material	Phosphorus (46%) Material	Potash (60%) Material	Phosphate Rock (33%) Material	Nitrogen (46%) Material	Phosphorus (46%) Material	Potash (60%) Material	Phosphate Rock (33%) Material
27 E. Central Africa	1975					149	168	75	120
	1985							193	566
	2000							417	
28 S. Africa	1975	166			8,714			162	705
	1985							489	1,295
	2000							1,105	148
29 Australia	1975	20						128	36
	1985							197	160
	2000							301	3,211
30 New Zealand	1975					20		169	1,173
	1985							232	1,287
	2000							326	1,457
31 India	1975	278						260	818
	1985							581	1,396
	2000							1,105	2,340
32 Pakistan	1975	489						8	57
	1985							50	132
	2000							105	231
33 Burma	1975						10		
	1985							29	66
	2000							69	138
34 Other Far East	1975					257	170	37	
	1985							112	370
	2000							205	537
35 South Korea	1975					77	257	105	357
	1985							239	952
	2000							411	1,262

336

TABLE 15.1 (cont.)

Region	Year	Commodity Exports				Commodity Imports			
		Nitrogen (46%) Material	Phosphorus (46%) Material	Potash (60%) Material	Phosphate Rock (33%) Material	Nitrogen (46%) Material	Phosphorus (46%) Material	Potash (60%) Material	Phosphate Rock (33%) Material
36 Malaya-Indonesia	1975					172	307	65	
	1985							150	576
	2000							301	846
37 Philippines-China (T)	1975							159	255
	1985	150						216	358
	2000							305	517
38 Japan	1975	77						1,425	2,191
	1985							1,694	2,675
	2000							2,182	3,553
World Totals	1975			6,681	33,916			6,681	33,916
	1985	2,519	4,123	10,906	48,544	2,519	4,123	10,906	48,544
	2000			15,277	63,450			15,277	63,450
Developing Countries	1975			1,227	17,300			1,722	4,925
	1985	2,057	1,755	1,454	25,688	2,499	2,666	4,019	11,033
	2000			6,290	27,458			7,205	14,613
Developed Countries	1975			5,454	16,615			4,958	28,990
	1985	461	2,369	9,453	22,856	20	1,454	6,884	37,512
	2000			8,987	31,923			8,074	48,837

337

TABLE 15.2. 1975 Interarea Net Trade Statistics for Nitrogen Fertilizer, Assuming (1) the Low Land Bounds and (2) Fertilizer Use Assumption A (trade flows in 1,000 metric tons; value of imports, c.i.f., in million dollars)

Importing Area	Exporting Areas (as numbered on the left)													Total Imports	Value of Imports
	1	2	3	4	5	6	7	8	9	10	11	12	13		
1 United States															
2 Canada															
3 Latin America															
4 EEC															
5 Other W. Europe															
6 E. Europe															
7 USSR															
8 Middle East						117								117	11
9 Africa					81			61		166	489			797	75
10 S. Africa-Oceania															
11 India-Pakistan															
12 Other E. Asia											278	77		355	33
13 Japan															
Total Exports					81	117		61		166	767	77		1,269	119

TABLE 15.3. 1975 Interarea Net Trade Statistics for Phosphate Fertilizers, Assuming (1) the Low Land Bounds and (2) Fertilizer Use Assumption A (trade flows in 1,000 metric tons; value of imports, c.i.f., in million dollars)

Importing Area	Exporting Areas (as numbered on the left)													Total Imports	Value of Imports
	1	2	3	4	5	6	7	8	9	10	11	12	13		
1 United States															
2 Canada															
3 Latin America	916								248					1,164	83
4 EEC															
5 Other W. Europe				313			14		2					329	21
6 E. Europe							1,126							1,126	71
7 USSR															
8 Middle East															
9 Africa								688						688	48
10 S. Africa-Oceania															
11 India-Pakistan															
12 Other E. Asia								529	215					744	55
13 Japan															
Total Exports	916			313			1,140	1,217	465					4,051	278

339

TABLE 15.4. 1975, 1985, and 2000 Interarea Net Trade Statistics for Potash, Assuming (1) the Low Land Bounds and (2) Fertilizer Use Assumption A (trade flows in 1,000 metric tons; value of imports, c.i.f., in million dollars)

Importing Area		Exporting Areas (as numbered on the left)													Total Imports	Value of Imports
		1	2	3	4	5	6	7	8	9	10	11	12	13		
1 United States	1975															
	1985															
	2000															
2 Canada	1975															
	1985															
	2000															
3 Latin America	1975		168			166				220					554	21
	1985		1,166					440							1,606	64
	2000	540				380									920	31
4 EEC	1975															
	1985															
	2000															
5 Other W. Europe	1975				488		606	1,356							2,450	86
	1985						37	2,874							2,911	107
	2000						1,812								1,812	61
6 E. Europe	1975				399			225							624	19
	1985							1,303							1,303	44
	2000				786			1,562							2,348	70
7 USSR	1975															
	1985															
	2000															
8 Middle East	1975							61							61	2
	1985							194							194	6
	2000							393							393	12
9 Africa	1975					28			128						156	6
	1985							417							417	16
	2000					595									595	22

TABLE 15.4 (cont.)

Importing Area		Exporting Areas (as numbered on the left)													Total Imports	Value of Imports
		1	2	3	4	5	6	7	8	9	10	11	12	13		
10 S. Africa-Oceania	1975		269							190					459	18
	1985		429					51		439					919	36
	2000		326							1,406					1,732	67
11 India-Pakistan	1975								23	245					268	10
	1985		44						396	191					631	24
	2000									1,210					1,210	46
12 Other E. Asia	1975		264							102					366	15
	1985		746												746	30
	2000		411							879					1,290	51
13 Japan	1975		1,425												1,425	55
	1985		1,694												1,694	66
	2000		2,182												2,182	85
Total Exports	1975		2,126		887		606	1,642	151	757					6,363	232
	1985		4,079			194	37	5,279	396	630					10,421	393
	2000	540	2,919		786	975	1,812	1,955		3,495					12,482	445

341

TABLE 15.5. 1975, 1985, and 2000 Interarea Net Trade Statistics for Phosphate Rock, Assuming (1) the Low Land Bounds and (2) Fertilizer Use Assumption A (trade flows in 1,000 metric tons; value of imports, c.i.f., in million dollars)

Importing Area		Exporting Areas (as numbered on the left)													Total Imports	Value of Imports
		1	2	3	4	5	6	7	8	9	10	11	12	13		
1 United States	1975															
	1985															
	2000															
2 Canada	1975	1,213													1,213	10
	1985	1,482													1,482	12
	2000	631		1,305											1,936	15
3 Latin America	1975	447								728					1,175	5
	1985	2,571								1,927					4,498	30
	2000	1,325								879					2,204	6
4 EEC	1975							6,327		4,862					11,189	44
	1985							2,396		10,299					12,695	52
	2000							15,339							15,339	59
5 Other W. Europe	1975							1,448		4,189					5,637	29
	1985							1,633		5,636					7,269	37
	2000							1,897		7,161					9,058	47
6 E. Europe	1975							6,137		709					6,846	4
	1985							9,663		985					10,648	5
	2000							12,730	1,405						14,135	7
7 USSR	1975															
	1985															
	2000															
8 Middle East	1975							419							419	0
	1985							580							580	0
	2000								998						998	5
9 Africa	1975								120						120	1
	1985								1,387						1,387	8
	2000															

342

TABLE 15.5 (cont.)

Importing Area		1	2	3	4	5	6	7	8	9	10	11	12	13	Total Imports	Value of Imports
10 S. Africa-Oceania	1975	625		180					36	1,074					1,915	17
	1985	1,857							35	850					2,742	26
	2000			1,457						3,360					4,817	41
11 India-Pakistan	1975								875						875	7
	1985								1,527						1,527	12
	2000									2,570					2,570	17
12 Other E. Asia	1975								612						612	6
	1985								2,322						2,322	22
	2000									3,301					3,301	27
13 Japan	1975								2,191						2,191	23
	1985	2,675													2,675	31
	2000									3,553					3,553	34
Total	1975	2,285		180				14,331	3,834	11,562					32,192	146
	1985	8,585						14,272	5,271	19,697					47,825	235
	2000	1,956		2,762				29,966	2,403	20,824					57,911	258

343

ume, 1.3 million tons, is distributed among the Middle East, Africa, and Other East Asia.

Since lack of information precluded the possibility for differentiation of price among regions, exports are a function of availability and location of productive capacity. Interarea trade in nitrogen and phosphates is not allowed for projections beyond 1975. Implications of that constraint for domestic production and plant capacity are presented and discussed later in this chapter.

As indicated in Table 15.6, known world potash reserves are present

TABLE 15.6. Geographic Region Reserves of Potash and Phosphate Rock (million metric tons)

Region	Potash (K_2O)	Phosphate Rock (material)
United States	400.0	4,747.0
Canada	6,400.0	
Mexico		
Caribbean		11,538.0
Central America		
Northern S. America		60.6[a]
Brazil	28.0	61.0
Eastern S. America	12.0	1,421.0
Southern S. America	2.4	7.0
EEC	10,306.4	
Ireland-UK	127.0	
Scandinavia		
Spain-Portugal	363.0	
Austria-Switzerland		
Northern E. Europe	12,700.0	
Yugoslavia		
Other E. Europe		
USSR	17,237.0	5,907.0
Greece-Turkey		
UAR		87.5
Iran-Iraq		
Other Middle East	2,000.0	920.0
Northern Africa		3,540.0[b]
Western Africa		459.0
West Central Africa	95.0	43.0
Ethiopia-Sudan	50.0	
East Central Africa		
S. Africa		374.0
Australia		70.0
New Zealand		78.7[c]
India-Ceylon		
Pakistan		
Burma		
Other Far East		
South Korea		
Malaya-Indonesia		
Philippines-China		
Japan		

[a] Includes islands of Curacao and Aruba.
[b] Includes Spanish Sahara.
[c] Reserves of Oceania. Nauru, Makatea, and Christmas Islands. They are assumed adequate to maintain current Australian production levels until 1985. After that year, they are assumed exhausted.

in only 13 of the 38 geographic regions studied. Those regions are distributed among 9 of the 13 trade areas. The Republic of South Africa-Oceania, India-Pakistan, Other East Asia, and Japan, being completely devoid of known reserves, never appear as exporting regions.

In 1975, 6.4 million tons of potash exports are projected relative to a 1965 level of 2.5 to 2.7 million tons. The major exporters are Canada, 2.1 million tons; the USSR, 1.6 million tons; the EEC, 0.9 million tons; Africa, 0.8 million tons; and East Europe, 0.6 million tons. Over half of total potash fertilizer imports are received by Other West Europe and Japan. Projected 1975 imports by those two regions are 3.9 million tons.

By 1985 potash trade among trade areas is projected to rise to 10.4 million tons. As a result of increased domestic demands, areas with limited potash production capacity reduce their exports and utilize capacity to satisfy domestic demand. Countries with large plant capacity fill the markets vacated by lesser exporters. On that basis, Canada and the USSR supply 90 percent of the 1985 total potash export market.

In 2000, 12.5 million tons of potash move in interarea trade. The elimination of plant investment costs from the 2000 analysis[3] results in considerable redistribution of potash exports among potential exporters. For instance, in 1985 Canada's projected exports are 4.1 million tons and those of the USSR are 5.3 million tons. In 2000, exports of Canada and the USSR are 2.9 and 2 million tons, respectively.

The distribution of world phosphate rock, as indicated in Table 13.11, is much less general than that of potash. Only 6 of the 13 trade areas have known reserves of phosphate rock of commercial grade: the United States, Latin America, the USSR, the Middle East, Africa, and South Africa-Oceania. In 1965 they export some 30 million tons of rock. In 1975 their exports are 32.2 million tons. Exports from the USSR, the Middle East, and Africa account for 29.7 million tons of that total. The major consumers, developed nations, import 37.5 million tons or 77 percent of the total rock moving in trade in 1985.

By 1985 trade in phosphate rock used in the production of fertilizer reaches 47.8 million tons, a projected 1975–85 increase of 32.7 percent. The entire quantity consists of exports from four areas: the United States, the USSR, the Middle East, and Africa. Forty-one percent of total exports are from Africa, and developed areas receive 88 percent of total exports. The largest importers are the EEC, Other West Europe, and East Europe who receive 12.7, 7.3, and 10.7 million tons, respectively.

The assumed obsolescence of existing phosphate rock production plants by 2000 results in considerable redistribution of phosphate rock exports among potential exporters. Of the 47.8 million tons traded in 1985, approximately 8.6 million tons or 18 percent are U.S. exports. Trade in phosphate rock for fertilizer use totaled 57.9 million tons in

---

3. An adjustment equivalent to the assumption of universally equal plant construction costs.

2000; of that total, 2 million tons, or 3.4 percent, are supplied by the United States. Rock exports from the USSR rise from 14.3 to 30 million tons between 1985 and 2000. The major importers in 2000 are the EEC, 15.3 million tons; Other West Europe, 9.1 million tons; and East Europe, 14.1 million tons. The largest developing world importers are Other East Asia, 3.3 million tons, and India-Pakistan, 2.6 million tons.

FERTILIZER IMPORT COSTS AND PLANT CAPACITY CONSIDERATIONS

Value of fertilizer and phosphate rock imports are presented in conjunction with the import-export data in Tables 15.2 to 15.5. Tables 15.7 and 15.8 contain plant capacity conditions projected in the analysis.

Review of the 1975 import value and plant capacity statistics reveals a situation where value of net fertilizer and phosphate rock trade is $1.4 billion, with imports to developing areas valued at $442 million. The kinds of fertilizer imported by developing countries cause their value to be out of proportion when compared to world volume and value data. Their major imports, nitrogen and phosphate fertilizers, have much higher per unit cost than potash and phosphate rock which are dominant among developed country imports.

The most striking feature of the plant utilization tables is the quantities of excess capacity for fertilizer production. Even in the case of the highest utilization level, phosphate rock, 24.2 percent of capacity remains unused. In the case of nitrogen with the lowest percent utilization, only 53.6 percent of actual capacity is utilized. Such underutilization is massive, especially given the efficiency assumption employed in the analysis.[4] In short, excess capacity for production abounds in the fertilizer industry, given 1975 projections.

Since potash and phosphate rock were the only fertilizer commodities for which interregional trade was allowed beyond 1975, the cost of fertilizer and phosphate rock imports remained near the 1975 level at $1.6 billion. However, large quantities of additional funds ($1.6 billion) are required to provide capital necessary to add fertilizer plant capacity in developing nations. The cost of adding nitrogen plant capacity in developing regions is $400 per ton. In 1985 the average cost (c.i.f.) of nitrogen received by developing countries is $94.13 per ton of 46 percent material. Given those costs, where foreign currency is required for construction of fertilizer plant capacity and fertilizer imports must be purchased in foreign currencies, considerable support exists for the suggestion that "where the urgency of the food problem is great and capital scarce, it would . . . seem that priority should be attached to importing fertilizer rather than producing it locally."[5] Indeed, with such vast

---

4. Production efficiency of 80 percent for developing and 90 percent for developed countries is assumed for nitrogen and phosphate fertilizers. Universal efficiency of 80 percent is assumed for potash and phosphate rock.

5. United Nations, Food and Agriculture Organization. *Fertilizer: A World Report on Production and Consumption.* Rome, Italy, 1952.

TABLE 15.7. 1975, 1985, and 2000 Nitrogen and Phosphate Fertilizer Plant Capacity Utilization Relative to Estimated 1975 Capacity and Quantity and Investment Cost of Capacity Added, Assuming (1) the Low Land Bounds and (2) Fertilizer Use Assumption A

Trade Area	Year	1975 Capacity[a] (1,000 metric tons nutrient)		Capacity Utilized				Capacity Added (1,000 metric tons nutrient)		Investment Cost (mil $)	
				Quantity (1,000 metric tons nutrient)		Percent					
		Nitrogen	Phosphate	Nitrogen	Phosphate	Nitrogen	Phosphate	Nitrogen	Phosphate	Nitrogen	Phosphate
1 United States	1975	11,314	7,243	7,716	6,713	68.2	92.7	none	none		
	1985			11,314	7,243	100.0	100.0	932	1,354	267	169
	2000			11,314	7,243	100.0	100.0	10,140	5,958	2,900	745
2 Canada	1975	1,276	942	250	445	19.6	47.3	none	none		
	1985			447	543	35.0	57.7	none	none		
	2000			781	710	61.2	75.4	none	none		
3 Latin America	1975	2,454	944	1,474	650	60.1	68.8	none	none		
	1985			1,806	775	73.6	82.1	1,487	1,332	593	233
	2000			2,265	944	92.3	100.0	3,650	2,475	1,459	434
4 EEC	1975	8,142	4,278	3,612	4,103	44.4	95.9	none	none		
	1985			5,034	4,278	61.8	100.0	none	376		47
	2000			6,973	4,278	85.6	100.0	none	1,346		168
5 Other W. Europe	1975	3,837	2,092	2,315	2,067	60.3	98.8	none	none		
	1985			3,100	2,092	80.8	100.0	38	573	11	72
	2000			3,656	2,092	95.3	100.0	793	1,229	277	154
6 E. Europe	1975	6,035	2,914	2,738	2,510	45.4	86.2	none	none		
	1985			4,328	2,611	71.7	89.6	none	1,293		161
	2000			6,035	2,765	100.0	94.9	850	2,418	243	302
7 USSR	1975	6,000	7,428	2,405	3,079	40.1	41.5	none	none		
	1985			3,602	3,100	60.0	41.7	none	none		
	2000			5,396	3,997	89.9	53.8	none	none		
8 Middle East	1975	1,114	1,166	929	1,166	83.4	100.0	none	none		
	1985			918	685	82.5	58.8	429	46	172	8
	2000			995	837	89.4	71.8	863	151	345	26

[a] J. R. Douglas, Jr., and E. A. Harre. TVA, Muscle Shoals, Alabama. Fertilizer Plant Capacity Estimates. Private Communication, 1968.

TABLE 15.7 (cont.)

Trade Area		1975 Capacity[a] (1,000 metric tons nutrient)		Capacity Utilized				Capacity Added (1,000 metric tons nutrient)		Investment Cost (mil $)	
				Quantity (1,000 metric tons nutrient)		Percent					
		Nitrogen	Phosphate	Nitrogen	Phosphate	Nitrogen	Phosphate	Nitrogen	Phosphate	Nitrogen	Phosphate
9 Africa	1975	316	574	144	574	45.5	100.0	none	none		
	1985			316	430	100.0	74.9	855	593	343	104
	2000			316	574	100.0	100.0	1,933	988	774	174
10 S. Africa-Oceania	1975	921	2,370	362	1,952	39.4	82.4	none	none		
	1985			531	2,194	57.7	92.6	354	62	102	8
	2000			669	2,325	72.6	98.1	1,291	468	369	58
11 India-Pakistan	1975	2,760	1,101	1,703	365	61.7	33.1	none	none		
	1985			1,858	636	67.3	57.8	none	none		
	2000			2,657	998	96.3	90.6	70	73	28	13
12 Other E. Asia	1975	967	305	965	255	99.8	83.5	none	none		
	1985			967	298	100.0	97.6	782	670	313	117
	2000			967	305	100.0	100.0	1,598	1,070	640	187
13 Japan	1975	3,058	1,284	1,212	803	39.6	62.6	none	none		
	1985			1,528	981	49.9	76.4	none	none		
	2000			2,172	1,284	71.0	100.0	none	19		2
Developed Nations	1975	40,583	28,550	20,611	21,672	50.8	75.9	none	none		
	1985			29,883	23,043	73.6	80.7	1,324	3,658	380	457
	2000			36,995	24,695	91.1	86.5	13,075	11,437	3,739	1,429
Developing Nations	1975	7,610	4,091	5,215	3,010	68.5	73.6	none	none		
	1985			5,865	2,825	77.1	69.1	3,553	2,642	1,421	462
	2000			7,200	3,658	94.6	89.4	8,115	4,757	3,246	834
96-Nation Total	1975	48,193	32,641	25,826	24,682	53.6	75.6	none	none		
	1985			35,748	25,868	74.2	79.3	4,877	6,300	1,801	919
	2000			44,195	28,353	91.7	86.9	21,190	16,194	6,985	2,263

TABLE 15.8. 1975, 1985, and 2000 Potash Fertilizer and Phosphate Rock Plant Capacity Utilization Relative to 1975 Capacity and Quantity and Investment Cost of Capacity Added, Assuming (1) the Low Land Bounds and (2) Fertilizer Use Assumption A

Trade Area		1975 Capacity (1,000 metric tons nutrient)		Capacity Utilized				Capacity Added (1,000 metric tons nutrient)		Investment Cost (mil $)	
				Quantity (1,000 metric tons nutrient)		Percent					
		Potash	Phosphate Rock	Potash	Phosphate Rock	Potash	Phosphate Rock	Potash	Phosphate Rock	Potash	Phosphate Rock
1 United States	1975	3,250	37,572	3,250	22,652	100.0	60.3	1,476	none	84	
	1985			3,250	35,231	100.0	93.8	3,824	none	218	
	2000			3,250	37,572	100.0	100.0	8,784	4,181	501	192
2 Canada	1975	6,408	none	1,624		25.3		none	none		
	1985			3,012		47.0		none	none		
	2000			2,413		37.7		none	none		
3 Latin America	1975	220	650	220	650	100.0	100.0	none	none		
	1985			220	650	100.0	100.0	none	none		
	2000			220	650	100.0	100.0	1,662	9,270	95	426
4 EEC	1975	4,461	none	4,461		100.0		none	none		
	1985			4,461		100.0		125	none	7	
	2000			4,461		100.0		1,614	none	92	
5 Other W. Europe	1975	302	none	302		100.0		none	none		
	1985			302		100.0		none	none		
	2000			302		100.0		2,025	none	115	
6 E. Europe	1975	2,400	none	2,400		100.0		none	none		
	1985			2,400		100.0		none	none		
	2000			2,400		100.0		1,761	none	100	
7 USSR	1975	7,500	25,000	2,891	25,000	38.5	100.0	none	none		
	1985			5,890	25,000	78.5	100.0	none	none		
	2000			4,593	25,000	61.2	100.0	none	19,955		918

[a] Reserves of phosphate rock associated with the specified capacity are assumed exhausted in 1985.

TABLE 15.8 (cont.)

Trade Area		1975 Capacity (1,000 metric tons nutrient)		Capacity Utilized				Capacity Added (1,000 metric tons nutrient)		Investment Cost (mil $)	
				Quantity (1,000 metric tons nutrient)		Percent					
		Potash	Phosphate Rock	Potash	Phosphate Rock	Potash	Phosphate Rock	Potash	Phosphate Rock	Potash	Phosphate Rock
8 Middle East	1975	360	7,200	124	7,200	34.5	100.0	none	none		
	1985			360	7,200	100.0	100.0	none	none		
	2000			195	5,396	54.2	74.9	none	none		
9 Africa	1975	600	23,000	600	14,400	100.0	62.6	none	none		
	1985			600	23,000	100.0	100.0	none	none		
	2000			600	11,023	100.0	47.9	2,257	15,322	129	705
10 S. Africa-Oceania	1975	none	3,750		3,750		100.0	none	none		
	1985				3,750		100.0	none	none		
	2000				Exhausted[a]			none	2,480		114
11 India-Pakistan	1975	none	none					none	none		
	1985							none	none		
	2000							none	none		
12 Other E. Asia	1975	none	none					none	none		
	1985							none	none		
	2000							none	none		
13 Japan	1975	none	none					none	none		
	1985							none	none		
	2000							none	none		
Developed Nations	1975	24,321	66,322	14,928	51,402	61.4	77.5	1,476	none	84	
	1985			19,314	63,981	79.4	96.5	3,949	none	225	
	2000			17,419	62,572	71.6	94.3	14,183	26,616	808	1,224
Developing Nations	1975	1,180	30,850	944	22,250	80.0	72.1	none	none		
	1985			1,180	30,850	100.0	100.0	none	none		
	2000			1,015	17,069	86.0	55.3	3,919	24,592	224	1,131
96-Nation Total	1975	25,501	97,172	15,872	73,652	62.2	75.8	1,476	none	84	
	1985			20,494	94,831	80.4	97.6	3,949	none	225	
	2000			18,434	79,641	72.3	81.9	18,102	51,208	1,032	2,855

quantities of excess capacity in developed nations, the extension of credit for purchase of fertilizer rather than construction of fertilizer plants may be a "best" policy.

The 2000 plant capacity addition statistics (Tables 15.7 and 15.8) and import-export statistics (Tables 15.1 to 15.5) provide a succinct summary of the fertilizer industry situation projected for the next several decades. It is one where, unless the rate of introduction of fertilizer technology is advanced through some revolution in farmer education and extension practice, fertilizer manufacturers will go sorely wanting for markets, existing capacities will be grossly underutilized, and the critical decision point will be whether or not to improve the position of developing countries through provision of capital in the form of fertilizer or fertilizer plants, assuming both or either is desirable.[6]

---

6. It might well be that research funds to study such things as institutional structures, plant genetics, and agricultural technology are more desirable than either.

CHAPTER 16

The Perspective and Context—
Some Further Dimensions Considered

THE INTERREGIONAL FLOWS of cereals and fertilizers presented in Chapters 14 and 15 indicate the 96-nation food and fertilizer situation projected to evolve over time under specified assumptions. Those assumptions include (a) short-run inflexibility of fertilizer plant capacities, (b) cereal production projections constrained by the low land bounds, and (c) requirement that U.S. cereal exports to developing countries move in U.S. flag vessels whose transportation costs are high.

In addition, they and all preceding chapters of the book exclude Mainland China. The purpose of this chapter is to present results which are produced when these conditions are modified. The intent, here as throughout the book, is to indicate the extent to which such analyses of the world food situation are subject to variation depending on the assumptions underlying a particular analysis.

ADDITION OF MAINLAND CHINA CONSIDERED

As indicated earlier, the inadequacy of available data made inclusion of Mainland China in the general analysis impossible. To determine the implications of this omission, we developed estimates of 1975 and 1985 net cereal import requirements of Mainland China and introduced them to the model.

The net import requirements introduced were estimated in a crude fashion by using historical data on cereal import statistics for Mainland China contained in Table 16.1 and assuming increases in exports from 1963 base levels at 2.3 percent per year for wheat, 2.1 percent per year for rice, and 1.7 percent per year for other grains. The resulting projected import levels by commodity class are presented in Table 16.2.

The results produced when the model was expanded to include Mainland China and the import demands presented in Table 16.2 were added are presented in Tables 16.3 to 16.7, together with the corresponding results for earlier solutions excluding Mainland China.

Wheat is the cereal class dominating projected imports of Mainland China. Tables 16.3 and 16.4 illustrate the projected inter*area*

TABLE 16.1. Cereal Imports to Mainland China, 1951–65

Year	Thousand Metric Tons of Cereal		
	Wheat[a]	Rice	Other Grains
1951	0	20	0
1952	2	16	0
1953	14	5	0
1954	38	1	0
1955	0	157	0
1956	50	92	2
1957	41	106	29
1958	127	12	31
1959	40	61	0
1960	104	29	0
1961	3,573	67	1,458
1962	4,438	7	1,002
1963	4,920	97	38
1964	5,119	58	548
1965	5,905	102	179

SOURCE: U.S. Department of Agriculture, *World Trade in Selected Agricultural Commodities 1951–65, Volume II, Food and Feed Grains*, Foreign Agricultural Economic Report 45, 1968.

[a] Including wheat flour expressed as wheat equivalents.

TABLE 16.2. Import Data and Projections for Mainland China, 1975–85

	Avg. Imports	Assumed Growth Rate	Projected Imports	
			1975	1985
	(1961–65)	*(%/year)*	*(1,000 metric tons)*	
Wheat	4,955.0	2.3	6,537	8,022
Rice	66.2	2.1	85	103
Other Grains	645.0	1.7	796	931

trading patterns for wheat in 1975 and 1985 and provide comparative data for solutions with and without Mainland China. Tables 16.5 to 16.7[1] indicate the 1985 inter*regional* cereal flows and provide comparative statistics to indicate the significance of the addition of Mainland China to the analysis.

In 1975 the flow of wheat among the 13 trade areas totaled 44.9 million metric tons. In 1985 the comparable quantity was 69.5 million metric tons. The addition of China's wheat import requirements of 6.5 million metric tons to 1975 requirements is a 14.5 percent increase. The 1985 China requirement of 8,022 metric tons represents an 11.5 percent increase in total wheat and rye import requirements for that period. This large relative importance of China's projected import demand may in no small measure explain why major wheat exporters

---

1. In these and all other regional tables contained in this chapter, entries are included for only those regions which actually export and/or import the commodity to which the table relates.

TABLE 16.3. The Implications of Including Import Requirements of Mainland China for 1975 Interarea Net Trade in Wheat and Rye, Assuming (1) the Low Land Bounds and (2) Fertilizer Use Assumption A (1,000 metric tons)

Importing Area	Exporting Areas (as numbered on the left)														Total Imports
	1	2	3	4	5	6	7	8	9	10	11	12	13	14	
1 United States															
2 Canada															
3 Latin America	10,520ª *10,520*														10,520 *10,520*
4 EEC															
5 Other W. Europe	6,458 *6,458*														6,458 *6,458*
6 E. Europe	5,415 *5,415*			2,653 *2,653*											8,068 *8,068*
7 USSR															
8 Middle East	2,638 *4,626*		554 *554*				450 *450*			2,868 *880*					6,510 *6,510*
9 Africa	2,748 *2,748*									635 *635*					3,383 *3,383*
10 S. Africa-Oceania	200 *323*														200 *323*
11 India-Pakistan										3,740 *3,740*					3,740 *3,740*
12 Other E. Asia		*496*								2,065 *1,569*					2,065 *2,065*
13 Japan	4,004 *4,004*														4,004 *4,004*
14 Mainland China	*3,139.9*	*790.1*								*2,607*					*6,537*
15 World Totals	31,983 *37,233.9*	*1,286.1*	554 *554*	2,653 *2,653*			450 *450*			9,308 *9,431*					44,948 *51,608*

ª Only figures in italics include Mainland China.

354

TABLE 16.4. The Implications of Including Import Requirements of Mainland China for 1985 Interarea Net Trade in Wheat and Rye, Assuming (1) the Low Land Bounds and (2) Fertilizer Use Assumption A (1,000 metric tons)

Importing Area	Exporting Areas (as numbered on the left)														Total Imports
	1	2	3	4	5	6	7	8	9	10	11	12	13	14	
1 United States															
2 Canada															
3 Latin America	15,191[a] *15,191*									488 *488*					15,679 *15,679*
4 EEC															
5 Other W. Europe	6,406 *6,406*														6,406 *6,406*
6 E. Europe	3,581 *3,581*			6,426 *6,426*	171 *171*										10,178 *10,178*
7 USSR	7,756 *7,756*														7,756 *7,756*
8 Middle East	5,510 *5,510*	2,264 *2,669*	654 *654*							3,125 *3,100*	380				11,933 *11,933*
9 Africa	4,594 *4,594*									870 *870*					5,464 *5,464*
10 S. Africa-Oceania	550 *550*														550 *550*
11 India-Pakistan										3,913 *3,913*					3,913 *3,913*
12 Other E. Asia		629 *629*								2,034 *2,059*	25				2,688 *2,688*
13 Japan	4,897 *4,897*														4,897 *4,897*
14 Mainland China	*6,760*	*857*									*405*				*8,022*
15 World Totals	48,485 *55,245*	2,893 *4,155*	654 *654*	6,426 *6,426*	171 *171*					10,430 *10,430*	405 *405*				69,464 *77,486*

[a] Only figures in italics include Mainland China.

355

TABLE 16.5. The Implications of Including Import Requirements of Mainland China for 1985 Interregional Net Trade in Wheat and Rye, Assuming (1) the Low Land Bounds and (2) Fertilizer Use Assumption A (1,000 metric tons)

Importing Area	Exporting Areas (as numbered in Table 12.1)							World Totals
	1	2	3	10	14	29	31	
4 Caribbean	786[a] *786*							786 *786*
5 Central America	545 *545*							545 *545*
6 Northern S. America	7,007 *7,007*							7,007 *7,007*
7 Brazil	5,322 *5,322*							5,322 *5,322*
8 Western S. America	1,531 *1,531*							1,531 *1,531*
9 Southern S. America						488 *488*		488 *488*
11 Ireland-United Kingdom	4,537 *4,537*							4,537 *4,537*
12 Scandinavia	1,539 *1,539*							1,539 *1,539*
13 Spain-Portugal	330 *330*							330 *330*
15 Northern E. Europe	3,581 *3,581*			6,426 *6,426*	171 *171*			10,178 *10,178*
18 USSR	7,756 *7,756*							7,756 *7,756*
19 Greece-Turkey	1,904 *1,904*	2,004 *2,004*						3,908 *3,908*
20 United Arab Republic	3,606 *3,606*							3,606 *3,606*
21 Iran-Iraq						1,552 *1,932*	380	1,932 *1,932*
22 Other Middle East		260 *665*	654 *654*			1,573 *1,168*		2,487 *2,487*

[a] Only figures in italics include Mainland China.

TABLE 16.5 (cont.)

Importing Area	Exporting Areas (as numbered in Table 12.1)							World Totals
	1	2	3	10	14	29	31	
23 Northern Africa	4,021 *4,021*							4,021 *4,021*
24 Western Africa	573 *573*							573 *573*
25 West Central Africa						176 *176*		176 *176*
26 Ethiopia-Sudan						158 *158*		158 *158*
27 East Central Africa						536 *536*		536 *536*
28 S. Africa	392 *392*							392 *392*
30 New Zealand	158 *158*							158 *158*
32 Pakistan						3,913 *3,913*	495 *495*	4,408 *4,408*
33 Burma						25	25	25 *25*
34 Other Far East						234 *234*		234 *234*
35 South Korea		629 *629*						629 *629*
36 Malaya-Indonesia						541 *541*		541 *541*
37 Philippines-China (T)						1,259 *1,259*		1,259 *1,259*
38 Japan	4,897 *4,897*							4,897 *4,897*
39 Mainland China	6,760	857					405	8,022
World Totals	48,485 *55,245*	2,893 *4,155*	654 *654*	6,426 *6,426*	171 *171*	10,430 *10,430*	900 *900*	69,959 *77,981*

357

TABLE 16.6. The Implications of Including Import Requirements of Mainland China for 1985 Interregional Net Trade in Rice, Assuming (1) the Low Land Bounds and (2) Fertilizer Use Assumption A (1,000 metric tons)

Importing Area	Exporting Areas (as numbered in Table 12.1)											World Totals
	1	3	6	7	13	20	27	29	31	33	38	
2 Canada	55ª											55
	55											*55*
4 Caribbean	346											346
	346											*346*
5 Central America	79											79
	79											*79*
8 Western S. America	149	13										162
	162											*162*
9 Southern S. America				63								63
				63								*63*
10 EEC	8											13
	13				*5*							*13*
11 Ireland-United Kingdom	133											133
	133											*133*
12 Scandinavia	46											46
	46											*46*
14 Austria-Switzerland	78											78
	78											*78*
15 Northern E. Europe	360											360
	360											*360*
16 Yugoslavia	54											54
	54											*54*
17 Other East Europe	61											61
	61											*61*
18 USSR	714											714
	714											*714*
19 Greece-Turkey						131						131
	4	*37*			*52*	*38*						*131*

ª Only figures in italics include Mainland China.

358

TABLE 16.6 (cont.)

Importing Area	Exporting Areas (as numbered in Table 12.1)

Importing Area	1	3	6	7	13	20	27	29	31	33	38	World Totals
21 Iran-Iraq						59						59
						59						*59*
22 Other Middle East						226						226
						226						*226*
23 Northern Africa	47											47
					47							*47*
24 Western Africa		24	164	477								665
		24	*164*	*477*								*665*
25 West Central Africa				101								101
				101								*101*
28 S. Africa	88											88
	88											*88*
30 New Zealand								5				5
	5											*5*
32 Pakistan						124	97		71	1,757		2,049
						217	*102*		*71*	*1,659*		*2,049*
34 Other Far East											197	197
											197	*197*
35 South Korea											1,222	1,222
											1,222	*1,222*
36 Malaya-Indonesia								132		1,967	602	2,701
								137		*1,962*	*602*	*2,701*
37 Philippines-China (T)											2,820	2,820
											2,820	*2,820*
39 Mainland China										*103*		*103*
World Totals	2,171	37	164	641	52	540	97	137	71	3,724	4,841	12,475
	2,269	*37*	*164*	*641*	*52*	*540*	*102*	*137*	*71*	*3,724*	*4,841*	*12,578*

359

TABLE 16.7. The Implications of Including Import Requirements of Mainland China for 1985 Interregional Net Trade in Other Grains Assuming (1) the Low Land Bounds and (2) Fertilizer Use Assumption A (1,000 metric tons)

Importing Area	Exporting Areas (as numbered in Table 12.1)							World Totals
	1	3	9	18	29	34	36	
1 Canada	2,168[a]							2,168
	2,168							2,168
4 Caribbean	409							409
	409							409
5 Central America	1,109							1,109
	1,109							1,109
6 Northern S. America	1,215							1,215
	1,215							1,215
7 Brazil			2,430					2,430
			2,430					2,430
8 Western S. America	773							773
	773							773
11 Ireland-United Kingdom	3,841							3,841
	3,841							3,841
13 Spain-Portugal	60							60
	60							60
14 Austria-Switzerland	920							920
	920							920
16 Yugoslavia	2,040							2,040
	2,040							2,040
19 Greece-Turkey	4,330			665				4,995
	4,330			665				4,995
20 United Arab Republic	2,361							2,361
	2,361							2,361
21 Iran-Iraq				1,785				1,785
				1,785				1,785

[a] Only figures in italics include Mainland China.

TABLE 16.7 (cont.)

Importing Area	Exporting Areas (as numbered in Table 12.1)							World Totals
	1	3	9	18	29	34	36	
22 Other Middle East	1,054 *1,054*							1,054 *1,054*
23 Northern Africa	1,997 *1,997*							1,997 *1,997*
24 Western Africa	2,441 *2,441*							2,441 *2,441*
25 West Central Africa	117 *725*		608					725 *725*
26 Ethiopia-Sudan	298 *298*	882 *882*						1,180 *1,180*
27 East Central Africa			1,168 *1,776*		1,135 *527*			2,303 *2,303*
28 S. Africa	853 *853*							853 *853*
31 India					534 *701*	2,639 *2,472*	677 *677*	3,850 *3,850*
32 Pakistan					468 *909*	441		909 *909*
33 Burma						2 *2*		2 *2*
35 South Korea	1,000 *1,000*							1,000 *1,000*
37 Philippines-China (T)	323					1,381 *1,058*		1,381 *1,381*
38 Japan	6,823 *6,823*							6,823 *6,823*
39 Mainland China						931		931
World Totals	33,809 *34,740*	882 *882*	4,206 *4,206*	2,450 *2,450*	2,187 *2,137*	4,463 *4,463*	677 *677*	48,624 *49,555*

361

such as Canada have eagerly pursued the establishment of good relations with China. Any nation whose import demands represent in excess of 10 percent of the projected world trade area requirements merits the close attention of major exporters.

Review of either the wheat trade solutions presented in Chapter 15 or the statistics in Tables 16.3 and 16.4 quickly reveals those countries projected to meet China's import demands: Canada and the United States.[2] The addition of China causes a projected increase of 6.8 million metric tons in U.S. wheat exports and 1.3 million metric tons in those of Canada.

The 1975 analysis excluding Mainland China resulted in projected excess wheat and rye production capacity in the United States, Canada, Yugoslavia, and Other East Europe. The failure of Yugoslavia and Other East Europe to enter the expanded solutions indicates their inability to compete with Canada and the United States for China's market under the conditions of the model. This situation also occurs in the expanded 1985 solution.

The 1985 interregional trade data presented in Tables 16.5 to 16.7 indicate the impact of China's addition on export levels and trade patterns for cereals in each of the three classes. The adjustments in interregional flows which occur when China is added are minimal. This indicates the stable nature of trade flows among regions (under the conditions of the model) relative to China's participation in the market. A brief glance at the world totals in each of Tables 16.5 to 16.7 reveals the small number of regions whose export levels change, given China's introduction. The reason is simple. Only a few regions have projected excess capacity for production in each cereal class.

THE 2000 HIGH LAND CASE

As indicated earlier, low and high cropland area upper bounds were developed for use in our studies. Due to the slow rate at which cropland area has and is projected to expand, even the low land bounds tend to be unrestrictive in areas with expansion potential. Consequently, utilization of the high land bounds in any particular analysis tends to have a relatively small impact on total world production. Therefore, detailed analyses employing the high land bound assumption were not conducted. However, results were produced for the 2000 high land case to indicate the effect of introducing the higher land bound in that case where the effect was largest.

Raising the bounds affected projected cropland area in only 3 of

---

2. The 123,000 metric tons supplied by South Africa-Oceania are merely a change in the flow of exports from Australia. Prior to the addition of China, Australia exported 123,000 tons to South Africa. When China was introduced, those 123,000 tons were exported to China by Australia, and the United States met South Africa's equivalent import requirements.

TABLE 16.8. Cropland Area and Cereal Production Implications of Assuming the High Cropland Upper Bound: The 2000 Case

Country Affected	Cropland Area Given		Change in Cropland Area	Cereal Production Given		Change in Cereal Production
	Low Bound	High Bound		Low Bound	High Bound	
	(1,000 ha)			*(1,000 metric tons)*		
Iran	7,587	9,415	1,828	7,009	8,696	1,687
Israel	273	325	52	649	781	132
Jordan	399	437	38	199	217	18
Syria	2,082	2,720	638	1,086	1,418	332
India	161,789	169,395	7,606	132,865	139,488	6,623
Pakistan	26,040	28,088	2,048	26,775	28,881	2,106
Thailand	8,478	10,025	1,547	17,101	19,084	1,983
South Vietnam	4,880	5,949	1,069	6,026	7,456	1,430
Philippines	7,765	9,460	1,695	5,475	6,836	1,361

the 13 trade areas in 2000. Cropland area in the Middle East, India-Pakistan, and Other East Asia increases 2.56, 9.65, and 4.31 million hectares, respectively. Expressed in percent, the increases are 7.6 percent in the case of the Middle East, 5.1 percent in India-Pakistan, and 6.8 percent in Other East Asia. Nine countries are affected. They are Iran, Israel, Jordan, Syria, India, Pakistan, Thailand, South Vietnam, and the Philippines. The actual effect on cropland area and cereal production in each country is indicated in Table 16.8. The high to low land bound ratio of total interregional trade volume for each cereal grain class (0.91 for wheat and rye, including wheat substituted for rice; 1.00 for rice; and 0.97 for other grains) provides a succinct summary of total trade impact caused by raising the land bound. It is very small in aggregate even though it appears, and indeed is, very significant for the countries and regions whose production is increased.

The actual adjustments it does cause are indicated for each region and cereal class in Tables 16.9 to 16.11. The impact may be summarized simply as one where major developed nation producers of the classes wheat and other grains experience reduction in volume of exports, and imports fall for the nine countries whose production increases. Adjustments[3] in interregional flow patterns are minimal.

The section can be appropriately concluded with a restatement of two observations made earlier: (a) Increases in extensive (area expansion) and intensive (multiple cropping) land use are indeed modest. (b) Relative to available land, there exists vast potential for expansion of cropland area. Our data for South America are a quantitative example. Quantification of other land areas suited to but yet unused in food production would most certainly indicate further expansion potential. The potential in Africa is one example.

---

3. The increase in wheat imports (including wheat substituted for rice) to Malaya-Indonesia appears to be an anomaly. It occurs due to a cost-minimizing adjustment which shifts wheat for rice substitution among regions.

TABLE 16.9. 2000 Interregional Net Trade Statistics for Wheat and Rye Including Wheat Substituted for Rice under Each of the High and Low Land Bounds (high land in italics)

Importing Area	Exporting Areas (as numbered in Table 12.1)							World Totals
	1	2	3	10	14	29	31	
3 Mexico	10 *10*							10 *10*
4 Caribbean	1,202 *1,202*							1,202 *1,202*
5 Central America	933 *933*							933 *933*
6 Northern S. America	11,305 *11,305*							11,305 *11,305*
7 Brazil	8,043 *8,043*							8,043 *8,043*
8 Western S. America	2,547 *2,547*							2,547 *2,547*
9 Southern S. America	1,969 *1,969*		252 *252*					2,221 *2,221*
11 Ireland-United Kingdom	4,707 *4,708*							4,707 *4,708*
12 Scandinavia	2,052 *2,052*							2,052 *2,052*
13 Spain-Portugal	57 *57*							57 *57*
15 Northern E. Europe	1,872 *1,872*			10,070 *10,070*	584 *584*			12,526 *12,526*
18 USSR	21,043 *21,043*							21,043 *21,043*
19 Greece-Turkey	3,323 *276*	4,933 *7,980*						8,256 *8,256*
20 United Arab Republic	5,828 *5,828*							5,828 *5,828*
21 Iran-Iraq		2,685				1,702 *2,970*	*277*	4,387 *3,247*
22 Other Middle East		4,327 *3,965*						4,327 *3,965*

TABLE 16.9 (cont.)

Importing Area	Exporting Areas (as numbered in Table 12.1)							World Totals
	1	2	3	10	14	29	31	
23 Northern Africa	7,063 *7,063*							7,063 *7,063*
24 Western Africa	1,368 *1,368*							1,368 *1,368*
25 West Central Africa	309 *309*							309 *309*
26 Ethiopia-Sudan		278 *278*						278 *278*
27 East Central Africa						946 *946*		946 *946*
28 S. Africa	973 *973*							973 *973*
30 New Zealand	228 *228*							228 *228*
31 India						727		727
32 Pakistan		3,287				8,178 *5,874*	1,103	11,465 *6,977*
33 Burma		11					*11*	11 *11*
34 Other Far East	4,336	1,562						4,336 *1,562*
35 South Korea	2,734 *2,734*							2,734 *2,734*
36 Malaya-Indonesia	1,323	3,283 *3,184*				1,763		4,606 *4,947*
37 Philippines-China (T)	4,423 *108*	1,836						4,423 *1,944*
38 Japan	6,196 *6,196*							6,196 *6,196*
World Totals	93,844 *80,824*	18,804 *18,805*	252 *252*	10,070 *10,070*	584 *584*	11,553 *11,553*	1,391	135,107 *123,479*

TABLE 16.10. 2000 Interregional Net Trade Statistics for Rice under Each of the High and Low Land Bounds (high land in italics)

Importing Area	Exporting Areas (as numbered in Table 12.1)											World Totals
	1	3	6	7	13	20	27	29	31	33	38	
1 Canada	73											73
	73											*73*
4 Caribbean	346											346
	346											*346*
5 Central America	86											86
	86											*86*
8 Western S. America	180											180
	180											*180*
9 Southern S. America				117								117
				117								*117*
10 EEC												
11 Ireland United Kingdom	152											152
	152											*152*
12 Scandinavia	51											51
	51											*51*
14 Austria-Switzerland	85											85
	85											*85*
15 Northern E. Europe	395											395
	395											*395*
16 Yugoslavia	62											62
	62											*62*
17 Other East Europe	77											77
	77											*77*
18 USSR	907											907
	907											*907*
19 Greece-Turkey					18	87		39				144
						14				*130*		*144*
21 Iran-Iraq										311		311
										66		*66*

TABLE 16.10 (cont.)

Importing Area	Exporting Areas (as numbered in Table 12.1)											World Totals
	1	3	6	7	13	20	27	29	31	33	38	
22 Other Middle East						274						274
						347						*347*
23 Northern Africa	25				27							52
	7				*45*							*52*
24 Western Africa	411		238	254								903
	429		*238*	*236*								*903*
25 West Central Africa				15				115				130
				33				*96*				*129*
28 S. Africa	133											133
	133											*133*
30 New Zealand								7				7
								7				*7*
31 India								2		2,838		2,840
32 Pakistan										1,387		1,387
										3,870		*3,870*
34 Other Far East											630	630
35 South Korea											655	655
											655	*655*
36 Malaya-Indonesia											2,463	2,463
								59		*470*	*1,594*	*2,123*
37 Philippines-China (T)											4,403	4,403
											5,903	*5,903*
World Totals	2,983		238	386	45	361		162		4,536	8,152	16,863
	2,983		*238*	*386*	*45*	*361*		*162*		*4,536*	*8,152*	*16,863*

TABLE 16.11. 2000 Interregional Net Trade Statistics for Other Grains under Each of the High and Low Land Bounds (high land in italics)

Importing Area	Exporting Areas (as numbered in Table 12.1)								World Totals
	1	3	9	13	18	29	34	36	
1 Canada	5,335 / *5,335*								5,335 / *5,335*
3 Mexico	494 / *494*								494 / *494*
4 Caribbean	777 / *777*								777 / *777*
5 Central America	2,337 / *2,337*								2,337 / *2,337*
6 Northern S. America	2,626 / *2,626*								2,626 / *2,626*
7 Brazil	2,325 / *2,325*		4,215 / *4,215*						6,540 / *6,540*
8 Western S. America	1,611 / *1,611*								1,611 / *1,611*
11 Ireland-United Kingdom	371 / *371*								371 / *371*
13 Spain-Portugal									
14 Austria-Switzerland	139 / *139*								139 / *139*
16 Yugoslavia	3,186 / *3,186*								3,186 / *3,186*
19 Greece-Turkey	7,086 / *6,784*				915 / *1,217*				8,001 / *8,001*
20 United Arab Republic	4,499 / *4,499*								4,499 / *4,499*
21 Iran-Iraq					3,550 / *3,248*				3,550 / *3,248*

TABLE 16.11 (cont.)

Importing Area	1	3	9	13	18	29	34	36	World Totals
					Exporting Areas (as numbered in Table 12.1)				
22 Other Middle East	1,651								1,651
	1,457								*1,457*
23 Northern Africa	4,041								4,041
	4,041								*4,041*
24 Western Africa	7,364								7,364
	7,364								*7,364*
25 West Central Africa	1,685								1,685
	1,685								*1,685*
26 Ethiopia-Sudan	2,777								2,777
	2,777								*2,777*
27 East Central Africa	3,808			1,032		1,650			6,490
	3,462			*1,032*		*1,996*			*6,490*
28 S. Africa	3,977								3,977
	3,977								*3,977*
31 India						346	8,374		8,763
							7,056		*7,099*
32 Pakistan	1,687						1,325		1,687
	261								*1,586*
33 Burma							42		42
							42		*42*
35 South Korea	1,879								1,879
	1,879								*1,879*
37 Philippines-China (T)	3,539								3,539
	3,157								*3,157*
38 Japan	7,952								7,952
	7,952								*7,952*
World Totals	71,146		4,215	1,032	4,465	1,996	8,416	43	91,313
	68,496		*4,215*	*1,032*	*4,465*	*1,996*	*8,423*	*43*	*88,670*

INFLUENCE OF U.S. TRANSPORTATION POLICY
IN RESPECT TO GOVERNMENT-SPONSORED SHIPMENTS ON THE
COMPETITIVENESS OF U.S. CEREAL GRAIN AND RICE EXPORTS

Under the Cargo Preference Act of 1954 at least 50 percent of all U.S. government-sponsored cereal exports must move in U.S. vessels. Since the rates on U.S. flag vessels are much higher than those on foreign flag vessels, this constitutes a considerable cost increase to countries importing grain under government-sponsored programs. Comparison of rates on the shipment of wheat from New Orleans to Calcutta illustrates the type of cost increase this imposes. The calculated shipping rate per ton for that voyage is $12.58 (see Table 13.15) for a foreign flag vessel. If, as indicated by Hutchinson,[4] U.S. flag rates are at least double those on foreign flag vessels, the requirement constitutes a 50 percent increase in per ton cost of the shipment.[5] When related to the U.S. wheat export price of $64.10 used in our analyses, that increase ($6.29) is shown to represent an 8.2 percent increase over the delivered price of $76.68, given shipment via foreign flag vessel.

Throughout the analyses conducted under the "most probable" conditions, the U.S. ocean rates to developing country ports were set at 1.5[6] times foreign flag rates. To determine the effect of the high U.S. vessel transportation rate specified in our analyses and to indicate the nature of its potential influence on the competitiveness of U.S. cereal exports, an analysis was conducted, given a reduction of the U.S. rates to the level of those charged on foreign flag vessels. The analysis was conducted using the projected regional requirements in 1985, given the low cropland upper bounds. The results are presented in Tables 16.12 to 16.14.

The statistical comparisons presented in each of Tables 16.12 to 16.14 indicate that U.S. rice exports rise only 97,000 tons but exports of wheat and rye increase 2.9 million tons or 106 million bushels and other grain exports rise 4.9 million tons or 194 million bushels. Both are substantial increases. The increased wheat and rye exports flow to countries in the Middle East and Other East Asia trade areas. The increased exports of other grains go to countries in Other East Asia, Africa, and the Middle East. The small increase in rice exports is spread among countries in Africa and the Middle East.

It is interesting to note that with the exception of 97,000 tons of rice markets secured at the expense of East Central Africa and 882,000 tons of Mexican other grain exports which are replaced by U.S. exports,

---

4. T. Q. Hutchinson. *Heavy Grain Exports in Voyage Chartered Ships: Rates and Volume.* U.S. Department of Agriculture, ERS-Marketing Research Report 812, 1968.
5. Average increase in per ton cost equals ($12.58 + $25.16)/2 − $12.58 = $6.29.
6. Specification of the factor was based on the assumptions that (a) all U.S. exports to developing countries are government sponsored, and (b) the U.S. flag rates are double those on foreign flag vessels.

TABLE 16.12. A Comparison of 1985 Interregional Patterns of Trade in Wheat and Rye, Given Alternate United States Transportation Rates to Developing Countries

Importing Area	Exporting Areas (as numbered in Table 12.1)							World Totals
	1	2	3	10	14	29	31	
4 Caribbean	786[a]							786
	786							*786*
5 Central America	545							545
	545							*545*
6 Northern S. America	7,007							7,007
	7,007							*7,007*
7 Brazil	5,322							5,322
	5,322							*5,322*
8 Western S. America	1,531							1,531
	877		*654*					*1,531*
9 Southern S. America						488		488
						488		*488*
11 Ireland-United Kingdom	4,537							4,537
	4,537							*4,537*
12 Scandinavia	1,539							1,539
	1,539							*1,539*
13 Spain-Portugal	330							330
	330							*330*
15 Northern E. Europe	3,581			6,426	171			10,178
	3,581			*6,426*	*171*			*10,178*
18 USSR	7,756							7,756
	7,756							*7,756*
19 Greece-Turkey	1,904	2,004						3,908
	3,908							*3,908*
20 United Arab Republic	3,606							3,606
	3,606							*3,606*
21 Iran-Iraq						1,552	380	1,932
						1,552	*380*	*1,932*
22 Other Middle East		260	654			1,573		2,487
	1,464					*1,023*		*2,487*

TABLE 16.12 (cont.)

Importing Area	Exporting Areas (as numbered in Table 12.1)							World Totals
	1	2	3	10	14	29	31	
23 Northern Africa	4,021 *4,021*							4,021 *4,021*
24 Western Africa	573 *573*							573 *573*
25 West Central Africa						176 *176*		176 *176*
26 Ethiopia-Sudan						158 *158*		158 *158*
27 East Central Africa						536 *536*		536 *536*
28 S. Africa	392					*392*		392 *392*
30 New Zealand	158					*158*		158 *158*
32 Pakistan						3,913 *3,913*	495 *495*	4,408 *4,408*
33 Burma							25 *25*	25 *25*
34 Other Far East						234 *234*		234 *234*
35 South Korea	*629*	629						629 *629*
36 Malaya-Indonesia						541 *541*		541 *541*
37 Philippines-China (T)						1,259 *1,259*		1,259 *1,259*
38 Japan	4,897 *4,897*							4,897 *4,897*
World Totals	48,485 *51,378*	2,893	654 *654*	6,426 *6,426*	171 *171*	10,430 *10,430*	900 *900*	69,959 *69,959*

ᵃ The italicized entries are those which result when United States rates to developing countries are equal those of other foreign flag vessels. Other entries indicate trading patterns which result when United States rates are 1.5 times those on foreign flag vessels.

TABLE 16.13. A Comparison of 1985 Interregional Patterns of Trade in Rice, Given Alternate United States Transportation Rates to Developing Countries

Importing Area	Exporting Areas (as numbered in Table 12.1)											World Totals
	1	3	6	7	13	20	27	29	31	33	38	
1 Canada	55[a] *55*											55 *55*
4 Caribbean	346 *346*											346 *346*
5 Central America	79 *79*											79 *79*
8 Western S. America	149 *125*	13 *37*										162 *162*
9 Southern S. America				63 *63*								63 *63*
10 EEC	8				5 *13*							13 *13*
11 Ireland-United Kingdom	133 *133*											133 *133*
12 Scandinavia	46 *46*											46 *46*
14 Austria-Switzerland	78 *39*				*39*							78 *78*
15 Northern E. Europe	360 *360*											360 *360*
16 Yugoslavia	54 *54*											54 *54*
17 Other East Europe	61 *27*					*34*						61 *61*
18 USSR	714 *714*											714 *714*
19 Greece-Turkey	*131*					131						131 *131*

373

TABLE 16.13 (cont.)

Importing Area	\[Exporting Areas (as numbered in Table 12.1)\] 1	3	6	7	13	20	27	29	31	33	38	World Totals
21 Iran-Iraq						59 / *59*						59 / *59*
22 Other Middle East						226 / *226*						226 / *226*
23 Northern Africa					47 / *47*							47 / *47*
24 Western Africa	/ *112*	24	164 / *164*	477 / *389*								665 / *665*
25 West Central Africa				101 / *101*								101 / *101*
28 S. Africa	88 / *88*											88 / *88*
30 New Zealand								5 / *5*				5 / *5*
32 Pakistan						124 / *221*	97		71 / *71*	1,757 / *1,757*		2,049 / *2,049*
34 Other Far East											197 / *197*	197 / *197*
35 South Korea											1,222 / *1,222*	1,222 / *1,222*
36 Malaya-Indonesia								132 / *132*		1,967 / *1,967*	602 / *602*	2,701 / *2,701*
37 Philippines-China (T)											2,820 / *2,820*	2,820 / *2,820*
World Totals	2,171 / *2,268*	37 / *37*	164 / *164*	641 / *641*	52 / *52*	540 / *540*	97	137 / *137*	71 / *71*	3,724 / *3,724*	4,841 / *4,841*	12,475 / *12,475*

374

ª The italicized entries are those which result when United States rates to developing countries are equal those of other foreign flag vessels. Other entries indicate trading patterns which result when United States rates are 1.5 times those on foreign flag vessels.

TABLE 16.14. A Comparison of 1985 Interregional Patterns of Trade in Other Grains, Given Alternate United States Transportation Rates to Developing Countries

Importing Area	Exporting Areas (as numbered in Table 12.1)							World Totals
	1	3	9	18	29	34	36	
1 Canada	2,168ᵃ / *2,168*							2,168 / *2,168*
4 Caribbean	409 / *409*							409 / *409*
5 Central America	1,109 / *1,109*							1,109 / *1,109*
6 Northern S. America	1,215 / *1,215*							1,215 / *1,215*
7 Brazil			2,430 / *2,430*					2,430 / *2,430*
8 Western S. America	773 / *773*							773 / *773*
11 Ireland-United Kingdom	3,841 / *3,841*							3,841 / *3,841*
13 Spain-Portugal	60 / *60*							60 / *60*
14 Austria-Switzerland	920 / *920*							920 / *920*
16 Yugoslavia	2,040 / *2,040*							2,040 / *2,040*
19 Greece-Turkey	4,330 / *4,995*			665				4,995 / *4,995*
20 United Arab Republic	2,361 / *2,361*							2,361 / *2,361*
21 Iran-Iraq	*1,406*			1,785		379		1,785 / *1,785*

ᵃ The italicized entries are those which result when United States rates to developing countries are equal those of other foreign flag vessels. Other entries indicate trading patterns which result when United States rates are 1.5 times those on foreign flag vessels.

TABLE 16.14 (cont.)

Importing Area	Exporting Areas (as numbered in Table 12.1)							World Totals
	1	3	9	18	29	34	36	
22 Other Middle East	1,054 *1,054*							1,054 *1,054*
23 Northern Africa	1,997 *1,997*							1,997 *1,997*
24 Western Africa	2,441 *2,441*							2,441 *2,441*
25 West Central Africa	117 *725*		608					725 *725*
26 Ethiopia-Sudan	298 *1,180*	882						1,180 *1,180*
27 East Central Africa	*846*		1,168 *923*		1,135 *534*			2,303 *2,303*
28 S. Africa	853		*853*					853 *853*
31 India					534	2,639 *3,173*	677 *677*	3,850 *3,850*
32 Pakistan					468	441 *909*		909 *909*
33 Burma						2 *2*		2 *2*
35 South Korea	1,000 *1,000*							1,000 *1,000*
37 Philippines-China (T)	*1,381*					1,381		1,381 *1,381*
38 Japan	6,823 *6,823*							6,823 *6,823*
World Totals	33,809 *38,744*	882	4,206 *4,206*	2,450	2,137 *534*	4,463 *4,463*	677 *677*	48,624 *48,624*

TABLE 16.15. Adjustments in Projected 1985 U.S. Cropland Utilization, Given Reduction in U.S. Shipping Rates to Developing Countries

Unit	1965 Area	Projected 1985 Area	Projected Utilization		Projected Area Not Used		Area Upper Bound
			US = Foreign Flag Rates	US = 1.5x Foreign Flag Rates	US = Foreign Flag Rates	US = 1.5x Foreign Flag Rates	
1,000 Hectares	84,035	102,331	88,692	86,747	13,639	15,584	125,000
1,000 Acres	207,650	252,860	219,158	214,352	33,702	38,508	308,875

the entire market gain of the United States, given the rate adjustment, came as a result of increased competitiveness with other developed nation exporters. This, as was the case in results presented elsewhere in the book, suggests the general competitiveness of developing nation cereal grain exports.

The information presented in Table 16.15 illustrates the effect of the changes in export levels, given ocean rate reductions, on projected cropland utilization. The total cropland area effect, given the decrease in ocean rates to developing countries, is an increase of 1.9 million hectares or 4.8 million acres in required cropland.

This analysis suggests that adjustment of U.S. shipping policy and consequently shipping rates respecting developing countries would increase the competitiveness of U.S. cereal exports with those of other developed countries while having minimal negative impact on developing nation exports. Such policy adjustments would at the same time reduce the cost of developing nation food imports. It would seem, therefore, that unless the United States wishes (a) to subsidize its shipping industry at the expense of developing nations and (b) reduce its competitiveness with other exporters of grain to developing countries some shipping policy adjustment may be desirable.

IMPLICATIONS OF EXISTING POTASH AND PHOSPHATE ROCK PLANT CAPACITIES FOR DISTRIBUTION OF PRODUCTION AND TRADE IN THOSE COMMODITIES

In analyses conducted for 1975 and 1985, existing potash and phosphate rock plant installations were utilized before capacity additions were allowed. This was accomplished by levying investment costs against activities increasing potash and phosphate rock plant capacity. To determine the extent to which this affected the distribution of production and trade, the 1985 analysis was conducted with the investment cost requirement removed.

Tables 16.16 and 16.17 contain the 1985 interarea results for both

TABLE 16.16. The Implications of Existing Regional Plant Capacities for 1985 Interarea Net Potash Trade Statistics, Assuming (1) the Low Land Bounds and (2) Fertilizer Use Assumption A (1,000 metric tons)

Importing Area	Exporting Areas (as numbered on the left)													Total Imports
	1	2	3	4	5	6	7	8	9	10	11	12	13	
1 United States														
2 Canada														
3 Latin America	*219*	1,166ᵃ					440							1,606 *219*
4 EEC														
5 Other W. Europe						37 *1,494*	2,874							2,911 *1,494*
6 E. Europe				553			1,303 *750*							1,303 *1,303*
7 USSR														
8 Middle East							194 *194*							194 *194*
9 Africa					290		417							417 *290*
10 S. Africa-Oceania		429 *232*					51		439 *686*					919 *918*
11 India-Pakistan		44						396	191 *630*					631 *630*
12 Other E. Asia		746 *239*							*507*					746 *746*
13 Japan		1,694 *1,694*												1,694 *1,694*
Totals	219	4,079 *2,165*		553	290	37 *1,494*	5,279 *944*	396	630 *1,823*					10,420 *7,488*

ᵃ The results for the modified analysis are in italics. They are those which result when investment costs of added plant capacity are removed. Or restated, they are the results produced when all regions have zero initial plant capacity and equal per unit costs of plant capacity additions.

TABLE 16.17. The Implications of Existing Regional Plant Capacities for 1985 Interarea Net Phosphate Rock Trade Statistics, Assuming (1) the Low Land Bounds and (2) Fertilizer Use Assumption A (1,000 metric tons)

Importing Area	1	2	3	4	5	6	7	8	9	10	11	12	13	Total Imports
						Exporting Areas (as numbered on the left)								
1 United States			719											719
2 Canada	1,482		*1,482*											1,482 / *1,482*
3 Latin America	2,571								1,927					4,498
4 EEC							2,396 / *12,695*		10,299					12,695 / *12,695*
5 Other W. Europe							1,633 / *1,633*		5,636 / *5,636*					7,269 / *7,269*
6 E. Europe							9,663 / *9,663*	*985*	985					10,648 / *10,648*
7 USSR														
8 Middle East							580							580
9 Africa								1,387 / *821*						1,387 / *821*
10 S. Africa-Oceania	1,857		*1,161*					35	850					2,742 / *1,161*
11 India-Pakistan								1,527	*1,527*					1,527 / *1,527*
13 Other E. Asia								2,322	*2,322*					2,322 / *2,322*
13 Japan	2,675								2,675					2,675 / *2,675*
Totals	8,585		3,362				14,272 / *23,991*	5,271 / *1,806*	19,697 / *12,160*					47,825 / *41,319*

a The results for the modified analysis are in italics. They are those which result when investment costs of added plant capacity are removed. Or restated, they are the results produced when all regions have zero initial plant capacity and equal per unit costs of plant capacity additions.

TABLE 16.18. 1985 Interregional Potash and Phosphate Rock Import-Export Levels under Alternate Plant Capacity Cost Assumptions

Region	Potash Exports	Potash Imports	Phosphate Rock Exports	Phosphate Rock Imports
1 United States	4,079,081 *219,207ᵃ*		8,583,899	1,481,532 *1,481,532*
2 Canada	*2,164,728*			746,676
3 Mexico		219,207 *219,207*	31,818	627,078 *627,078*
4 Caribbean		309,277 *309,277*		628,860 *628,860*
5 Central America		291,445 *291,445*		600,406
6 Northern South America		346,330 *346,330*	*1,738,917*	1,000,970
7 Brazil	390,656	346,372		1,000,970
8 Western South America	148,765 *766,914*		205,110 *2,159,232*	
9 Southern South America		242,852 *210,518*		1,130,643
10 EEC	*552,602*			12,694,644 *12,694,644*
11 Ireland-UK		1,474,010		2,829,606 *2,829,606*
12 Scandinavia		947,908 *947,908*		1,633,266 *1,633,266*
13 Spain-Portugal	57,639 *290,495*			1,845,042 *1,845,042*
14 Austria-Switzerland		546,275 *546,275*		961,680 *961,680*
15 Northern East Europe	36,562 *1,494,183*			5,284,359
16 Yugoslavia		552,602 *552,602*		985,434 *985,434*
17 Other East Europe		750,793 *750,793*		4,378,860
18 USSR	5,278,935 *944,331*		14,272,426 *14,327,910*	
19 Greece-Turkey		141,103 *141,103*		846,956

ᵃ The results for the modified analysis are in italics. They are those which result when investment costs of added plant capacity are removed. Or restated, they are the results produced when all regions have zero initial plant capacity and equal per unit costs of plant capacity additions.

TABLE 16.18 (cont.)

Region	Potash		Phosphate Rock	
	Exports	Imports	Exports	Imports
20 United Arab Republic		86,247	2,723,555	1,806,528
21 Iran-Iraq	481,924	52,435		197,142
22 Other Middle East	86,247	290,495	3,012,506	
23 Northern Africa			18,128,271	5,636,328
24 Western Africa		126,175	1,586,187	
25 West Central Africa	438,564	615,620		17,061
26 Ethiopia-Sudan	384,247	1,528,016		821,094
27 East Central Africa		193,462	6,524,123	565,884
28 S. Africa		489,445		1,295,412
29 Australia		196,919	125,895	160,469
30 New Zealand		231,915		1,286,574
31 India		580,860		1,395,630
32 Pakistan		49,525		131,595
33 Burma		28,838		66,042
34 Other Far East		111,971		369,801
35 South Korea		238,738		952,479
36 Malaya-Indonesia		150,390		576,162
37 Philippines-China (T)		216,051		357,812
38 Japan		1,694,075		2,674,602
World Totals	10,905,717	9,052,999	48,543,772	32,318,933

cases.[7] The results indicate that potash exports, for the solution in which investment costs were levied, were largely a function of available plant capacity. With investment costs removed, Canadian potash exports fell 1.9 million tons and those of the USSR fell 4.3 million tons. The respective 1.5 and 1.2 million ton increases in potash exports from Eastern Europe and Africa indicate the locational advantage of their deposits relative to 1985 international requirements and deposits in Canada and the USSR.

Phosphate rock trade among areas declines 6.5 million tons when consideration of plant investment costs is removed. That is an indication of the long-run feasibility of nations tending toward increased use of domestic sources of rock, given adequate capital to develop reserves.

Adjustments also occur in patterns of trade in phosphate rock, given removal of investment costs. Exports from Latin America and the USSR increase 3.4 and 9.7 million tons, respectively. Withdrawal of the United States from the phosphate rock export market suggests the absence of any locational advantage for U.S. phosphate rock reserves.

Tables 16.16 and 16.17 illustrate the potash and phosphate rock trading adjustments among trade areas. The data contained in Table 16.18 present the regional dimensions of the adjustments caused by the removal of plant investment costs. Comparison of the results contained in those tables with domestic fertilizer use and plant capacity data presented in Tables 13.5 to 13.10, 15.7, and 15.8 provides a ready explanation of why countries, such as Canada which have plant capacities far in excess of domestic requirements for potash and/or phosphate rock, are facing and will likely continue to face problems of tremendous overcapacity. These conditions will prevail unless great strides are made in respect to increased fertilizer use and unless new additions to total world capacity are minimal.

---

7. The case including plant investment costs and the case in which they are excluded.

APPENDIX

Table A-1. Population and Per Capita Income Projections Presented in the National Totals and by Development Class and Trade Area

| Region | Year | Projected Population | | Indices of Projected Income Per Capita (1960 = 100) |
		Number (1,000's)	% 96-Nation Total	
United States	1960	180,676	8	100
	1975	219,390	7	121
	1985	254,403	7	127
	2000	316,376	7	129
Canada	1960	17,909	1	100
	1975	23,446	1	121
	1985	28,360	1	125
	2000	37,444	1	122
Latin America	1960	207,559	9	100
	1975	318,341	11	105
	1985	419,085	12	103
	2000	614,143	13	102
EEC	1960	171,696	8	100
	1975	187,742	6	151
	1985	197,337	6	183
	2000	211,782	5	224
Other West Europe	1960	126,986	6	100
	1975	136,154	5	139
	1985	140,936	4	170
	2000	147,555	3	222
East Europe	1960	115,254	5	100
	1975	130,492	4	162
	1985	140,053	4	195
	2000	152,826	3	239
USSR	1960	214,400	10	100
	1975	260,800	9	170
	1985	296,332	8	202
	2000	353,099	8	234

Region	Year	Projected Population Number (1,000's)	% 96-Nation Total	Indices of Projected Income Per Capita (1960 = 100)
Middle East	1960	100,136	5	100
	1975	148,091	5	117
	1985	191,338	5	117
	2000	268,196	6	114
Africa	1960	193,490	9	100
	1975	280,127	10	107
	1985	366,926	10	105
	2000	554,972	12	103
South Africa-Oceania	1960	28,509	1	100
	1975	39,592	1	113
	1985	50,240	1	110
	2000	70,763	2	109
India-Pakistan	1960	535,224	24	100
	1975	752,756	26	100
	1985	907,370	26	102
	2000	1,160,009	25	109
Other East Asia	1960	232,306	10	100
	1975	343,550	12	108
	1985	445,339	13	108
	2000	645,668	14	106
Japan	1960	93,210	4	100
	1975	106,174	4	165
	1985	114,615	3	201
	2000	122,400	3	256
Developed Countries	1960	948,640	43	100
	1975	1,103,790	37	149
	1985	1,222,276	34	173
	2000	1,412,245	30	201
Developing Countries	1960	1,268,715	57	100
	1975	1,842,865	63	105
	1985	2,330,058	66	105
	2000	3,242,979	70	106
96-Nation Total	1960	2,217,355	100	100
	1975	2,946,655	100	121
	1985	3,552,334	100	128
	2000	4,655,233	100	135

TABLE A-2. Population and Income Projections Expressed as Geographic Region Aggregates

Region	Year	Projected Population (1,000's)	Indices of Projected Real Income Per Capita (1960=100)
United States	1960	180,676	100
	1975	219,390	121
	1985	254,403	127
	2000	316,376	129
Canada	1960	17,909	100
	1975	23,446	121
	1985	28,360	125
	2000	37,444	122
Mexico	1960	34,988	100
	1975	57,603	109
	1985	80,811	101
	2000	127,703	100
Caribbean	1960	15,574	100
	1975	22,051	106
	1985	27,850	107
	2000	38,922	108
Central America	1960	11,788	100
	1975	18,722	112
	1985	25,344	110
	2000	38,163	106
Northern S. America	1960	23,706	100
	1975	38,440	103
	1985	52,868	102
	2000	82,275	101
Brazil	1960	70,459	100
	1975	110,792	100
	1985	145,412	100
	2000	211,480	100
Eastern S. America	1960	18,250	100
	1975	28,027	106
	1985	37,223	102
	2000	55,093	100
Southern S. America	1960	32,794	100
	1975	42,706	109
	1985	49,577	113
	2000	60,498	115
EEC	1960	171,696	100
	1975	187,742	151
	1985	197,337	183
	2000	211,782	224
Ireland-UK	1960	55,342	100
	1975	58,013	137
	1985	59,175	172
	2000	60,856	240
Scandinavia	1960	20,072	100
	1975	21,927	136
	1985	22,945	165
	2000	24,177	219

Region	Year	Projected Population (1,000's)	Indices of Projected Real Income Per Capita (1960=100)
Spain-Portugal	1960	39,129	100
	1975	42,906	139
	1985	45,073	164
	2000	48,129	197
Austria-Switzerland	1960	12,443	100
	1975	13,308	154
	1985	13,743	188
	2000	14,393	235
Northern E. Europe	1960	60,598	100
	1975	68,670	149
	1985	73,885	174
	2000	80,876	205
Yugoslavia	1960	18,402	100
	1975	21,151	173
	1985	22,537	216
	2000	24,425	274
Other East Europe	1960	36,254	100
	1975	40,671	177
	1985	43,631	221
	2000	47,525	279
USSR	1960	214,400	100
	1975	260,800	170
	1985	296,332	202
	2000	353,099	234
Greece-Turkey	1960	36,145	100
	1975	51,016	121
	1985	63,998	123
	2000	87,256	119
United Arab Republic	1960	25,952	100
	1975	40,153	113
	1985	53,651	109
	2000	77,911	100
Iran-Iraq	1960	27,182	100
	1975	40,300	117
	1985	51,320	120
	2000	69,568	124
Other Middle East	1960	10,857	100
	1975	16,622	112
	1985	22,369	113
	2000	33,461	114
Northern Africa	1960	28,009	100
	1975	43,103	112
	1985	57,668	111
	2000	84,096	107
Western Africa	1960	71,059	100
	1975	109,502	103
	1985	147,507	102
	2000	236,939	100

TABLE A-2 (cont.)

Region	Year	Projected Population (1,000's)	Indices of Projected Real Income Per Capita (1960=100)
West Central Africa	1960	22,878	100
	1975	29,756	111
	1985	37,197	106
	2000	53,746	103
Ethiopia-Sudan	1960	31,770	100
	1975	42,570	104
	1985	52,980	102
	2000	72,562	100
East Central Africa	1960	39,774	100
	1975	55,196	110
	1985	71,574	109
	2000	107,629	106
South Africa	1960	15,822	100
	1975	23,401	107
	1985	30,911	101
	2000	46,335	100
Australia	1960	10,315	100
	1975	13,002	124
	1985	15,373	130
	2000	19,176	135
New Zealand	1960	2,372	100
	1975	3,189	107
	1985	3,956	104
	2000	5,252	100
India-Ceylon	1960	442,646	100
	1975	616,151	100
	1985	737,612	102
	2000	933,509	111
Pakistan	1960	92,578	100
	1975	136,605	101
	1985	169,758	100
	2000	226,500	100
Burma	1960	22,325	100
	1975	30,870	114
	1985	38,704	113
	2000	53,696	106
Other Far East	1960	46,138	100
	1975	69,513	113
	1985	87,149	114
	2000	121,295	109
South Korea	1960	24,665	100
	1975	38,075	124
	1985	49,197	127
	2000	67,418	126
Malaya-Indonesia	1960	101,159	100
	1975	143,780	100
	1985	185,559	100
	2000	268,203	100

387

TABLE A-2 (cont.)

Region	Year	Projected Population (1,000's)	Indices of Projected Real Income Per Capita (1960=100)
Philippines-China (T)	1960	38,019	100
	1975	61,312	106
	1985	84,730	107
	2000	135,056	106
Japan	1960	93,210	100
	1975	106,174	165
	1985	114,615	201
	2000	122,400	256

SOURCE: L. L. Blakeslee. "An Analysis of Projected World Food Production and Demand in 1970, 1985, and 2000." Unpublished Ph.D. thesis, Ames: Iowa State University Library, 1967.

TABLE A-3. Geographic Region Production-Requirement Comparisons of Cereal Grains for 1975 (1,000 metric tons)

Region	Domestic Requirements	Domestic Production		Requirements Less Production	
		Low Land Bounds	High Land Bounds	Low Land Bounds	High Land Bounds
United States					
Wheat, rye	18,435	61,245		−42,810	
Rice	619	2,703		−2,084	
Other grain	126,819	190,563		−63,744	
Canada					
Wheat, rye	4,509	20,312		−15,803	
Rice	45	0		45	
Other grain	12,873	12,188		685	
Mexico					
Wheat, rye	2,355	2,911		−556	
Rice	285	322		−37	
Other grain	9,135	10,003		−868	
Caribbean					
Wheat, rye	652	0		652	
Rice	636	372		264	
Other grain	738	517		221	
Central America					
Wheat, rye	429	26		403	
Rice	283	254		29	
Other grain	2,263	1,752		511	
Northern S. America					
Wheat, rye	5,259	182		5,077	
Rice	593	682		−89	
Other grain	2,257	1,647		610	
Brazil					
Wheat, rye	4,558	663		3,895	
Rice	4,992	5,553		−561	
Other grain	13,172	12,361		811	

TABLE A-3 (cont.)

Region	Domestic Requirements	Domestic Production Low Land Bounds	Domestic Production High Land Bounds	Requirements Less Production Low Land Bounds	Requirements Less Production High Land Bounds
Eastern S. America					
Wheat, rye	1,453	331		1,122	
Rice	531	467		64	
Other grain	1,766	1,379		387	
Southern S. America					
Wheat, rye	7,832	8,465		—633	
Rice	299	269		30	
Other grain	4,939	9,077		—4,138	
EEC					
Wheat, rye	35,977	38,631		—2,654	
Rice	615	590		25	
Other grain	37,929	34,780		3,149	
Ireland-UK					
Wheat, rye	9,481	4,583		4,898	
Rice	121	0		121	
Other grain	16,562	9,907		6,655	
Scandinavia					
Wheat, rye	2,923	1,944		979	
Rice	42	0		42	
Other grain	6,655	11,209		—4,554	
Spain-Portugal					
Wheat, rye	6,270	5,816		454	
Rice	389	444		—55	
Other grain	5,276	4,855		421	
Austria-Switzerland					
Wheat, rye	2,072	1,947		125	
Rice	72	0		72	
Other grain	2,724	1,583		1,141	
Northern E. Europe					
Wheat, rye	26,225	18,158		8,067	
Rice	331	0		331	
Other grain	13,003	13,155		—152	
Yugoslavia					
Wheat, rye	4,587	5,988		—1,401	
Rice	65	17		48	
Other grain	8,059	6,950		1,109	
Other East Europe					
Wheat, rye	9,209	11,063		—1,854	
Rice	134	87		47	
Other grain	15,298	15,913		—615	
USSR					
Wheat, rye	68,728	69,179		—451	
Rice	869	278		591	
Other grain	41,442	41,977		—535	
Greece-Turkey					
Wheat, rye	12,352	10,754		—1,598	
Rice	203	126		77	
Other grain	8,335	5,253		3,082	

Region	Domestic Requirements	Domestic Production		Requirements Less Production	
		Low Land Bounds	High Land Bounds	Low Land Bounds	High Land Bounds
United Arab Republic					
Wheat, rye	4,558	2,096		2,462	
Rice	1,070	1,565		—495	
Other grain	3,828	2,593		1,235	
Iran-Iraq					
Wheat, rye	6,107	5,265	5,265	842	842
Rice	859	878	878	—19	—19
Other grain	2,989	2,064	2,064	925	925
Other Middle East					
Wheat, rye	2,784	1,174	1,408	1,610	1,376
Rice	166	0	0	166	166
Other grain	1,477	716	826	761	651
Northern Africa					
Wheat, rye	5,212	2,901		2,311	
Rice	50	17		33	
Other grain	3,303	2,450		853	
Western Africa					
Wheat, rye	436	0		436	
Rice	1,808	1,397		411	
Other grain	8,424	7,571		853	
West Central Africa					
Wheat, rye	187	40		147	
Rice	172	105		67	
Other grain	2,201	1,876		325	
Ethiopia-Sudan					
Wheat, rye	465	352		113	
Rice	0	0		0	
Other grain	6,681	6,137		544	
East Central Africa					
Wheat, rye	543	169		374	
Rice	1,108	1,222		—114	
Other grain	8,224	7,487		737	
South Africa					
Wheat, rye	1,414	1,214		200	
Rice	72	4		68	
Other grain	5,018	5,625		—607	
Australia					
Wheat, rye	2,995	12,427		—9,432	
Rice	28	150		—122	
Other grain	1,582	3,837		—2,255	
New Zealand					
Wheat, rye	445	322		123	
Rice	4	0		4	
Other grain	104	156		—52	
India					
Wheat, rye	17,835	17,115	17,115	720	720
Rice	46,978	47,197	47,197	—219	—219
Other grain	28,418	26,374	26,374	2,044	2,044

TABLE A-3   (cont.)

Region	Domestic Requirements	Domestic Production		Requirements Less Production	
		Low Land Bounds	High Land Bounds	Low Land Bounds	High Land Bounds
Pakistan					
Wheat, rye	7,657	4,637	4,637	3,020	3,020
Rice	15,555	14,338	14,338	1,217	1,217
Other grain	1,797	1,277	1,277	520	520
Burma					
Wheat, rye	79	47		32	
Rice	4,679	7,744		—3,065	
Other grain	128	143		—15	
Other Far East					
Wheat, rye	189	0	0	189	189
Rice	12,729	13,802	15,066	—1,073	—2,337
Other grain	138	2,645	2,645	—2,507	—2,507
South Korea					
Wheat, rye	813	317		496	
Rice	3,992	3,288		704	
Other grain	1,886	1,312		574	
Malaya-Indonesia					
Wheat, rye	416	0		416	
Rice	13,828	12,379		1,449	
Other grain	3,324	4,059		—735	
Philippines-China (T)					
Wheat, rye	1,008	77	77	931	931
Rice	6,856	5,829	5,829	1,027	1,027
Other grain	2,170	1,728	1,728	442	442
Japan					
Wheat, rye	5,414	1,410		4,004	
Rice	11,426	14,444		—3,018	
Other grain	6,017	905		5,112	

TABLE A-4.   Geographic Region Production-Requirement Comparisons of Cereal Grains for 1985 (1,000 metric tons)

Region	Domestic Requirements	Domestic Production		Requirements Less Production	
		Low Land Bounds	High Land Bounds	Low Land Bounds	High Land Bounds
United States					
Wheat, rye	21,242	83,286		—62,044	
Rice	724	3,198		—2,474	
Other grain	147,604	243,867		—96,263	
Canada					
Wheat, rye	5,457	22,607		—17,150	
Rice	55	0		55	
Other grain	15,613	13,445		2,168	

Region	Domestic Requirements	Domestic Production Low Land Bounds	Domestic Production High Land Bounds	Requirements Less Production Low Land Bounds	Requirements Less Production High Land Bounds
Mexico					
Wheat, rye	3,226	3,882		—654	
Rice	390	428		—38	
Other grain	12,582	13,465		—883	
Caribbean					
Wheat, rye	786	0		786	
Rice	788	443		345	
Other grain	941	532		409	
Central America					
Wheat, rye	574	29		545	
Rice	382	302		80	
Other grain	3,050	1,941		1,109	
Northern S. America					
Wheat, rye	7,208	202		7,006	
Rice	791	957		—166	
Other grain	3,063	1,848		1,215	
Brazil					
Wheat, rye	5,983	661		5,322	
Rice	6,552	7,193		—641	
Other grain	17,288	14,858		2,430	
Eastern S. America					
Wheat, rye	1,897	367		1,530	
Rice	701	540		161	
Other grain	2,316	1,544		772	
Southern S. America					
Wheat, rye	9,099	8,612		487	
Rice	349	287		62	
Other grain	5,676	9,883		—4,207	
EEC					
Wheat, rye	37,724	44,152		—6,428	
Rice	660	650		10	
Other grain	42,237	46,128		—3,891	
Ireland-UK					
Wheat, rye	9,597	5,060		4,537	
Rice	133	0		133	
Other grain	17,489	13,650		3,839	
Scandinavia					
Wheat, rye	3,012	1,474		1,538	
Rice	46	0		46	
Other grain	7,250	13,343		—6,093	
Spain-Portugal					
Wheat, rye	6,386	6,056		330	
Rice	416	469		—53	
Other grain	5,962	5,903		59	
Austria-Switzerland					
Wheat, rye	2,100	2,272		—172	
Rice	78	0		78	
Other grain	2,896	1,976		920	

Region	Domestic Requirements	Domestic Production Low Land Bounds	Domestic Production High Land Bounds	Requirements Less Production Low Land Bounds	Requirements Less Production High Land Bounds
Northern E. Europe					
Wheat, rye	28,972	18,794		10,178	
Rice	360	0		360	
Other grain	14,286	16,243		−1,957	
Yugoslavia					
Wheat, rye	4,696	7,508		−2,812	
Rice	73	19		54	
Other grain	9,175	7,135		2,040	
Other East Europe					
Wheat, rye	9,626	12,053		−2,427	
Rice	149	88		61	
Other grain	17,348	17,719		−371	
USSR					
Wheat, rye	75,784	68,027		7,757	
Rice	1,031	317		714	
Other grain	49,265	51,716		−2,451	
Greece-Turkey					
Wheat, rye	15,448	11,541		3,908	
Rice	251	119		132	
Other grain	10,500	5,505		4,995	
United Arab Republic					
Wheat, rye	6,029	2,423		3,606	
Rice	1,415	1,956		−541	
Other grain	5,061	2,700		2,361	
Iran-Iraq					
Wheat, rye	7,765	5,834	6,218	1,931	1,547
Rice	1,098	1,040	1,111	58	−13
Other grain	3,923	2,138	2,244	1,785	1,679
Other Middle East					
Wheat, rye	3,701	1,214	1,451	2,487	2,250
Rice	226	0	0	226	226
Other grain	1,881	829	948	1,052	933
Northern Africa					
Wheat, rye	6,921	2,901		4,020	
Rice	64	17		47	
Other grain	4,451	2,453		1,998	
Western Africa					
Wheat, rye	573	0		573	
Rice	2,312	1,648		664	
Other grain	11,351	8,912		2,439	
West Central Africa					
Wheat, rye	223	48		175	
Rice	206	105		101	
Other grain	2,650	1,924		726	
Ethiopia-Sudan					
Wheat, rye	578	420		158	
Rice	0	0		0	
Other grain	8,245	7,066		1,179	

TABLE A-4   (cont.)

Region	Domestic Requirements	Domestic Production Low Land Bounds	Domestic Production High Land Bounds	Requirements Less Production Low Land Bounds	Requirements Less Production High Land Bounds
East Central Africa					
Wheat, rye	718	182		536	
Rice	1,354	1,460		—106	
Other grain	10,778	8,478		2,300	
South Africa					
Wheat, rye	1,846	1,454		392	
Rice	92	5		87	
Other grain	6,609	5,755		854	
Australia					
Wheat, rye	3,520	13,951		—10,431	
Rice	34	172		—138	
Other grain	1,868	4,006		—2,138	
New Zealand					
Wheat, rye	556	398		158	
Rice	5	0		5	
Other grain	129	195		—66	
India					
Wheat, rye	21,598	22,498	22,498	—900	—900
Rice	56,821	56,894	56,894	—73	—73
Other grain	34,115	30,265	30,265	3,850	3,850
Pakistan					
Wheat, rye	9,473	5,065	5,065	4,408	4,408
Rice	19,261	17,213	17,213	2,048	2,048
Other grain	2,230	1,321	1,321	909	909
Burma					
Wheat, rye	99	74		25	
Rice	5,862	9,587		—3,725	
Other grain	160	158		2	
Other Far East					
Wheat, rye	234	0	0	234	234
Rice	15,926	15,732	17,471	194	—1,545
Other grain	174	4,639	4,639	—4,465	—4,465
South Korea					
Wheat, rye	1,060	430		630	
Rice	5,189	3,967		1,222	
Other grain	2,444	1,444		1,000	
Malaya-Indonesia					
Wheat, rye	541	0		541	
Rice	17,806	15,104		2,702	
Other grain	4,277	4,955		—678	
Philippines-China (T)					
Wheat, rye	1,358	99	99	1,259	1,259
Rice	9,329	6,508	6,911	2,821	2,418
Other grain	3,100	1,718	1,905	1,382	1,195
Japan					
Wheat, rye	6,042	1,144		4,898	
Rice	12,208	17,050		—4,842	
Other grain	6,949	126		6,823	

TABLE A-5. Geographic Region Production-Requirement Comparisons of Cereal Grains for 2000 (1,000 metric tons)

Region	Domestic Requirements	Domestic Production		Requirements Less Production	
		Low Land Bounds	High Land Bounds	Low Land Bounds	High Land Bounds
United States					
Wheat, rye	26,359	122,558		−96,199	
Rice	904	3,888		−2,984	
Other grain	183,809	334,790		−150,981	
Canada					
Wheat, rye	7,202	26,007		−18,805	
Rice	73	0		73	
Other grain	20,578	15,243		5,335	
Mexico					
Wheat, rye	5,086	5,338		−252	
Rice	615	606		9	
Other grain	19,842	19,347		495	
Caribbean					
Wheat, rye	1,027	0		1,027	
Rice	1,070	549		521	
Other grain	1,329	553		776	
Central America					
Wheat, rye	847	33		814	
Rice	575	371		204	
Other grain	4,543	2,208		2,335	
Northern S. America					
Wheat, rye	11,534	230		11,304	
Rice	1,203	1,443		−240	
Other grain	4,759	2,133		2,626	
Brazil					
Wheat, rye	8,701	658		8,043	
Rice	9,528	9,916		−388	
Other grain	25,142	18,602		6,540	
Eastern S. America					
Wheat, rye	2,759	419		2,340	
Rice	1,035	647		388	
Other grain	3,386	1,775		1,611	
Southern S. America					
Wheat, rye	11,067	8,846		2,221	
Rice	427	310		117	
Other grain	6,814	11,030		−4,216	
EEC					
Wheat, rye	40,435	50,507		−10,072	
Rice	727	729		−2	
Other grain	48,119	64,018		−15,899	
Ireland-UK					
Wheat, rye	9,800	5,092		4,708	
Rice	152	0		152	
Other grain	18,742	18,373		369	
Scandinavia					
Wheat, rye	3,131	1,081		2,050	
Rice	51	0		51	
Other grain	8,084	16,026		−7,942	

Region	Domestic Requirements	Domestic Production		Requirements Less Production	
		Low Land Bounds	High Land Bounds	Low Land Bounds	High Land Bounds
Spain-Portugal					
Wheat, rye	6,560	6,503		57	
Rice	452	498		—46	
Other grain	6,885	7,919		—1,034	
Austria-Switzerland					
Wheat, rye	2,161	2,746		—585	
Rice	85	0		85	
Other grain	3,109	2,971		138	
Northern E. Europe					
Wheat, rye	32,542	20,016		12,526	
Rice	395	0		395	
Other grain	15,911	21,585		—5,674	
Yugoslavia					
Wheat, rye	4,864	9,738		—4,874	
Rice	83	21		62	
Other grain	10,523	7,337		3,186	
Other East Europe					
Wheat, rye	10,188	12,937		—2,749	
Rice	167	89		78	
Other grain	19,914	20,724		—810	
USSR					
Wheat, rye	87,848	66,804		21,044	
Rice	1,274	367		907	
Other grain	60,991	65,456		—4,465	
Greece-Turkey					
Wheat, rye	20,826	12,643		8,183	
Rice	330	113		217	
Other grain	13,935	5,934		8,001	
United Arab Republic					
Wheat, rye	8,547	2,718		5,829	
Rice	2,006	2,368		—362	
Other grain	7,172	2,674		4,498	
Iran-Iraq					
Wheat, rye	10,493	6,105	7,245	4,388	3,248
Rice	1,502	1,192	1,437	310	65
Other grain	5,662	2,111	2,413	3,551	3,249
Other Middle East					
Wheat, rye	5,496	1,243	1,531	4,253	3,965
Rice	347	0	0	347	347
Other grain	2,620	970	1,164	1,650	1,456
Northern Africa					
Wheat, rye	9,940	2,901		7,039	
Rice	93	17		76	
Other grain	6,499	2,459		4,040	
Western Africa					
Wheat, rye	883	0		883	
Rice	3,392	2,004		1,388	
Other grain	18,212	10,850		7,362	

Region	Domestic Requirements	Domestic Production		Requirements Less Production	
		Low Land Bounds	High Land Bounds	Low Land Bounds	High Land Bounds
West Central Africa					
Wheat, rye	305	60		245	
Rice	298	105		193	
Other grain	3,680	1,995		1,685	
Ethiopia-Sudan					
Wheat, rye	786	509		277	
Rice	0	0		0	
Other grain	11,192	8,416		2,776	
East Central Africa					
Wheat, rye	1,100	198		902	
Rice	1,854	1,812		42	
Other grain	16,351	9,861		6,490	
South Africa					
Wheat, rye	2,762	1,790		973	
Rice	138	5		133	
Other grain	9,902	5,924		3,978	
Australia					
Wheat, rye	4,372	15,927		—11,555	
Rice	42	205		—163	
Other grain	2,327	4,325		—1,998	
New Zealand					
Wheat, rye	744	517		227	
Rice	7	0		7	
Other grain	173	263		—90	
India					
Wheat, rye	28,566	30,722	31,946	—2,156	—3,380
Rice	74,565	68,843	72,578	5,722	1,987
Other grain	43,819	35,056	36,720	8,763	7,099
Pakistan					
Wheat, rye	12,640	5,250	5,663	7,390	6,977
Rice	25,699	20,237	21,829	5,462	3,870
Other grain	2,975	1,288	1,389	1,687	1,586
Burma					
Wheat, rye	135	123		12	
Rice	8,062	12,599		—4,537	
Other grain	222	180		42	
Other Far East					
Wheat, rye	309	0	0	309	309
Rice	21,758	17,101	20,507	4,657	1,251
Other grain	242	8,660	8,667	—8,418	—8,425
South Korea					
Wheat, rye	1,448	564		884	
Rice	7,099	4,594		2,505	
Other grain	3,346	1,468		1,878	
Malaya-Indonesia					
Wheat, rye	808	0		808	
Rice	25,786	19,523		6,263	
Other grain	6,162	6,208		—46	

TABLE A-5 (cont.)

Region	Domestic Requirements	Domestic Production		Requirements Less Production	
		Low Land Bounds	High Land Bounds	Low Land Bounds	High Land Bounds
Philippines-China (T)					
Wheat, rye	2,075	132	132	1,943	1,943
Rice	14,507	7,624	8,604	6,883	5,903
Other grain	5,168	1,629	2,010	3,539	3,158
Japan					
Wheat, rye	6,708	511		6,197	
Rice	12,907	21,061		−8,154	
Other grain	8,031	79		7,952	

TABLE A-6. Country Inland Transportation Rates for Average Freight Commodities (dollars per metric ton)

	Country	Rate
1	United States	$ 3.19
2	Canada	8.44
3	Mexico	5.37
4	Cuba	4.15 (13)
5	Dominican Republic	3.60 (7)
6	Haiti	3.60 (7)
7	Jamaica	3.60
8	British Honduras	4.15 (13)
9	Costa Rica	7.77
10	El Salvador	4.15 (13)
11	Guatemala	4.15 (13)
12	Honduras	1.72
13	Nicaragua	4.15
14	Panama	4.15 (13)
15	Trinidad, Tobago	3.60 (7)
16	Colombia	2.27
17	Venezuela	2.27 (16)
18	Brazil	2.49
19	Bolivia	4.43
20	Ecuador	7.09
21	Peru	4.47
22	Argentina	4.54 (23)
23	Uruguay	4.54
24	Chile	9.33
25	Paraguay	2.77
26	Belgium, Luxembourg	3.51
27	West Germany	5.26
28	France	6.84
29	Netherlands	4.27
30	Italy	11.65
31	Ireland	17.72
32	United Kingdom	5.96
33	Denmark	12.34
34	Norway	4.17
35	Sweden	6.04

Country	Rate
36 Finland	$ 5.26
37 Spain	5.46
38 Portugal	7.76
39 Austria	5.58
40 Switzerland	5.53
41 East Germany	5.21 (43)
42 Poland	5.21 (43)
43 Czechoslovakia	5.21
44 Yugoslavia	7.87
45 Hungary	7.87 (44)
46 Rumania	7.87 (44)
47 Bulgaria	7.87 (44)
48 USSR[a]	5.60
49 Greece	13.49
50 Turkey	6.56
51 United Arab Republic	4.13 (61)
52 Iran	11.07
53 Iraq	6.62
54 Israel	3.42
55 Jordan	3.42 (54)
56 Lebanon	3.42 (54)
57 Syria	7.29
58 Cyprus	3.42 (54)
59 Morocco	4.13 (61)
60 Algeria	4.13 (61)
61 Tunisia	4.13
62 Libya	4.13 (61)
63 Ghana	5.42
64 Guinea	5.42 (63)
65 Ivory Coast	5.42 (63)
66 Liberia	5.42 (63)
67 Nigeria	11.24
68 Senegal	5.42 (63)
69 Sierra Leone	25.01
70 Togo	5.42 (63)
71 Angola	6.02
72 Cameroun	4.88 (73)
73 Congo (Kinshasa), Rwanda, Burundi	4.88
74 Ethiopia	10.55
75 Sudan	9.58
76 Kenya[b]	8.46
77 Tanganyika	8.46 (76)
78 Uganda	8.46 (76)
79 Malagasy Republic	8.46 (76)
80 Malawi, Rhodesia, Zambia	5.59
81 South Africa	3.40
82 Australia	6.58
83 New Zealand	7.94
84 India	4.51
85 Ceylon	3.42 (54)
86 Pakistan	5.08
87 Burma	6.23
88 Cambodia	7.14
89 Thailand	5.21
90 South Vietnam	2.68

TABLE A-6 (cont.)

Country	Rate
91 South Korea	$ 2.21
92 Fed. of Malaya	5.56
93 Indonesia	5.15
94 Philippines	5.14
95 China (T)	5.56 (92)
96 Japan	5.18

SOURCE: H. Sampson, ed. *World Railways, 1961–62,* 7th ed. London: Sampson-Lows, World Railways, Ltd., 1963.
NOTES: The rates are based on 1960 data.
Where lack of data prevented determination of the rate for a particular country, data for a country with similar characteristics were used. In such cases, the number of the country whose data were used appears in parentheses to the right of the rate specified.
[a] The USSR rate is set at 70 percent of the U.S. rate of 1.59 cents average (1959–61) revenue per ton mile.
[b] Kenya rate is developed from data for East Africa contained in H. Sampson, ed. *World Railways, 1961–62,* 7th ed. London: Sampson-Lows, World Railways, Ltd., 1963.

TABLE A-7. Country Reserves of Potash and Phosphate Rock (million metric tons)

Country	Potash (K_2O)	Phosphate Rock (material)
United States	400	4,747
Canada	6,400	
Mexico		1,154
Cuba		
Dominican Republic		
Haiti		
Jamaica		
British Honduras		
Costa Rica		
El Salvador		
Guatemala		
Honduras		
Nicaragua		
Panama		
Trinidad, Tobago		
Colombia		2
Venezuela[a]		59
Brazil	28	61
Bolivia		
Ecuador		
Peru	12	1,421
Argentina		
Uruguay		
Chile	[b]	1
Paraguay		
Belgium, Luxembourg		
West Germany	9,979	
France	318	
Netherlands		
Italy	[c]	

400

TABLE A-7   (cont.)

Country	Potash (K$_2$O)	Phosphate Rock (material)
Ireland		
United Kingdom	127	
Denmark		
Norway		
Sweden		
Finland		
Spain	363	
Portugal		
Austria		
Switzerland		
East Germany	12,700	
Poland		
Czechoslovakia		
Yugoslavia		
Hungary		
Rumania		
Bulgaria		
USSR	17,237	5,907
Greece		
Turkey		
United Arab Republic		87
Iran		
Iraq		
Israel	1,000	650
Jordan	1,000	270
Lebanon		
Syria		
Cyprus		
Morocco[d]		1,550
Algeria		830
Tunisia		1,160
Libya		
Ghana		
Guinea		
Ivory Coast		
Liberia		
Nigeria		
Senegal		339
Sierra Leone		
Togo		120
Angola		43
Cameroun		
Congo (Kinshasa), Rwanda, Burundi	95	
Ethiopia	50	
Sudan		
Kenya		
Tanganyika		
Uganda		323
Malagasy Republic		
Malawi, Rhodesia, Zambia		51
South Africa		70
Australia[e]		92
New Zealand		
India		
Ceylon		
Pakistan		

Country	Potash (K₂O)	Phosphate Rock (material)
Burma		
Cambodia		
Thailand		
South Vietnam		
South Korea		
Fed. of Malaya		
Indonesia		
Philippines		
China (T)		
Japan		

SOURCES: The British Sulphur Corporation, Ltd. *A World Survey of Phosphate Deposits*, 2nd ed. London, 1964; The British Sulphur Corporation, Ltd. *A World Survey of Potash*, London, 1966; J. R. Douglas, Jr., and E. A. Harre. TVA, Muscle Shoals, Alabama. *Fertilizer Plant Capacity Estimates*. Private communication, 1968.

[a] Includes islands of Curacao and Aruba.

[b] Quantity adequate for present production level.

[c] Assumes reserves adequate to maintain 1972 K₂O capacity production level.

[d] Includes Spanish Sahara.

[e] Australian production is based on reserves on Australian Pacific island possessions (Oceania and Nauru Islands, Makatea Islands, and Christmas Islands). Reserves are assumed adequate for continuation of present production levels through 1985. Subsequent projections assume exhaustion of reserves and, therefore, zero production.

TABLE A-8.   Estimated 1975 Nitrogen, Phosphate, Potash, and Phosphate Rock Plant Capacities by Geographic Region (1,000 metric tons)

Region	Nitrogen	Phosphate	Potash	Phosphate Rock
United States	11,314	7,243	3,250	37,572
Canada	1,276	942	6,408	
Mexico	469	480		35
Caribbean	27	10		
Central America		28		
Northern S. America	1,007	61		100
Brazil	246	147		100
Eastern S. America	49	43	123	400
Southern S. America	655	176	97	15
EEC	8,142	4,278	4,461	
Ireland-UK	1,374	818		
Scandinavia	958	553		
Spain-Portugal	1,262	590	302	
Austria-Switzerland	242	131		
Northern E. Europe	2,938	1,263	2,400	
Yugoslavia	555	664		
Other East Europe	2,542	987		
USSR	6,000	7,428	7,500	25,000
Greece-Turkey	435	693		
United Arab Republic	297	121		3,500
Iran-Iraq	321	97		
Other Middle East	60	255	360	3,700

TABLE A-8   (cont.)

Region	Nitrogen	Phosphate	Potash	Phosphate Rock
Northern Africa	291	497		21,000
Western Africa		27		2,000
West Central Africa			300	
Ethiopia-Sudan			300	
East Central Africa	25	50		
South Africa	366	613		600
Australia	555	1,195		3,150
New Zealand		562		
India	2,317	902		
Pakistan	443	199		
Burma	30			
Other Far East	24			
South Korea	374	149		
Malaya-Indonesia	139			
Philippines-China (T)	400	157		
Japan	3,058	1,283		

SOURCE: J. R. Douglas, Jr., and E. A. Harre. TVA, Muscle Shoals, Alabama. *Fertilizer Plant Capacity Estimates.* Private communication, 1968.

TABLE A.9.   Country Plant Capacity for Production of Nitrogen, Phosphorus, Potash, and Phosphate Rock, and Specified Quantity of That Capacity Allocated to Fertilizer Production (1,000 metric tons)

Country	Total Nutrient[a] Capacity				Fertilizer Production Capacity			
	PR	N	P	K	PR	N	P	K
United States	37,572	14,142	8,923	3,421	37,572	11,314	7,243	3,250
Canada		1,595	1,160	6,745		1,276	942	6,408
Mexico	35	521	494		35	469	480	
Cuba		30	10			27	10	
Dominican Republic								
Haiti								
Jamaica								
British Honduras								
Costa Rica								
El Salvador			29				28	
Guatemala								
Honduras								
Nicaragua								
Panama								
Trinidad, Tobago		320				288		
Colombia		365	22			329	21	
Venezuela	100	434	41		100	391	40	
Brazil	100	273	151		100	246	147	
Bolivia								
Ecuador								
Peru	400	55	44	123	400	49	43	123
Argentina		266	13			239	13	
Uruguay			29				28	
Chile	15	462	139	97	15	416	135	97
Paraguay								

Country	Total Number[a] Capacity				Fertilizer Production Capacity			
	PR	N	P	K	PR	N	P	K
Belgium, Luxembourg		735	1,187			588	1,095	
West Germany		3,134	1,202	2,400		2,507	1,109	2,280
France		3,275	1,211	1,900		2,620	1,117	1,805
Netherlands		1,469	316			1,175	291	
Italy		1,565	722	396		1,252	666	376
Ireland		33	222			26	205	
United Kingdom		1,685	665			1,348	613	
Denmark		29	148			23	137	
Norway		741	161			593	149	
Sweden		143	150			114	138	
Finland		285	140			228	129	
Spain		1,254	514	318		1,003	474	302
Portugal		323	126			258	116	
Austria		221	131			177	121	
Switzerland		82	11			66	10	
East Germany		600	321	2,400		480	318	2,400
Poland		2,099	718			1,679	711	
Czechoslovakia		974	236			779	234	
Yugoslavia		694	670			555	664	
Hungary		861	216			689	214	
Rumania		1,232	182			986	180	
Bulgaria		1,084	599			867	593	
USSR	25,000	7,500	7,500	7,500	25,000	6,000	7,428	7,500
Greece		459	453			413	441	
Turkey		24	259			22	252	
United Arab Republic	3,500	330	124		3,500	297	121	
Iran		303	100			273	97	
Iraq		54				49		
Israel	1,700	26	252	360	1,700	23	245	360
Jordan	2,000				2,000			
Lebanon			10				10	
Syria		41				37		
Cyprus								
Morocco	16,000		228		16,000		223	
Algeria	1,000	270	16		1,000	243	16	
Tunisia	4,000	53	264		4,000	48	258	
Libya								
Ghana								
Guinea								
Ivory Coast								
Liberia								
Nigeria								
Senegal	1,000		28		1,000		27	
Sierra Leone								
Togo	1,000				1,000			
Angola								
Cameroun								
Congo (Kinshasa), Rwanda, Burundi				300				300
Ethiopia				300				300
Sudan								
Kenya			1				1	
Tanganyika								

TABLE A-9   (cont.)

Country	Total Nutrient[a] Capapcity				Fertilizer Production Capacity			
	PR	N	P	K	PR	N	P	K
Uganda			5				5	
Malagasy Republic								
Malawi, Rhodesia, Zambia		28	45			25	44	
South Africa	600	406	627		600	366	613	
Australia	3,150	694	1,238		3,150	555	1,195	
New Zealand			582				562	
India		2,413	927			2,171	902	
Ceylon		162				146		
Pakistan		492	205			443	199	
Burma		33				30		
Cambodia								
Thailand		27				24		
South Vietnam								
South Korea		416	153			374	149	
Fed. of Malaya		41				37		
Indonesia		113				102		
Philippines		97	101			87	98	
China (T)		347	60			312	58	
Japan		3,823	1,330			3,058	1,283	
World Totals	97,172	59,133	35,411	26,260	97,172	48,193	32,640	25,501

SOURCE: J. R. Douglas, Jr., and E. A. Harre. TVA, Muscle Shoals, Alabama. *Fertilizer Plant Capacity Estimates*. Private communication, 1968.
[a] PR—Phosphate Rock, N—Nitrogen, P—Phosphate (P_2O_5), K—Potash (K_2O).

TABLE A-10. 1975 Interarea Net Volume and Pattern of Trade in Wheat and Rye, Rice, Other Grains, and Total Cereal Exports, Assuming (1) the Low Land Bounds and (2) Fertilizer Use Assumption A (1,000 metric tons)

Importing Area	Commodity Exported	Exporting Areas (as numbered on the left)													Total Imports
		1	2	3	4	5	6	7	8	9	10	11	12	13	
1 United States	Wheat, rye														
	Rice													45	45
	Other grains	685													685
	Total cereals	685												45	730
2 Canada	Wheat, rye														
	Rice														
	Other grains														
	Total cereals														
3 Latin America	Wheat, rye	10,520													10,520
	Rice													359	359
	Other grains	1,729													1,729
	Total cereals	12,249												359	12,608
4 EEC	Wheat, rye														
	Rice												27		27
	Other grains	3,150													3,150
	Total cereals	3,150											27		3,177
5 Other W. Europe	Wheat, rye	6,458													6,458
	Rice											98	137		235
	Other grains	8,220													8,220
	Total cereals	14,678										98	137		14,913
6 E. Europe	Wheat, rye	5,415			2,653										8,068
	Rice												427		427
	Other grains	1,109													1,109
	Total cereals	6,524			2,653								427		9,604
7 USSR	Wheat, rye												592		592
	Rice														
	Other grains														
	Total cereals												592		592

406

TABLE A-10 (cont.)

Importing Area	Commodity Exported	\multicolumn{13}{c}{Exporting Areas (as numbered on the left)}													Total Imports
		1	2	3	4	5	6	7	8	9	10	11	12	13	
8 Middle East	Wheat, rye	2,638		551				450			2,868				6,510
	Rice												244		244
	Other grains	1,207		3,012				535			1,036		213		6,003
	Total cereals	3,845		3,566				985			3,904		457		12,767
9 Africa	Wheat, rye	2,748									635				3,383
	Rice			37									474		511
	Other grains	855		1,182							740		544		3,321
	Total cereals	3,603		1,219							1,375		1,018		7,215
10 S. Africa-Oceania	Wheat, rye	200													200
	Rice												67	4	71
	Other grains														
	Total cereals	200											67	4	271
11 India-Pakistan	Wheat, rye										3,740				3,740
	Rice												1,096		1,096
	Other grains										1,084		1,482		2,566
	Total cereals										4,824		2,578		7,402
12 Other E. Asia	Wheat, rye										2,065				2,065
	Rice													2,111	2,111
	Other grains														
	Total cereals										2,065			2,111	4,176
13 Japan	Wheat, rye	4,004													4,004
	Rice														
	Other grains	5,111													5,111
	Total cereals	9,115													9,115

NOTE: Volume of cereal grain trade is the sum of projected trade in wheat and rye, rice, and other grains.

TABLE A-11. 1985 Interarea Net Volume and Pattern of Trade in Wheat and Rye, Rice, Other Grains, and Total Cereal Exports, Assuming (1) the Low Land Bounds and (2) Fertilizer Use Assumption A (1,000 metric tons)

Importing Area	Commodity Exported	Exporting Areas (as numbered on the left)													Total Imports
		1	2	3	4	5	6	7	8	9	10	11	12	13	
1 United States	Wheat, rye														
	Rice	55													55
	Other grains	2,168													2,168
	Total cereals	2,223													2,223
2 Canada	Wheat, rye														
	Rice														
	Other grains														
	Total cereals														
3 Latin America	Wheat, rye	15,191									488				15,679
	Rice	574													574
	Other grains	3,506													3,506
	Total cereals	19,271									488				19,759
4 EEC	Wheat, rye	8				5									13
	Rice														
	Other grains														
	Total cereals	8				5									13
5 Other W. Europe	Wheat, rye	6,406													6,406
	Rice	257													257
	Other grains	4,821													4,821
	Total cereals	11,484													11,484
6 E. Europe	Wheat, rye	3,581			6,426	171									10,178
	Rice	475													475
	Other grains	2,040													2,040
	Total cereals	6,096			6,426	171									12,693
7 USSR	Wheat, rye	7,756													7,756
	Rice	714													714
	Other grains														
	Total cereals	8,470													8,470

TABLE A-11 (cont.)

Importing Area	Commodity Exported	Exporting Areas (as numbered on the left)													Total Imports
		1	2	3	4	5	6	7	8	9	10	11	12	13	
8 Middle East	Wheat, rye	5,510	2,264								3,125	380			11,933
	Rice	7,745						2,450							10,195
	Other grains														
	Total cereals	13,255	2,261					2,450			3,125	380			22,128
9 Africa	Wheat, rye	4,594									870				5,464
	Rice			766		47									813
	Other grains	4,853		2,658							1,135				8,646
	Total cereals	9,447		3,424		47					2,005				14,923
10 S. Africa-Oceania	Wheat, rye	550													550
	Rice	88													88
	Other grains	853													853
	Total cereals	1,491													1,491
11 India-Pakistan	Wheat, rye										3,913				3,913
	Rice								124	97			1,757		1,978
	Other grains										1,002		3,757		4,759
	Total cereals								124	97	4,915		5,514		10,650
12 Other E. Asia	Wheat, rye		629								2,034	25			2,688
	Rice										132			4,841	4,973
	Other grains	1,000													1,000
	Total cereals	1,000	629								2,166	25		4,841	8,661
13 Japan	Wheat, rye	4,897													4,897
	Rice														
	Other grains	6,823													6,823
	Total cereals	11,720													11,720

NOTE: Volume of cereal grain trade is the sum of projected trade in wheat and rye, rice, and other grains.

409

TABLE A-12. 2000 Interarea Net Volume and Pattern of Trade in Wheat and Rye, Rice, Other Grains, and Total Cereal Exports, Assuming (1) the Low Land Bounds and (2) Fertilizer Use Assumption A (1,000 metric tons)

Importing Area	Commodity Exported	Exporting Areas (as numbered on the left)													Total Imports
		1	2	3	4	5	6	7	8	9	10	11	12	13	
1 United States	Wheat, rye														
	Rice														
	Other grains														73
	Total cereals														5,335
															5,408
2 Canada	Wheat, rye														
	Rice	73													73
	Other grains	5,335													5,335
	Total cereals	5,408													5,408
3 Latin America	Wheat, rye	25,497													25,497
	Rice	1,125													1,125
	Other grains	10,170													10,170
	Total cereals	36,792													36,792
4 EEC	Wheat, rye														
	Rice														
	Other grains														
	Total cereals														
5 Other W. Europe	Wheat, rye	6,817													6,817
	Rice	288													288
	Other grains	510													510
	Total cereals	7,615													7,615
6 E. Europe	Wheat, rye	1,872			10,070	584									12,526
	Rice	534													534
	Other grains	3,186													3,186
	Total cereals	5,592			10,070	584									16,246
7 USSR	Wheat, rye	21,043													21,043
	Rice														
	Other grains	907													907
	Total cereals	21,950													21,950

TABLE A-12 (cont.)

Importing Area	Commodity Exported	Exporting Areas (as numbered on the left)													Total Imports
		1	2	3	4	5	6	7	8	9	10	11	12	13	
8 Middle East	Wheat, rye	9,078	11,872								1,702				22,652
	Rice	73	73			18					38		311		513
	Other grains	13,236						4,465							17,701
	Total cereals	22,387	11,945			18		4,465			1,740		311		40,866
9 Africa	Wheat, rye	8,168	278								902				9,348
	Rice	1,008		507		27					115	44			1,701
	Other grains	19,675				1,032					1,650				22,357
	Total cereals	28,851	278	507		1,059					2,667	44			33,406
10 S. Africa-Oceania	Wheat, rye	1,201													1,201
	Rice	133													133
	Other grains	3,977													3,977
	Total cereals	5,311													5,311
11 India-Pakistan	Wheat, rye										7,390				7,390
	Rice		3,287								1,517		4,225		9,029
	Other grains	1,687									346		8,417		10,450
	Total cereals	1,687	3,287								9,253		12,642		26,869
12 Other E. Asia	Wheat, rye	3,136	819												3,955
	Rice	9,680	2,475											8,152	20,307
	Other grains	5,418													5,418
	Total cereals	18,234	3,294											8,152	29,680
13 Japan	Wheat, rye	6,196													6,196
	Rice														
	Other grains	7,952													7,952
	Total cereals	14,148													14,148

NOTE: Volume of cereal grain trade is the sum of projected trade in wheat and rye, rice, and other grains.

411

INDEX